Lecture Notes in Computer Science 2411

Edited by G. Goos, J. Hartmanis, and J. van Leeuwen

W0245813

Springer-Verlag Berlin Heidelberg GmbH

Fabio Paternò (Ed.)

Human Computer Interaction with Mobile Devices

4th International Symposium, Mobile HCI 2002
Pisa, Italy, September 18-20, 2002
Proceedings

Series Editors

Gerhard Goos, Karlsruhe University, Germany
Juris Hartmanis, Cornell University, NY, USA
Jan van Leeuwen, Utrecht University, The Netherlands

Volume Editor

Fabio Paternò
Consiglio Nazionale delle Ricerche, ISTI
Via G. Moruzzi,1, 56124 Pisa, Italy
E-mail: fabio.paterno@cnuce.cnr.it

Cataloging-in-Publication Data applied for

Die Deutsche Bibliothek - CIP-Einheitsaufnahme

Human computer interaction with mobile devices : 4th international symposium ; proceedings / Mobile HCI 2002, Pisa, Italy, September 18 - 20, 2002. Fabio Paternò (ed.).

(Lecture notes in computer science ; Vol. 2411)

ISBN 978-3-540-44189-2

CR Subject Classification (1998): H.5.2, H.5.3, H.5, H.4, C.2, I.2.1

ISSN 0302-9743

ISBN 978-3-540-44189-2 ISBN 978-3-540-45756-5 (eBook)

DOI 10.1007/978-3-540-45756-5

This work is subject to copyright. All rights are reserved, whether the whole or part of the material is concerned, specifically the rights of translation, reprinting, re-use of illustrations, recitation, broadcasting, reproduction on microfilms or in any other way, and storage in data banks. Duplication of this publication or parts thereof is permitted only under the provisions of the German Copyright Law of September 9, 1965, in its current version, and permission for use must always be obtained from Springer-Verlag Berlin Heidelberg GmbH
Violations are liable for prosecution under the German Copyright Law.

http://www.springer.de

© Springer-Verlag Berlin Heidelberg 2002

Originally published by Springer-Verlag Berlin Heidelberg New York in 2002

Typesetting: Camera-ready by author, data conversion by Steingräber Satztechnik GmbH, Heidelberg
Printed on acid-free paper SPIN: 10873714 06/3142 5 4 3 2 1 0

Preface

It was a wager, but it has worked. Mobile HCI used to be a workshop, often held in conjunction with other events. Instead, this time it was a true symposium. I felt there was a need for this change because of the ever-increasing interest prompted by the issues involved in interactive mobile systems and the lack of specific events focusing on such aspects. Although there are events addressing the broad area of ubiquitous computing, they tend to concentrate on other topics. For example, a paper on criteria for the design of interactive mobile phone applications would be considered inappropriate for such events, whereas it would certainly be relevant for Mobile HCI.

The response to the symposium has been positive in terms of submissions and participation. The contributions, especially the long papers, were selected carefully by the International Program Committee. The result is a set of interesting and stimulating papers that address such important issues as location awareness, design criteria for PDAs, context-dependent systems, innovative case studies, usability evaluation in small devices, and novel interfaces for mobile devices. The interest shown in the symposium has truly been worldwide: we have authors from 16 countries on three continents. There is a good balance of contributions from academia and industry. The final program of the symposium included two technical invited speakers (Brad Myers from Carnegie Mellon University and Luca Passani from Openwave), 18 full papers and 32 short papers, as well as a session with a representative of the European Commission to present and discuss their future programs in this area, and a number of interactive demos that allow participants to have direct experience of innovative results.

More generally, we can note that recent years have seen the introduction of many types of computers and devices (e.g., cellphones, PDAs, etc.) and the availability of this wide range of devices has become a fundamental challenge for designers of interactive software systems. Users wish to be able to seamlessly access information and services regardless of the device they are using, even when the system or the environment changes dynamically. To this end, computer-based applications need to run on a wide spectrum of devices. These challenges are addressed in research projects such as the CAMELEON IST Project (http://giove.cnuce.cnr.it/cameleon.html), which I coordinate. The project's main point is to develop methods and tools able to support the design and development of highly usable context-sensitive interactive software systems with the support of models that allow designers to better manage the increasing complexity of design. The resulting applications should behave like chameleons! They should be able to change their forms depending upon the types of devices utilized by users to perform their tasks and the surrounding environment.

Last, but not least, let me thank all those who helped to organize the symposium, in particular, the International Program Committee and Carmen Santoro who managed the symposium website and chaired the tutorial and workshop program.

July 2002 Fabio Paternò

Program Committee

Stephen Brewster – University of Glasgow, UK
Luca Chittaro – Universiy of Udine, Italy
Keith Cheverst – University of Lancaster, UK
Joelle Coutaz – University of Grenoble, France
Nicholas Graham – Queen's University, Canada
Mary Czerwinski – Microsoft Research, USA
Alan Dix – vfridge limited and University of Lancaster, UK
Mark Dunlop – University of Strathclyde, UK
Phil Gray – University of Glasgow, UK
Nicholas Graham – Queen’s University, Canada
Tom Gross – Fraunhofer FIT, Germany
Laurence Nigay – University of Grenoble, France
Dan Olsen – Brigham Young University, USA
Reinhardt Opperman – Fraunhofer FIT, Germany
Fabio Paternò (Chair) – ISTI CNR, Italy
Angel Puerta – RedWhale Software, USA
Matthias Rauterberg – University of Eindhoven, The Netherlands
Daniel Salber – IBM T.J. Watson Research Center, USA
Carmen Santoro – ISTI CNR, Italy
Albrecht Schmidt – Universität Karlsruhe, Germany
Mathias Schneider-Hufschmidt – Siemens, Germany
Ahmed Seffah – Concordia University, Canada
Constantine Stephanidis – FORTH, Greece
Kaisa Vaananen – Nokia, Finland
Jean Vanderdonckt – University of Louvain, Belgium
Bruno von Niman – Ericsson Enterprise, Sweden

Sponsoring Organizations

ERCIM
IFIP TC13
ACM-SIGCHI
ACM SIGCHI-Italy
ISTI-CNR

Table of Contents

Innovative Case Studies

Usability Evaluation in Small Devices

Novel User Interfaces for Mobile Devices

Short Papers

Mobile Devices for Control

Brad A. Myers

Human-Computer Interaction Institute,
Carnegie Mellon University,
School of Computer Science,
Pittsburgh, PA 15213, USA,
bam@cs.cmu.edu,
http://www.cs.cmu.edu/~bam

Abstract. With today's and tomorrow's wireless technologies, such as IEEE 802.11, BlueTooth, RF-Lite, and G3, mobile devices will frequently be in close, interactive communication. Many environments, including offices, meeting rooms, automobiles and classrooms, already contain many computers and computerized appliances, and the smart homes of the future will have ubiquitous embedded computation. When the user enters one of these environments carrying a mobile device, how will that device interact with the immediate environment? We are exploring, as part of the Pebbles research project, the many ways that mobile devices such as PalmOS Organizers or PocketPC / Windows CE devices, can serve as useful adjuncts to the "fixed" computers in the user's vicinity. This brings up many interesting research questions, such as how to provide a user interface that spans multiple devices that are in use at the same time? How will users and systems decide which functions should be presented and in what manner on what device? How can the user's mobile device be effectively used as a "Personal Universal Controller" to provide an easy-to-use and familiar interface to all of the complex appliances available to a user? How can communicating mobile devices enhance the effectiveness of meetings and classroom lectures? I will describe some preliminary observations on these issues, and discuss some of the systems that we have built to investigate them.

For more information, see http://www.pebbles.hcii.cmu.edu/.

1 Introduction

It has always been part of the vision of mobile devices that they would be in *continuous communication*. For example, the ParcTab small handheld devices [17], which were part of the original *ubiquitous computing* research project at Xerox PARC, were continuously communicating with the network using an infrared network. Mobile phones are popular because they allow people to stay in constant contact with others. However, the previous two or three generations of commercial handheld personal

F. Paternò (Ed.): Mobile HCI 2002, LNCS 2411, pp. 1–8, 2002.
© Springer-Verlag Berlin Heidelberg 2002

digital assistants (PDAs), such as the Apple Newton and the Palm Pilot, did not provide this capability, and only rarely communicated with other devices. For example, the Palm Pilot is designed to "HotSync" with a PC about once a day to update the information.

With the growing availability and popularity of new wireless technologies, such as IEEE 802.11, BlueTooth [3], RF-Lite [18], always-on two-way pagers, and email devices such as the Blackberry RIM, continuous communication is returning to commercial handhelds. What will be the impact of this on the user interfaces?

Another important observation is that most of people's time is spent in environments where there are already many computerized devices. Most offices have one or more desktop or laptop computers and displays. Many meeting rooms and classrooms have permanent or portable data projectors and PCs. Automobiles contain dozens of computers, and dashboards are likely to include LCD panels, sometimes replacing the conventional gauges. The more expensive airplane passenger seats provide individual LCD display screens for watching movies. Homes have televisions, PCs and many appliances with display screens and push buttons.

Our focus in the Pebbles project [5] is to look at how mobile devices will interoperate with each other and with other computerized devices in the users' environment. This brings up a number of interesting new research issues. For example:

- **How can the user interface be most effectively spread across all the devices that are available to the user?** If there is a large screen nearby, there may be no need for all the information to be crammed into the tiny screen of a PDA. When a PDA is near a PC, the PC's keyboard will often be an easier way to enter text than the PDA's input methods, but on the other hand, the PDA's stylus and touch screen may be a more convenient input device for drawing or selecting options for the PC than using a mouse. We call these situations *multi-machine user interfaces* since a person may be using multiple machines to complete the same task.
- **Can communicating mobile devices enhance the effectiveness of meetings and classroom lectures?** People at their seat may be able to use their PDAs to interact with the content displayed on the wall without having to physically take the keyboard and mouse away from the speaker. If there are multiple people in front of a large shared display, then mobile devices may be used for private investigation of the public information without disrupting the public displays. In classrooms, students may be able to answer questions using handhelds with the results immediately graded and summarized on the public display.
- **Can the user's mobile device be used to provide an easy-to-use and familiar interface to all of the complex appliances available to the user?** If the user has a mobile device with a high-quality screen and a good input method, why would a low-quality remote control be used for an appliance? Our preliminary studies suggest that users can operate a remote control on a PDA in one-half the time with one-half the errors as the manufacturers' original appliance interfaces [15]. Furthermore, allowing the remote to engage in a two-way communication with the appliances enables the creation of high-quality specialized devices that provide access to the disabled. For example, the INCITS V2 standardization effort [16] is

creating the Alternative Interface Access Protocol that will let people with visual difficulties use mobile Braille and speech devices to control household appliances.

The next sections provide a brief overview of how mobile devices can be used to *control* PCs and appliances. More information is available in the various publications about the Pebbles research project [2, 4-15]. See also the Pebbles web site for up-to-date information: http://www.pebbles.hcii.cmu.edu/.

2 Control of PCs

The first set of applications we created as part of the Pebbles project explores how mobile devices can be used to control a PC, in both group and individual settings.

The *Remote Commander* program [10] allows a Palm or PocketPC device to provide the keyboard and mouse input for a PC (see Figs. 1(a) and 1(b)). The input appears to applications running on the PC as if it came from the regular PC keyboard and mouse. The original concept was for participants in a meeting to use Remote Commander to interact with a public display. Remote Commander has also proven useful for system administrators to control "headless" computers that do not have keyboards and mice, such as servers and display computers in shops and museums.

Remote Commander has also helped people with certain neuromuscular disorders to use a computer more easily [11]. People with Muscular Dystrophy, for example, have difficulty with the larger movements required by conventional keyboards and mice, but can more easily make small movements to control a stylus on a PDA screen.

(a)

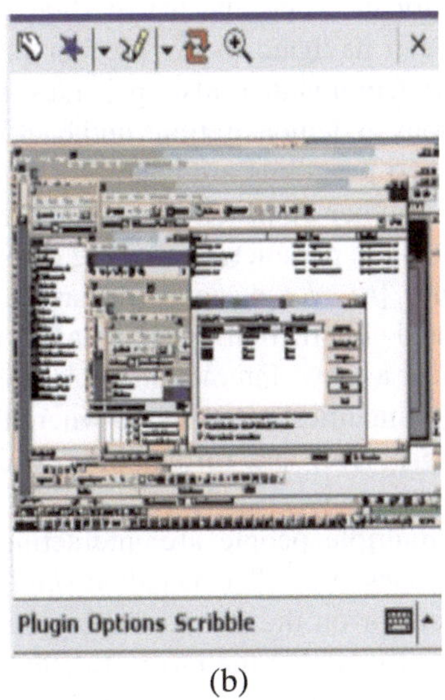

(b)

Fig. 1. Palm (a) and PocketPC (b) Remote Commander screens. The PocketPC version displays a PC's screen image.

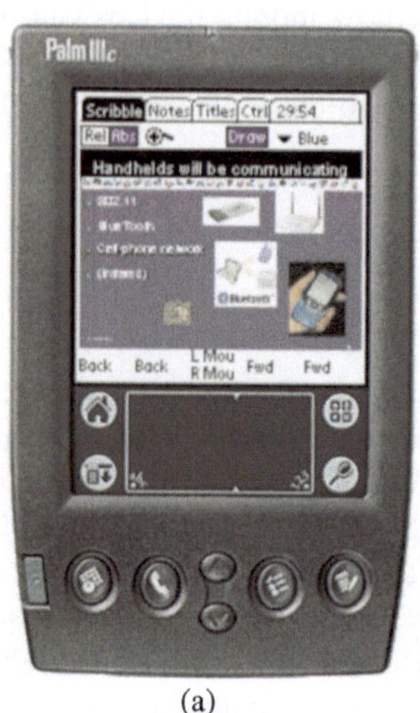

(a)

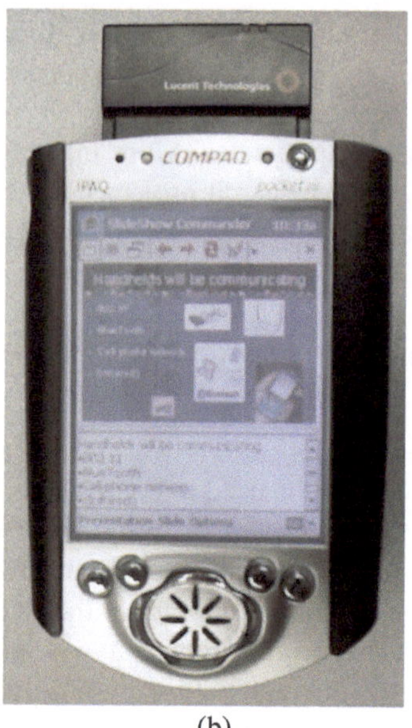

(b)

Fig. 2. SlideShow Commander screens for the Palm (a) and PocketPC (b).

The *SlideShow Commander* program [8] extends the idea of Remote Commander to provide more information on the handheld for controlling slide shows. When running a PowerPoint presentation on the PC, SlideShow Commander displays a thumbnail picture of the current slide on which the user can scribble with the stylus, as well as the notes for the slide, the list of slides, and other information (Figs. 2(a) and 2(b)). The user can navigate to the next or previous slide, or jump anywhere in the talk. SlideShow Commander also provides facilities to make it easier to switch from presentations to demonstrations and back.

These two programs are examples of using the mobile device for *interacting at a distance*. Another common way to interact at a distance is using a laser pointer. We have studied the parameters of using a laser pointer tracked by a camera as a computer input device [6]. We discovered that the beam wiggles about 10 pixels due to hand motion, and interactions using laser pointers tend to be slow. Therefore we investigated a new interaction technique called *semantic snarfing* [9] where the contents ("semantics") in the area where the beam is pointing are copied ("snarfed") to the mobile device, and further interaction takes place on the mobile device, where increased accuracy is possible.

When multiple people are interacting with the same shared display, many user interface issues arise. This is called *single-display groupware*. For example, if there is only one cursor on the shared display, how will users decide who is in control of the cursor? We found that the most effective strategy for such face-to-face sharing was to let whoever wanted to take control do so, but to impose a small timeout before the control was switched to prevent accidental overlapping [11].

In the context of a military environment, called the Command Post of the Future, we studied *private drill down of public information*. Here, multiple people are sharing public maps and other information displays, so it would be inappropriate for anyone to usurp the big displays for their private use. Instead, there is fluid transfer of information and control between the large public displays and each user's mobile device [4].

We also investigated uses for mobile wireless devices in a classroom. One application we have studied is instantaneous test taking. We have used PDAs in a second-level chemistry class with about 100 undergraduates to enable the instructor to ask multiple choice questions and get a bar graph of all the student's answers. This helps keep the students thinking about the material and allows the instructor to evaluate the students' level of understanding during a lecture. The students reported a strong preference for using the mobile wireless devices over non-computerized alternatives, such as raising their hands or using paper [2].

Most of the above situations involved multiple users. We also studied how *individuals* working alone might find a mobile device useful even when they had a regular PC available.

Most mobile devices are rechargeable, so it is reasonable for users to put them in a cradle beside the keyboard while at a PC. We studied how a PDA could be used as an extra input device for the non-dominant hand while in this configuration (see Fig. 3(a)). For example, a study showed that the users could scroll and select more quickly using their left hands to scroll with a PDA while their right hands were on the mouse, as shown in Fig 3(a) [7].

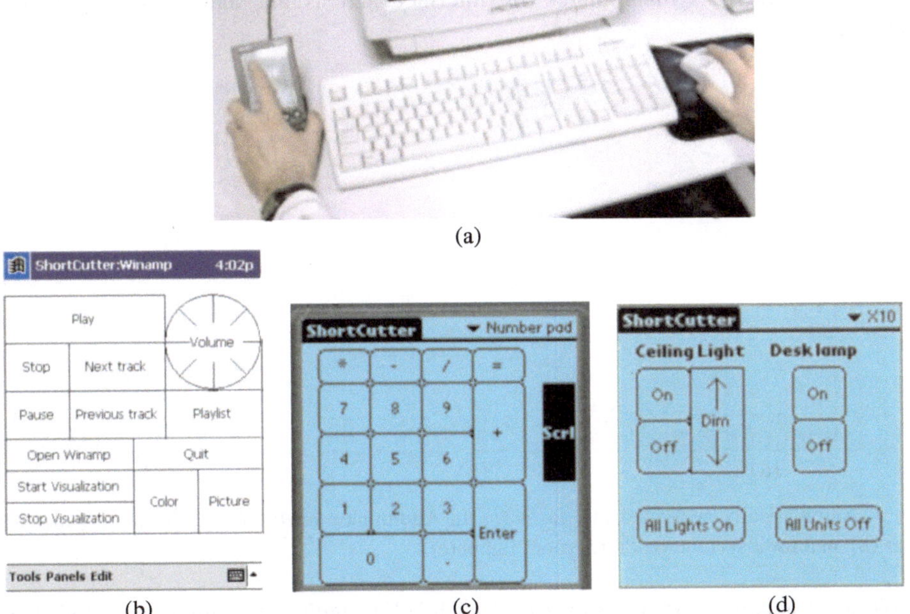

Fig. 3. PDA on left of a keyboard (a) makes it useful to use Shortcutter on a PocketPC (b) or Palm (c)(d) to control PC applications for an individual.

As a more general application of this concept, we created the *Shortcutter* program, which allows users to draw a panel of controls on the PocketPC (Fig. 3(b)) or Palm (Figs. 2(c)(d)), and use these panels to control any PC application [8]. The user might create buttons to perform the most common operations. For example, Fig. 3(b) shows a control panel for the Winamp media player.

3 Control of Appliances

A new area we are investigating is how to use mobile devices to control everyday home and office appliances, such as stereos, VCRs, room lights, copiers, etc. These are becoming more complex as embedded computers enable new kinds of functions, but as complexity increases, appliance user interfaces usually get harder to use [1]. Our concept is that each user would use their mobile device as a *personal universal controller* (PUC) that would allow the user to interact with all the appliances and services in the environment. A PUC could take many forms: an unimpaired user might have a handheld mobile device with a graphical user interface (GUI), whereas a blind user might have an interactive Braille surface or headset that supports speech recognition and speech output. When the user wants to control an appliance, the PUC would communicate with the appliance, download a specification of the appliance's functions, and then automatically generate a remote-control interface suited to the PUC device and the user. The PUC and the appliance would continue to exchange messages as the user manipulates the interface and as the state of the appliance changes.

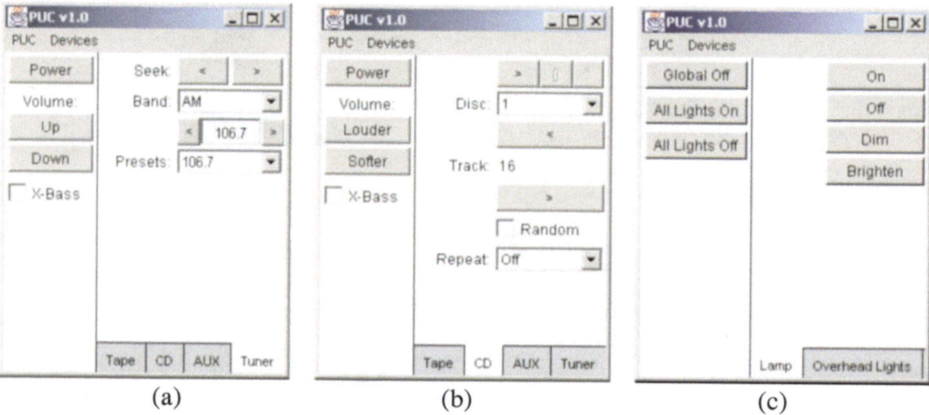

Fig. 4. Automatically generated interfaces for an Audiophase shelf stereo with its CD (a) and tuner (b); and for a system to control room lights (c).

We approached the PUC project by first hand-designing user interfaces, and then studying how well they performed [15]. We were encouraged by the results, which showed that for both simple and complex tasks, user were able to use our handheld interfaces in about ½ the time with ½ the errors of using the manufacturer's interfaces. Based on our user studies and hand-designs, we developed a set of requirements for the specification language [13]. We now are developing algorithms that will automatically generate high-quality graphical and speech user interfaces from the specifications [12, 14]. Fig. 4 shows some of the current interfaces that can be generated.

4 Looking Forward

Much of the research in the area of mobile human-computer interaction has focused on the user interfaces to the mobile devices themselves: their input methods and displays. It is important to also study the broader picture and look at how the devices will fit into the users' entire information and control space. As more and more electronics are computerized and are able to communicate, mobile devices can serve as a personal, portable focal point for interactions with the world. Let us work to have mobile devices *improve* the user interfaces for everything else, rather than just being additional complex gadgets that must also be mastered.

Acknowledgements

The Pebbles research project is supported by grants from DARPA, NSF, Microsoft, and the Pittsburgh Digital Greenhouse, and equipment grants from Symbol Technologies, Palm, Hewlett-Packard, Lucent, IBM, SMART Technologies, Inc., and TDK Systems Europe, LTD. This research was performed in part in connection with contract number DAAD17-99-C-0061 with the U.S. Army Research Laboratory. The National Science Foundation funded this work through a Graduate Research Fellowship, and under Grant No. IIS-0117658. The views and conclusions contained in this document are those of the authors and should not be interpreted as presenting the official policies or position, either expressed or implied, of the U.S. Army Research Laboratory, the National Science Foundation, or the U.S. Government unless so designated by other authorized documents. Citation of manufacturer's or trade names does not constitute an official endorsement or approval of the use thereof.

References

1. Brouwer-Janse, M.D., Bennett, R.W., Endo, T., van Nes, F.L., Strubbe, H.J., and Gentner, D.R. "Interfaces for consumer products: "how to camouflage the computer?"" in *CHI'1992: Human factors in computing systems*. 1992. Monterey, CA: pp. 287-290.
2. Chen, F., Myers, B., and Yaron, D., *Using Handheld Devices for Tests in Classes*. Carnegie Mellon University, School of Computer Science Technical Report no. CMU-CS-00-152 and Human Computer Interaction Institute Technical Report no. CMU-HCII-00-101, July, 2000. Pittsburgh, PA. http://www.cs.cmu.edu/~pebbles/papers/CMU-CS-00-152.pdf.

3. Haartsen, J., Naghshineh, M., Inouye, J., Joeressen, O.J., and Allen, W., "Bluetooth: Vision, Goals, and Architecture." *ACM Mobile Computing and Communications Review*, 1998. **2**(4): pp. 38-45. Oct. www.bluetooth.com.
4. Myers, B., Malkin, R., Bett, M., Waibel, A., Bostwick, B., Miller, R.C., Yang, J., Denecke, M., Seemann, E., Zhu, J., Peck, C.H., Kong, D., Nichols, J., and Scherlis, B., *Flexi-modal and Multi-Machine User Interfaces.* submitted for publication, 2002. http://www.cs.cmu.edu/~cpof/papers/cpoficmi02.pdf.
5. Myers, B.A., "Using Hand-Held Devices and PCs Together." *Communications of the ACM*, 2001. **44**(11): pp. 34-41.
6. Myers, B.A., Bhatnagar, R., Nichols, J., Peck, C.H., Kong, D., Miller, R., and Long, A.C. "Interacting At a Distance: Measuring the Performance of Laser Pointers and Other Devices," in *ACM CHI'2002 Conference Proceedings: Human Factors in Computing Systems.* 2002. Minn, MN: pp. 33-40.
7. Myers, B.A., Lie, K.P.L., and Yang, B.-C.J. "Two-Handed Input Using a PDA And a Mouse," in *Proceedings CHI'2000: Human Factors in Computing Systems.* 2000. The Hague, The Netherlands: pp. 41-48.
8. Myers, B.A., Miller, R.C., Bostwick, B., and Evankovich, C. "Extending the Windows Desktop Interface With Connected Handheld Computers," in *4th USENIX Windows Systems Symposium.* 2000. Seattle, WA: pp. 79-88.
9. Myers, B.A., Peck, C.H., Nichols, J., Kong, D., and Miller, R. "Interacting At a Distance Using Semantic Snarfing," in *ACM UbiComp'2001.* 2001. Atlanta, Georgia: pp. 305-314.
10. Myers, B.A., Stiel, H., and Gargiulo, R. "Collaboration Using Multiple PDAs Connected to a PC," in *Proceedings CSCW'98: ACM Conference on Computer-Supported Cooperative Work.* 1998. Seattle, WA: pp. 285-294. http://www.cs.cmu.edu/~pebbles.
11. Myers, B.A., Wobbrock, J.O., Yang, S., Yeung, B., Nichols, J., and Miller, R. "Using Handhelds to Help People with Motor Impairments," in *Fifth International ACM SIGCAPH Conference on Assistive Technologies; ASSETS'02.* 2002. Scotland: pp. To appear.
12. Nichols, J. "Informing Automatic Generation of Remote Control Interfaces with Human Designs," in *ACM CHI'2002 Extended Abstracts.* 2002. Minneapolis, Minnesota: pp. 864-865. http://www-2.cs.cmu.edu/~jeffreyn/papers/chi2002puc.pdf.
13. Nichols, J., Myers, B.A., Harris, T.K., Rosenfeld, R., Shriver, S., Higgins, M., and Hughes, J. "Requirements for Automatically Generating Multi-Modal Interfaces for Complex Appliances," in *Submitted for Publication.* 2002. http://www.cs.cmu.edu/~pebbles/papers/pucICMI.pdf.
14. Nichols, J., Myers, B.A., Higgins, M., Hughes, J., Harris, T.K., Rosenfeld, R., and Pignol, M. "Generating Remote Control Interfaces for Complex Appliances," in *Submitted for Publication.* 2002. http://www.cs.cmu.edu/~pebbles/papers/PebblesPUCuist.pdf.
15. Nichols, J.W. "Using Handhelds as Controls for Everyday Appliances: A Paper Prototype Study," in *ACM CHI'2001 Extended Abstracts.* 2001. Seattle, WA: pp. 443-444. http://www.cs.cmu.edu/~pebbles/papers/NicholsRemCtrlShortPaper.pdf.
16. V2 Working Group, *Universal Remote Console Specification (AIAP-URC) of the Alternate Interface Access Protocol (AIAP).* http://www.ncits.org/tc_home/v2.htm, 2002.
17. Want, R., Schilit, B.N., Adams, N., Gold, R., Petersen, K., Goldberg, D., Ellis, J.R., and Weiser, M., "An Overview of the ParcTab Ubiquitous Computing Experiment." *IEEE Personal Communications*, 1995. pp. 28-43. December. Also appears as Xerox PARC Technical Report CSL-95-1, March, 1995.
18. Zigbee Alliance, *Zigbee Working Group Web Page for RF-Lite.* 2002. http://www.zigbee.org/.

Building Usable Wireless Applications for Mobile Phones

Luca Passani

Tools Development Manager,
Openwave Systems
http://www.openwave.com

1 Introduction

Building usable WAP applications is not simple. Wireless devices have many limitations, and the average user of a WAP application is not technically oriented (and possibly not even used to the Internet). Finally, the interpretation of WML varies greatly between devices from different vendors. This poses an extra challenge to good usability.

This situation resembles the browser war we're still witnessing on the Web.

This paper looks into these issues in depth, and discusses methods for overcoming some of the problems when attempting to code or convert applications for use on a variety of different browsers.

The paper assumes familiarity with the WAP architecture and WML programming [1].

2 Limitations of Wireless Devices

Let's review the limitations of WAP devices here:

- Small screens: In general, WAP devices are tiny. Those accustomed to web browsers will find navigating with a WAP phone a real pain.
- Limited input facilities: Most wireless devices lack a keyboard that is anything like a traditional QWERTY PC keyboard. Simple, mass-market, consumer-class data input technology that does not depend on a keyboard has yet to be invented.
- Limited processor power and memory: WAP browsers are simple and unforgiving.
- Limited bandwidth: At this stage, WAP devices have very little bandwidth available when compared to PCs. In Europe, users can count on a speed of 9600 bps (bits per second) as of April 2000. The introduction of GPRS may improve the situation slightly by the end of 2000.

F. Paternò (Ed.): Mobile HCI 2002, LNCS 2411, pp. 9–20, 2002.
© Springer-Verlag Berlin Heidelberg 2002

- Lack of graphics: Or at least, very limited support for them. Icons and graphics can go a long way towards helping the user in complicated situations.
- Limited deck size: A deck can contain only a limited amount of information.

These limitations have serious implications on the way WAP applications are designed.

3 WAP Users

WAP users are not sitting in front of a PC. They are on the move, on their way to a meeting, or in a crowded train. Sometimes they're under pressure. Building usable WAP systems is not straightforward, and goal when doing so should be to make it as simple to use as possible. While this is true for any application, it's an absolute must in the context of WAP. WAP users are subject to many distracting events in the environment that surrounds them, and this adds to the input/output limitations of WAP phones described above.

To add to that, users of wireless devices will soon outnumber conventional Internet users by far. One implication of this is that, in general, you cannot assume that the users of your application are also Internet users.

4 WML Interoperability Issues

WML delivers content and user interfaces across very different kinds of devices. The various browser implementations render WML in different ways, and this affects the usability of applications. This paper will discuss this issue in detail.

A usable WAP application should never confuse users, in that users should ideally be able to find the most obvious operations intuitively - just one click away. Unfortunately, if you tweak your application to be more usable on a particular device, the chances are that usability will suffer on other devices.

4.1 Different Devices

Fine-tuning usability necessarily implies getting involved with the idiosyncrasies of each device you intend to support. Generally speaking though, applications developed for small displays tend to look and work fine on large displays. Applications developed for large displays tend to look and work very badly on small displays.

If, while developing applications, one targets the smallest devices, in most cases larger devices are handled satisfactorily too. PDA-like devices will be especially well covered, as they support hyperlinks and features handled by <do> elements (features that are problematic on very small screens) very well.

5 Building Usable Application

It is important to develop a set of concrete guidelines that you can apply when designing a WAP application. Here are the most important questions one should ask themselves:

- Is the application easy to learn?
- Is the application efficient to use?
- Are unusual operations easy to remember?
- Do users get stuck when there are errors?
- Are users likely to be frustrated by their attempts to use your application?

Answering questions like these will put one on a good track to working out how usable an application is. Let's see some general rules for building usable applications.

- **Top 20% of functionality**: When porting an existing application to WAP (an HTML page, for example), you should identify the main activities that users will be interested in using while on the road. Porting parts of the application that are not commonly used will be more expensive for the developer, and the overall usability of the system will degrade, because of the extra levels and navigation paths introduced. In the case of new applications, think of this rule as "refrain from implementing functions that the majority of users won't use on the road".
- **Rate user activities**: One should identify the main activities that the majority of the users will perform, and build applications in a way that will let users perform these activities in the fastest way possible. Most common activities should be intuitively available for all users.
- **Design it as a tree structure**: Lay out a hierarchical tree of activities. Users should enter the application at the root and be able to perform any of the available activities through some path starting at the root. Each level of the tree should be laid out according to the likely popularity of the activities it contains.
- **Minimize data entry**: Most phones only have a phone keypad. Applications should require textual data entry only when absolutely necessary. Similarly, users should not be required to remember codes or other information when using an application – Apps should remember things for them. In addition, the input mode of the terminal should be set to support the expected format for the data that users will enter. This can be achieved through input masks.
- **Text should be terse**: Short, polished, and informative text is vital to guide users.
- **Always implement 'back' functionality**: All users like to explore when confronted with a new application, and a 'back' function should be available to them at all times. But be careful: users should go back to a logical and consistent place in the application, which is not necessarily the previous card.
- **Consistency is very important**: Applications can often require users to perform the same activity (or very similar activities) in different parts of the

application. It is important that you deploy a consistent set of metaphors that will help users find their way around easily.

- **Push**: Real-time information is a key piece of functionality that will give extra value to WAP. Don't give up on the extra value your application can acquire through push.
- **Be prepared to test**: When deploying a WAP application that is even moderately complex, one should be ready to build prototypes as early as possible in the development process. Find a non-technical person and let them use your application on real phones.
- **Identifying Activities.** When thinking about the main functionality of your application, user activities should be rated into categories according to how often users perform them. For example:
    - Activities that most users perform most of the time
    - Activities most users perform occasionally
    - Activities that specialized users performs once in a while

We give some more detailed groupings below. Identifying those activities is the basis for breaking down the navigation flow and optimizing the navigation path required to perform the main activities.

- Think of the tasks required to achieve the object of each activity
- Order the tasks by importance
- Lay out your user interface

For each activity, one should understand how users expect to perform them, according to models users are familiar with. This could mean similarity with corresponding PC or phone functions, or with the way users perform the activity in their work.

Classifying activities in the following groups will help you map the design directly into WML:

- **Required activities** (compulsory activities for all users): These are activities that all of the application users will have to perform. One typical such activity is configuring your e-mail address and POP server when using an e-mail application for the first time. Configuring access to a different gateway is another example. It goes without saying that user-friendliness degrades unacceptably if operations like this are requested each time the user accesses an application. Required activities should be avoided at all costs. In many cases you can uniquely identify each device and recall user preferences and configuration parameters automatically. Use this possibility and avoid requiring users to log in each time.
- **Primary-path activities** (high use activities): These are the activities that 80% or more of your users are likely to perform. Such activities should be easily and intuitively available, without any learning curve or unnatural interaction path for the inexperienced user. Performing Primary-path activities should be as easy as possible (always one click away).
- **Secondary-path activities** (high use by a large segment of, but not all, users): These are activities that many users (but not the majority) perform

often. Access to secondary-path activities should be kept as simple as possible, without being an impediment for users who do not perform them.

- **Side-path activities** (activities used occasionally by most users): These are activities that 80% of users will perform, but only 20% of the time they use your application. Replying to e-mail is one such activity. It makes sense to implement access to side-path activities in a menu that is not immediately accessible.
- **Rare-path activities** (activities never used by most users): These activities are there to support power users. Rare-path activities are natural candidates for removal. It's better to have a simple system than one with a lot of obscure options.

The classification of activities into primary-path, secondary-path, side-path, and rare-path gives developers an indication of how they should implement navigation to the different kinds of activities.

6 Guidelines for Specific Browsers

Good usability requires that applications be customized for each specific browser or family of browsers. In the rest of the presentation, I will look at two widespread families of WAP browsers: Nokia and Openwave.

6.1 Guidelines for Nokia

What follows is an overview of the interface for the Nokia WAP browser (Fig 1):

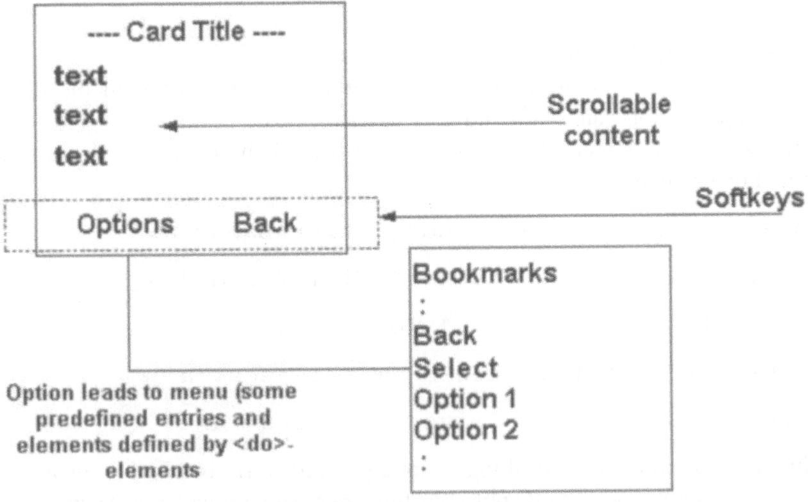

Fig 1: Entering Data on the Nokia browser

If you have an <input> or a <select> element in your code, you need to make it clear to the user how they can submit the data in order to move on to the next logical step.

The only sensible thing to do in these cases is to provide a link that will keep users going with one click:

```
<p>
    Name please:
    <input type="text" name="searchkey" value="" />
    <a href="search.asp">Submit data</a>
</p>
```

If you use <do> elements, the Nokia browser will interpret it by adding an extra element in the menu triggered by the left softkey. This is not good. Users will not immediately see this and so would be at a loss for what to do after they've entered the data.

6.1.1 Forms

There are two types of form: wizard forms and elective forms.

Wizard forms let users insert data one bit at the time. Each card contains an input element. It's a good idea to use wizards when you can, since they let users focus on entering data rather than navigation.

Forms implemented through a single card containing multiple input fields are called elective forms.

It goes without saying that on the Nokia browser you should provide a link to let users move from one card to the next, rather than using <do> elements.

6.1.2 Menu Navigation on the Nokia Browser

Menu navigation is a straightforward way to let your users access all the different parts of your application, laid out in a tree structure.

The best way to implement this for a Nokia browser is by building a menu as a list of anchors. For example:

```
<a href="#band" title="find">Artist/Band</a>
<a href="#song" title="songs">Title/Song</a>
<a href="#top" title="top">Top 20</a>
<a href="#new" title="new">New Releases</a>
<a href="#conc" title="live">Concerts</a>
```

Forcing users to do a lot of scrolling is not a good idea. If you need to display more than nine or ten items, you should split your links over several cards or decks.

6.1.3 Backward Navigation

The right softkey of the Nokia can only be used as a back button, and is labeled as such. By default, it has no action at all. Some sub-versions of the Nokia browser allow reprogramming of the back key with <do type="prev">. Unfortunately, this is not the case with all phones. The way to get the phone to do what you want is:

```
<card id="mycard" onenterforward="#nextcard"
      onenterbackward="http://logical_back">
```

Unfortunately, this has the side effect of spoiling the history stack.

Never provide a label for the <do type="prev"> element. The Nokia browser will remove the Back label from the right softkey and instead create a new entry in the menu accessed through the left softkey. This confuses users who expect to find the back key in a well-known position.

6.2 Guidelines for UP.Browser (Openwave Browser)

The Openwave browser has a different interface from the Nokia browser. The general idea behind the UP.Browser is that a <do> element is mapped directly to a softkey whenever possible.

Users will only ever be one click away from performing a primary activity. Other activities will also be reachable in a simple way. This is in sharp contrast with the Nokia browser.

<do type="accept"> elements are normally used to support primary activities. They are also called ACCEPT buttons. Other activities are supported through <do type="options"> elements (OPTION elements). See fig. 2.

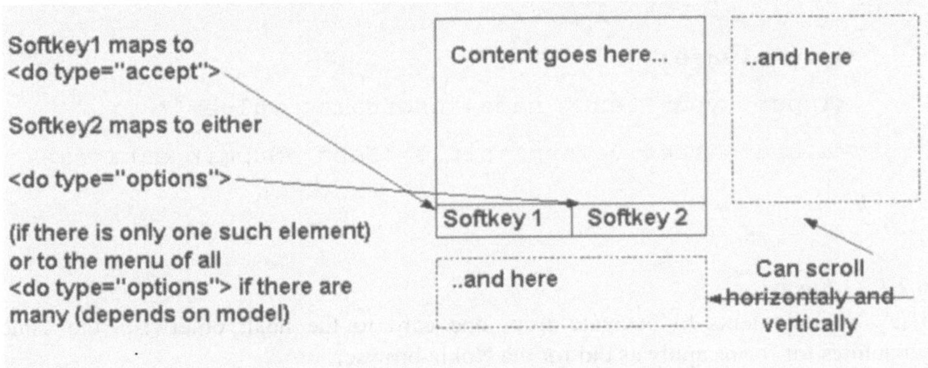

Fig 2: Interface for Openwave browser.

If you have a single primary-path activity, for example, the best you can do is to associate it with an ACCEPT button, which is always bound to a softkey (usually mapped to Softkey1). This way, your users are only one click away from the activity they are most likely to perform.

If you have a single <do type="option"> element (or OPTION button), this will be mapped to Softkey2. In the case of multiple OPTION buttons, Softkey2 will display the label menu, which leads to a pseudo-card that allows you to select the option from a simple list (similarly to Softkey1 on the Nokia browser).

Some UP.Browser phones support three softkeys. In that case, you can afford an ACCEPT task and two OPTION tasks without the need to go through an extra menu.

In the diagram, you can see softkey1 on the left and softkey2 on the right, but this is not always the case. Softkeys occupy different positions in different implementations of the UP.Browser.

Entering Data on the UP.Browser

Unlike the Nokia browser, the UP.browser does not require the presence of a link after an input element in order to be usable. A <do> element will work wonders there:

```
<do type="accept" label="Send">
   <go href="receive.asp" />
</do>
<p>
   name please:
   <input type="text" name="username" value="" />
</p>
```

The data can be posted with just one click (on Softkey1). If you insert a navigation link after the input element, as you should do with the Nokia browser, the mechanism will still work, but usability will degrade. The construct that follows will require no fewer than three clicks to post the data from the UP.Browser.

```
<p>
   name please:
   <input type="text" name="username" value="" />
   <a href="receive.asp" title="Send">Submit data</a>
</p>
```

6.2.1 Forms

Use <do> elements to navigate from one card to the next; otherwise, the same guidelines for forms apply as did for the Nokia browser.

6.2.2 Menu Navigation on the UP.Browser

If you implement menu navigation the same way as you would on the Nokia browser (with a bunch of links), the UP.Browser will work satisfactorily. Unfortunately, by doing so you'll lose a feature that the UP.Browser supports to facilitate navigation for slightly advanced users.

To demonstrate, the code below implements menu navigation in the optimal way for the UP.Browser:

```
<select>

   <option onpick="#band"
title="find">Artist/Band</option>

   <option onpick="#song"
title="songs">Title/Song</option>

   <option onpick="#top" title="top">Top 20</option>

   <option onpick="#new" title="new">New
Releases</option>

   <option onpick="#conc"
title="live">Concerts</option>

</select>
```

Figure 3 shows how the UP.Browser renders the code:

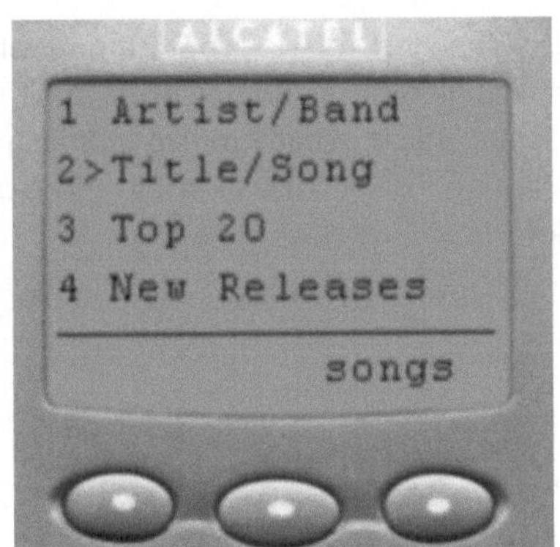

Fig 3: Menus rendered by the UP Browser

The numbers on the left of the screen are shortcuts. If you implement navigation menus this way, UP.Browser users will not be required to scroll to the menu item of choice, since they can press the number on their keypads and trigger the onpick event for the corresponding menu choice.

6.3 The Generic Approach

The basic idea behind the 'generic browser' is to identify a subset of WML that both the browsers used above interpret in a more or less usable way. Is Generic Navigation Good Enough? Generally speaking, the answer is no. The generic approach forces you to discard all of the really good things that each micro browser has to offer, and to settle for an alternative that proves to be mediocre in both environments, despite being useful in some cases.

For this reason, Openwave has always recommended developers to build multiple applications, one for each family of devices that needs to be supported.

Unfortunately, most developers will not go all the way and will stick to one application using generic mark-up.

6.4 Multi-serving WAP Applications

Multi-serving WAP applications according to the family of device is necessary in order to support an acceptable user experience. Unfortunately, most developers resist this idea because of the cost involved in building (and maintaining) multiple versions of their systems. For this reason, several companies have built and commercialized solutions and frameworks to abstract away differences in device families and allow developers to multi-serve without branching their code.

One example of this library is the Open Usability Interface (OUI), which has been contributed by Openwave to the developer community as open source under a Mozilla Public License.

OUI is simply a library that lets WAP programmers abstract away differences in WAP devices. When an actual request comes from a WAP device, the library will look at the device and decide there what kind of mark-up is best suited to optimize navigation and usability for the user. Just as a simple example, please consider the following snippet of code:

```
<%@ taglib uri="/WEB-INF/tld/oui.tld" prefix="oui" %>
<oui:wml>

 <oui:card id="start" title="Wireless World">
  <oui:p align="left" mode="nowrap">
   <oui:menu>
     <oui:menu_item href="ema.jsp" text="Email"
icon="envelope1" />
     <oui:menu_item href="fin.jsp" text="Finance"
icon="graph1" />
     <oui:menu_item href="ent.jsp" text="Entertainment"
icon="videocam" />
     <oui:menu_item href="spo.jsp" text="Sports"
icon="football" />
     <oui:menu_item href="new.jsp" text="News &amp;
weather" icon="partcloudy" />
     <oui:menu_item href="tra.jsp" text="Travel"
icon="plane" />
```

```
      <oui:menu_item href="sho.jsp" text="Shopping"
icon="dollarsign" />
      <oui:menu_item href="oth.jsp" text="Other"
icon="folder1" />
    </oui:menu>
   </oui:p>
  </oui:card>
</oui:wml>
```

The code above is similar to WML, but it's not WML. It is actually an abstraction, which is resolved at request time. The WML returned to different devices is different depending on the capabilities od the device. Figures 4,5 and 6 show how different WAP clients render the code above:

Fig 4: Nokia Phone

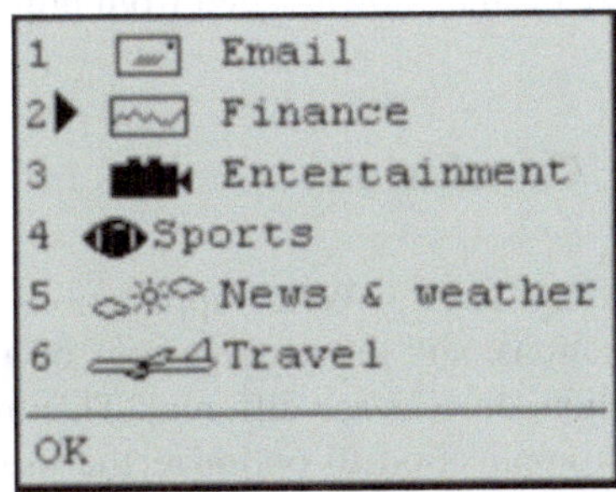

Fig 5: Openwave Mobile Browser textual

Fig 6: Openwave Mobile Browser, GUI

Here is a summary of the several OUI abstractions that represent developer intent. Those abstractions will render into the most usable WML constructs at request-time. Some of the abstractions are:

- Navigation paths:
  - Primary path (main operation for each card)
  - Secondary paths
  - Side paths
- Menus: numbered lists
- One-click calls
- Diminished latency thanks to page pre-loading
- Tables
- Automatic pagination for large documents
- Large forms: support for complex user input
- Composite pages
- In-line menus
- Back navigation: avoid trapping end users

OUI can be downloaded from http://oui.sourceforge.net

7 Conclusions

Building usable WAP applications is not simple. Wireless devices have many limitations, and the average user of a WAP application is not technically oriented. In addition, devices are different. This makes it necessary to build several versions of the same application to optimize the user experience on each class of devices. There is software which can diminish the cost of multi-serving applications. We saw one example (OUI).

References

[1] Arehart Charles, et. Al.: Professional WAP, Wrox Press Ltd. July 2000. ISBN: 1861004044

A Diary Study of Rendezvousing: Group Size, Time Pressure and Connectivity

Martin Colbert

Kingston University
Penrhyn Road, Kingston-upon-Thames,
Surrey, United Kingdom, KT1 2EE
+44 20 8547 2000
m.colbert@kingston.ac.uk

Abstract. This paper reports an initial analysis of a diary study of rendezvousing as performed by university students. The study's tentative findings are: (i) usability ratings for communication services are a little worse *during* a rendezvous (when at least one person is en route) than before (when none have yet departed); (ii) problems rendezvousing caused more stress when the rendezvousing group was large (6 or more participants) than when the group was small, but led to no more lost opportunity. Finding (i) is attributed to the desire for instant communication (which is stronger when users are under time pressure), and the constraints placed upon interaction (which are tighter in public spaces than in personal spaces). Finding (ii) is attributed to the suggestion that large rendezvous include more acquaintances (whose contact details may not be known) and different kinds of subsequent activity. If rendezvousers need anything, this study suggests that they need greater connectivity and service availability, rather than greater bandwidth. Implications for the design of position-aware communications services are discussed.

1 Setting Performance Goals for Mobile IP Services for the General Public

Industry plans to develop a wide range of multimedia, Internet-based (IP) services for the general public, and to deliver them over broadband, wireless networks to a range of mobile, context-aware devices. These services may include information seeking, shopping, personal communication, computer games, and audio/video entertainment.

The success of these plans rests on the assumption that the public will perceive such services to be valuable, because, for example, the services achieve performance benefits i.e. when a 'new' service is used, users achieve a higher quality of outcome, and/or incur lower levels of 'cost' so doing, than they did using the 'old' service. An initial step in the development of many future services, then, is to characterise "performance deficits" – trade-offs between quality and costs, that new services may be able to improve upon. The diagnosis of performance deficits may also indicate the kind of service that has the potential to make good the deficit, and support investigations into whether or not users would, indeed, value a service of that kind. If

F. Paternò (Ed.): Mobile HCI 2002, LNCS 2411, pp. 21–35, 2002.
© Springer-Verlag Berlin Heidelberg 2002

no deficits can be found, then the topic of research may change, and/or a less performance-oriented approach may be adopted.

2 Aims

This paper characterises some performance deficits associated with rendezvousing - the informal, geographical co-ordination of small groups of friends, family and team mates, such as "meeting a friend for lunch", and "collecting the kids from school". Rendezvousing is of general interest, because it is relevant to a wide range of possible services, such as personal organisation, navigation, information seeking, and communication. It is particularly relevant to position-aware services, because rendezvousing inherently involves a group of users moving towards a specific point in space and time, so position information seems particularly relevant.

This paper is an initial analysis of the interim results of a large diary study involving UK university students. Phase 1 of this study was conducted in Spring 2001. Ten diaries were selected for analysis here. Phases 2 and 3 are currently being conducted (Spring 2002). The aim of this initial analysis is to make an early contribution to the development of position-aware communication services. Some such services, and toolkits for their development, are already available (www.benefon.com, www.gate5.de, www.ericsson.com/mobilityworld).

This paper follows on from a previous diary study of rendezvousing, conducted in Spring 2000[1]. This pilot study (also based upon students) suggested that, if users *did* require an additional service to help them rendezvous, then a communication service was likely to be more frequently useful than, say, navigation or information services. The study found that approximately 5-10% of rendezvous caused a notable amount of stress and/or lost opportunity (rated at 4 or 5 out of 5, 5 = high). Given an average of 6 rendezvous per week, this proportion translates into a rate of about 1 'major problem' rendezvousing every 3 weeks – not obviously unacceptable, but also not obviously the best that could be achieved. The following reasons for rendezvousing problems were cited at least 10% of the time - transport delays, over-runs of previous activities, lack of information about other rendezvousers, lack of geographical information, promptness not being valued, and the spontaneous performance of additional tasks. The fact that rendezvousers very often had close personal relationships, and had met at that rendezvous point before, suggested that better communication was likely to be the most frequently useful at ameliorating the problem with such a range of causes. The previous study focussed upon the quality of outcomes, rather than 'costs incurred by users', because, with the benefit of hindsight, the conceptualisation and measurement of 'costs' was not the most appropriate. This paper complements previous work by focussing upon costs incurred by users, which are here measured in terms of usability ratings – satisfaction, effort, convenience, disruption, frustration, and social acceptability of communication services utilised. This paper also pursues the direction suggested by the previous study (position-aware communication), by attempting to identify the kind of rendezvous for which current communication services perform deficiently, and to suggest the kind of future communication service that has the potential to perform better. This paper, then, together with a previous one, prepare the ground for reporting of the full study.

The next Section outlines the design concerns considered in this analysis – choice of communication service, group size and time pressure (high task pacing, not necessarily under the user's control). A following Section describes the conduct of the diary study, summarises the results[1], and attempts to explain them. A final section considers the implications of the tentative findings for the design of position-aware messaging services.

3 Choice of Communication Service, Group Size and Time Pressure

The study reported here considers costs incurred by users associated with rendezvousing, in relation to choice of communication service, group size and time pressure. The potential importance of these concerns was suggested by diary entries from the pilot study (see Table 1).

The choice of communication service was studied, because the public already use various mobile communication services, and any new service will be introduced into this context. 'Competition' between traditional, fixed-access communication services has been the subject of user-centered research for many years. The perceived strengths and weaknesses of many of these services are summarised in [2]. For example, users like to telephone, because contributions are simultaneous and synchronous, voice tone help to express emotional content, and selected items of information can be conveyed quite rapidly. E-mail, in contrast, gives users time to reflect, but lacks the body language and contextual cues necessary for good emotional understanding. Wireless communication services are also beginning to be studied. For example, teenagers like to send text messages, despite the difficulties of entering the text, and of reading abbreviation-filled messages, because messages are sent almost immediately direct to the recipient's phone, the channel is private, and the service is cheap[3]. Deficits in rendezvousing performance are likely to emerge when the communication demands of the situation are not well satisfied by any existing service. Perhaps we should aim to develop a service whose strengths compensate for the weaknesses of existing alternatives.

Group size was studied, because it has been found to be a precondition for the use of some fixed-access services i.e. the user group must achieve a certain critical mass for a certain channel of communication to be adopted. For example, Whittaker compared successful and unsuccessful message databases in a large US corporation [4]. The successful (long-lived) databases concerned a topic relevant to a very large group, such as corporate issues, and for which other communications media were too difficult or unreliable. Interviewees said that they used a message database, because, although they didn't know who had the answer to a certain question, they thought that somebody using the database would. Databases that concerned a topic relevant to a smaller group, such as a specific project, were less successful, because potential users preferred to talk in person when they happened to meet, rather than use the database. Deficits in rendezvousing performance, then, may only emerge if, or until, a rendezvousing group exceeds a certain 'critical mass'.

[1] Statistical analysis of these interim results is not reported, so all conclusions are tentative.

Time pressure was studied, because it is a source of job stress, and a factor in many occupational health issues within the air traffic control, health care and manufacturing industries [5]. Time pressure, here, refers to the demand to perform tasks at a rate that exceeds a user's capacity to perform them, and that is not under the user's control. If time pressure also leads to stress in the context of rendezvousing, then it may also cause deficits in other aspects of rendezvousing performance, perhaps, user satisfaction, and disruption.

Table 1. Rendezvous Scenarios (based upon diary entries from a pilot study (Spring 2000))

Sport Club AGM. The organiser of a casual football team arranged for the team's Annual General Meeting to take place at "the pub in Paddington station" (a railway terminus in West London) - it was not a very formal AGM. Unfortunately, there are at least 3 venues in and around the station that fit this description. The manager's evening was constantly interrupted by various members of the squad telephoning him to ask, "Which pub?"

An Impromptu Invitation. Three flat-mates meet in a pub on a Friday evening. They decide to make a night of it, and go on to a club. Thinking it would be more fun if some of their friends came along, too, they text an invitation to two friends they think might be up for it (text is quick and cheap, and the pub is noisy anyway). One friend doesn't reply. The second friend declines, saying he's due to play pool elsewhere. Later, however, the second friend gets back to the flat-mates, saying his game is over, and he and his pool partner would like to join them, but where is the club? Actually in the club by this time, the flat-mates reply with rough directions, again by text, and the pool players turn up in their own time, after a few wrong turns.

Old Friends Get Back Together. A girl bumped into an old friend at Waterloo (another railway terminus, this time, in South London). The two had a coffee, got on famously and swapped telephone numbers so they could arrange a proper day out some other day. In due course, they arranged (via e-mail) a brunch at the old friend's house for a couple of days after that. When the day came, the girl rose late, walked to the end of the road and hailed a cab. When the driver asked "Where to?" she realised she didn't have any idea. The friends had known each other in Manchester, and the girl didn't have her friends new address in London. The cab circled, whilst the girl phoned her friend to get her address. Luckily she got through.

4 Diary Study

4.1 Method

The participants in this study were 5 male and 5 female students from the School of Computing and Information Systems, Kingston University.

Second and final year undergraduates registered for Human-Computer Interaction (HCI) modules on Computer Science and Software Engineering courses completed a diary as part of coursework exercises. The diary keeping procedure was identical for all students.

5 female students completed a diary and consented to its anonymous use here[2]. The diaries of 5 male students were then selected from a group of 46 possible diaries, to match the female participants as closely as possible in terms of age, marital status, number of children, ethnic origin and mobile phone ownership.

The keepers of the diaries had a mean age of 21 years, 8 months. 40% had been with the same partner for more than one year, and 60% were single. None had children, and 66% did more than 10 hours per week paid work in addition to their University studies. All participants owned a mobile telephone, and almost all had access to a fixed line telephone, plus a private e-mail account in addition to their University account (90% in each case). 50% used their mobile phone more than 10 times per week, and, of those with access to fixed-line phone, 40% used that more than 10 times per week. All participants lived within commuting distance of Kingston-upon-Thames, a suburb of southwest London, UK, and 80% of participants are non-White in ethnic origin. 80% owned a map of Greater London. None owned a GPS unit or a pager.

Participants kept diaries about their own rendezvousing behaviour for two, one week periods during the spring of 2001 - 12th-18th February (early in second Semester) and 11th-17th April (during Easter break).

At the outset of the study, all students were given an overview of future position-aware, computing and communications for mobile devices, and were introduced to the aims of the study and the obligations of diary keeping. To illustrate the kind of services that could be developed, participants examined fixed-access Web sites that provide map, transport and venue information (e.g. www.multimap.com, londontransport.co.uk). A possible future service was also described, which enabled each member of a small group to display on their mobile telephone the positions of other group members, superimposed upon an annotated map (see [1]).

One diary entry was made for each rendezvous event. Each entry comprised: (i) an open-ended, narrative description in the diary keeper's own words of what happened, and why; and (ii) the diary keeper's responses to a questionnaire, which asked for specific details of each rendezvous event. This questionnaire comprised 37 questions in total and breaks down as:

- Questions 1-6: the event (where, who, when, where, why of the rendezvous);
- Questions 7-11: outcomes (the additional stress and lost opportunity associated with attempts to meet at the time and place initially agreed);
- Questions 12-24: usage and usability of communication services *before* the rendezvous;
- Questions 25-37: usage and usability of communication services *during* the rendezvous.

Questionnaire responses were processed automatically by an Ocular Reading Machine, which generated a text file of responses, and a report of mean scores for each question. Recognition errors by the Reading Machine have not yet been ruled out by manual checking.

---

[2] The vast majority of diary keepers were male – this reflects the in-take to Computer Science and Software Engineering

4.2 Results

Task Type, Frequency, and Performance. Diary keepers took part in a total of 141 rendezvous - a rate of approximately 7 rendezvous per week/1 per day. The rendezvous were very often in locations at which diary keepers had rendezvoused before (61%), and included people with whom they had close relationships (immediate family 15%, close friends 64%, extended family 18%, acquaintances 17% and strangers 7%). On 44% of occasions, these rendezvous occurred as initially planned. Failure to meet as initially agreed very often resulted in stress and lost opportunity, but not very severe stress and lost opportunity. Failure to meet as initially agreed resulted in stress on 43% occasions, with a mean rating of 2.32 (1=low, 5 =high), and led to lost opportunities on 46% of occasions, with a mean rating of 2.11 (1=low, 5 =high). The usability ratings for the communication services used rendezvous were not bad (see Fig. 1). Effort, frustration and disruption are rated fairly low (1.6, 1.7, 1.7 respectively, 1=low) and satisfaction, convenience and acceptability are on the good side of middling (3.3, 3.3, 3.4 respectively, 5 = high).

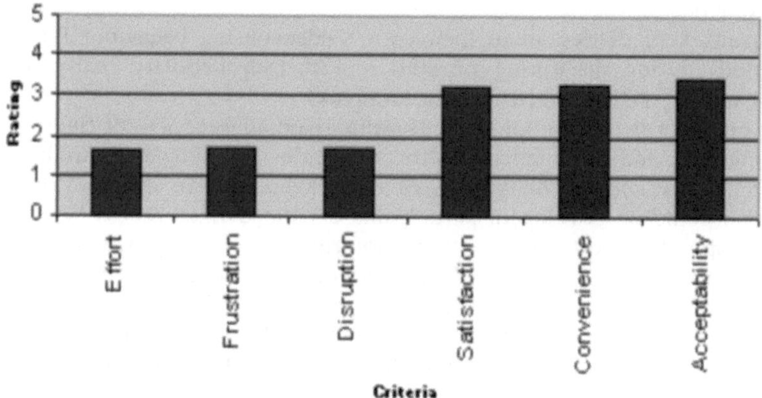

Fig. 1. Usability ratings for communication at any time before or during a rendezvous

The two most popular reasons for problems rendezvousing were overrunning of previous activities (cited on 36% of occasions), and disruption to mode of travel (26% of occasions) – if a rendezvous goes wrong, there is a fair chance the cause is either overrunning or travel disruption (see Fig. 2). On the other hand, a range of other reasons for also cited: success not valued (15%); the plan itself (11%); lack of travel information (9%); lack of information about others (8%); lack of geographical information (5%); and spontaneous performance of additional tasks (5%).

In general, these findings are similar to those reported for the previous pilot study. The similarity is reassuring, because diary keepers received little supervision, so there was little opportunity to check that diary keeping occurred systematically and as intended. However, the relative frequency of 'reasons for problems' is slightly different, probably due to interpretation errors in the pilot study. The questionnaire in the pilot study only elicited a brief free text description about reasons for problems, which was only categorised during subsequent analysis. In this study, the questionnaire asked respondents to tick a box for as many reasons as necessary from a list of nine options. The data in this study are probably the more accurate.

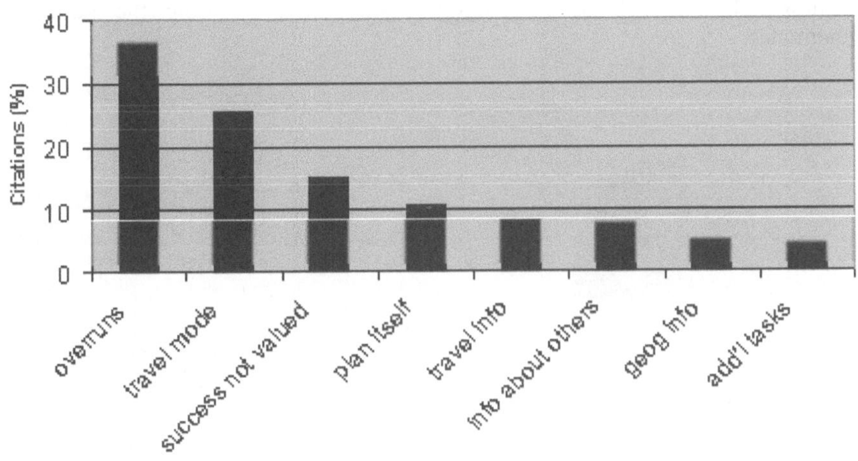

Fig. 2. Frequency of citation of reasons for failure to meet as initially agreed

Choice of Communication Service. Rendezvousers used various means of communication when rendezvousing. On average, they made 1.3 telephone calls per rendezvous, sent or received 0.8 text messages, sent or received 0.4 e-mails, and left 0.3 voicemails.

Diaries suggested that diary keepers exchanged position information about other rendezvousers under the following situations:

- another rendezvouser, 'B', has not shown up, does not answer the phone, or is making claims about their status that 'A' does not believe e.g. B says, "I'm half way – should be there in 10 minutes" but A suspects that "I have just left, and might make it in 15 minutes if the traffic is light" is nearer the truth;
- A is giving B directions to the rendezvous point over the telephone, or has been delayed him or herself, and wants to see whether B has been delayed as well, so that A and/or B are able to judge whether opportunities are about to be lost, and/or whether the original arrangements should be changed.
- A wants reassurance that everything is going to plan, so A gets B to confirm "B is having a safe trip";
- A and B are re-planning the rendezvous - agreeing a better time and place to meet, in the light of events that have unfolded since the original plan was made.

Groups Size. The mean size for a rendezvous was 3.6 i.e. the diary keeper, plus 2.6 other people. One-on-one rendezvous were more frequent than other sizes of rendezvous, but all sizes occurred roughly equally often. The size of the rendezvous was 2 (the diary keeper plus 1 other) on 33% of occasions, 3 on 20% of occasions, 4 on 13%, 5 on 16% and 6 or more on 18% (see Fig. 3).

Group size appears to be related to the diary keeper's familiarity with the rendezvous point – small and large rendezvous seem more likely to occur in familiar places than medium sized rendezvous (see Fig. 4 – notice the 'dip' for group sizes 4 and 5). Diaries suggested this may reflect subsequent activities that involve trying out 'new' places, such as a different restaurant or visiting the home of an acquaintance.

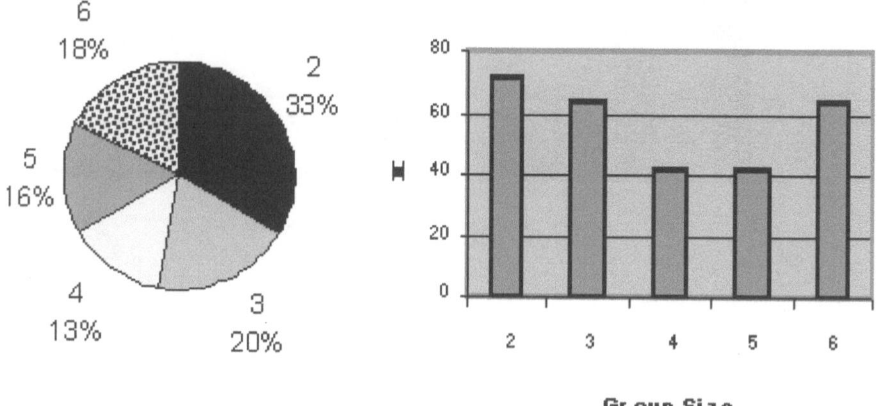

Fig. 3. Group Size

Fig. 4. Proportion of diary keepers who had rendezvoused at location before, by group size

Group size also appears to be related to the composition of the rendezvous in terms of the diary keeper's relationships with other rendezvousers. Acquaintances and strangers become involved when the group gets larger. This may partly explain the 'dip' in familiarity with location as group size increases. The location may not be familiar to the diary keeper, but it is familiar to the acquaintance. The diary keeper doesn't know the location, because he or she doe not know the acquaintance very well. Close friends are highly represented throughout, but particularly in medium-sized rendezvous (see Fig. 5). This may reflect the user population studied here – hanging out with their intimate circle is what students often do.

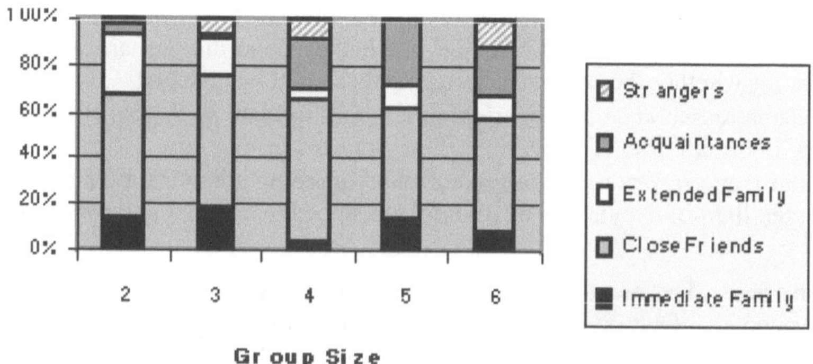

Fig. 5. The composition of a rendezvous for groups of different sizes

Choice of Communication Service and Group Size. The amount of communication appeared to change with group size - use of all services seems to increase with group size at first (until size equals 4), but then reduce as the rendezvousing group becomes even larger (see Fig. 6). This reduction may reflect the absence of contact details (note that inclusion of acquaintances and strangers increases with group size (see Fig.

5)), the fact that re-planning may not be a viable option for large group activities, and the fact that participants may travel to the rendezvous point in sub-groups (so one communication may in fact exchange information with a whole subgroup)(see Table 2). Indeed, smaller rendezvous sometimes occurred in advance of a larger one – a 'sub-group' may meet in advance to share a car, to take public transport together, or to ensure they met up successfully (say, in a pub), before meeting the whole group at the ultimate destination (say, a club, which may be full later on).

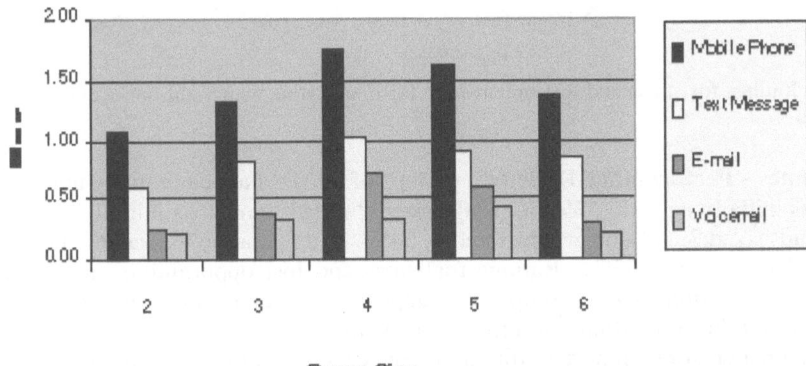

Fig. 6. Number of times a communication service was used before and during a rendezvous, by group size

Table 2. Rendezvous Scenarios (based upon diaries from the study reported here (Spring 2001))

The Holiday-Makers. A van load of holiday makers were late leaving home, and then hit heavy traffic as they began to pick up various other family members en route. One relative to be collected did not own a mobile phone, and was difficult to contact from the van anyway, as the van was travelling through the City of London (where lots of tall buildings to obscure the signals). They were getting later and later, but couldn't make sure the relative to be collected was ready to jump in when they came, or reassure him they were on their way. The holiday makers missed the ferry they had planned to catch, but, luckily, there was space on the next ferry, so they caught that. However, they didn't arrive at their host's house until passed midnight. It was all very stressful, but worked out alright in the end.

The Football Player. A football match was taking place at a ground the diary keeper's team hadn't played at before, in an unfamiliar part of town. The map distributed to the team in advance only showed the immediate vicinity of the ground. There was no route to the ground from familiar territory, and no indication where the car park was. The footballer didn't have the contact details for all his mates and was driving and navigating hard himself so as not to be late. Again, it was all very stressful, but, as it turned out, everybody arrived somehow and the delay did not affect the playing of the game.

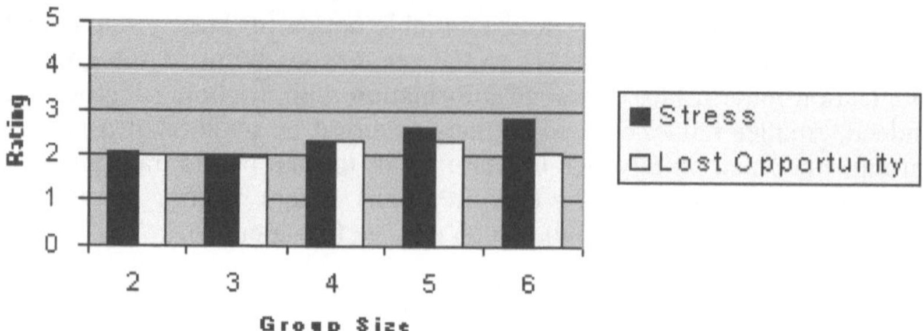

Fig. 7. Ratings for stress and lost opportunity for rendezvous which did not occur as originally planned

Outcomes - Performance Deficits? Groups of all sizes appear equally able to fail to meet as initially agreed. 53% of groups sized 2 failed to meet as initially agreed, 57% of groups sized 3, 58% of groups sized 4, 54% sized 5, and 56% sized 6. The average over all groups was 55%. Ratings for stress and lost opportunities are similar for small and medium sized groups, but appear to diverge for larger ones - stress increases but lost opportunity returns to the average (see Fig. 7).

The greater stress reported for large rendezvous is attributed to the subsequent group activity. Large rendezvous (6 or more) occurred for various reasons, including: playing sports (football), going on holiday (two families travel to, and stay at, a relative's house), hanging out with a crowd (going to pubs and clubs), and special occasions (birthdays/family reunions in restaurants). The rendezvous that returned high stress ratings (4 or 5, 5 = high) were the family holiday, and the football match (see Table 2). Both the holiday and football match are difficult to re-schedule for another time, are "all or nothing" - if the activity takes place without one of the participants, it is no longer the same event - and the lost opportunity affects many other people. Also, rendezvousers may not have the contact details for all participants in a large rendezvous, particularly acquaintances, strangers and the extended family.

Costs - Performance Deficits? Overall, the ratings for frustration, disruption, satisfaction, convenience, and social acceptability appear slightly worse (lower or higher, as applicable) for communication during a rendezvous than for communication before a rendezvous (see Fig. 8). Diaries suggested two, related reasons for this apparent difference.

Instant Communication. Communication becomes more frustrating and less satisfying during a rendezvous, because the need for instant communication increases as a rendezvous approaches, but network coverage and service availability is such that "en route to a rendezvous" is the kind of context that presents many barriers to instant communication. Diary keepers' negative comments frequently concerned network coverage (e.g. the phone doesn't work on the Underground, there was bad reception), and service availability (e.g. the other rendezvouser didn't own a phone, or had no e-mail where he was, the use of mobiles was not permitted in the hospital/aeroplane/university lab, or the device's battery had run out).

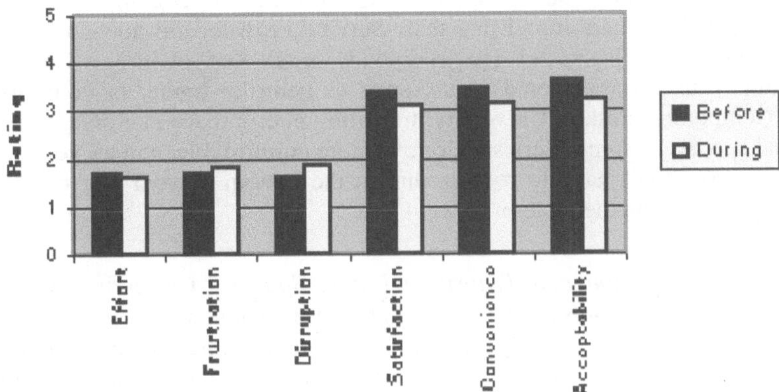

Fig. 8. Usability ratings for communication: before and during a rendezvous

Communication may be rated as less socially acceptable during a rendezvous, because rendezvousers who have instant communication sometimes feel the need to make a decision "right here, right now", and without contributions from others who do not have instant communication, although they feel bad for "leaving the other person out" or for "deciding without talking to them first". Also, users sometimes intentionally, or unintentionally, inhibit communication. For example, they may choose not to answer their phone, because they just don't want to anymore, turn the phone off accidentally, put it in silent mode and then not monitor it, or they may be on the phone to someone else. Users feel worse socially when they inhibited communication that needed to be instant (eg communication during the rendezvous it concerns), than communication that probably could have delayed, without any negative consequences (eg communication about the rendezvous before it has started).

Users seemed to like instant communication before a rendezvous as well. Diary keeper's positive comments most frequently concerned the speed with which services could be accessed, messages transmitted, and information grounded (i.e both parties knew the message had been received, read and understood). It was expected that users would take risks to achieve instant communication (e.g. to drive whilst talking on the phone). Other ways of achieving instant communication were surprising. For example, if one service was not available, some users would use another service immediately, rather than wait. And, if the mobile number was engaged, they would send a text message, or try them on their land-line, rather than try the mobile again a minute later. One user was also proud to have composed a text message whilst simultaneously talking on the phone.

Transiting Public Spaces. Communication may be more disruptive, and less convenient during a rendezvous, because at least one party is in, or moving through, public spaces when communication occurs. For example, users may be taking some form of transport (biking, in a car, on a train or bus), or walking towards the rendezvous point. When transiting public spaces, users may have to perform two tasks at once (walk and talk), and 'orient' to the spaces they are transiting through (perhaps, not disturbing other passengers, keeping an eye on your bag etc). Rendezvous points are often public spaces which, surprisingly, perhaps, are not necessarily good environments for communication. Some users had to go outside the building in which they had arranged to meet in order to get a signal or to hear what was being said. Others found that their Service Provider did not work at a location where somebody else's did (so everybody used that phone). Compared with inhabiting your own, personal space, such as being at home, in your office, or at 'your' (temporary) desk at University, transiting public space is more constraining. In personal spaces, as some users said, one is more comfortable, and all your services are to hand – you don't have to root around in the bottom of your bag for your phone, because it's lying on the table in front of you.

Instant Communication and Transiting Public Spaces. Currently, there are many technical, social, organisational and cognitive reasons why rendezvousers cannot always communicate instantly whilst at the same time make their way to the rendezvous point. To communicate instantly, users may have to stop transiting public spaces, or take detours. For example, a driver may have to pull over to the side of the road to answer the phone, or a visitor may have to walk to the nearest internet café to email someone to open the doors to the building he wishes to enter. To transit public spaces, users may have to stop or delay communication. For example, a user may have to cut short a phone call when their Underground train arrives. It's a pity, but instant communication "anywhere, anytime" for rendezvous isn't always a reality.

5 Implications for Position-Aware Communication

5.1 Connectivity and Availability, Not Bandwidth

Section 4 suggested two possible performance deficits. First, usability ratings for communications services were a little worse during a rendezvous than before, because the perceived need for instant communication is greater, and because transiting public spaces is a more constraining context for interaction than inhabiting personal spaces. Second, stress ratings may also be a little higher when the rendezvous is large, in part, because they involve more acquaintances. Weak connectivity, and non-availability of services were identified as the feature of technology that underlay these deficits. However, the motivation for this work was to seek applications for 3G networks i.e. greater bandwidth. Unfortunately, it is not obvious how greater bandwidth will improve performance - a broadband phone is no more likely to work on the underground or in a hospital than a narrowband phone. This study suggests that the work should change the kind of application it seeks to support - not multimedia applications (a typical 3G application), but applications that facilitate (almost) instant communication, despite weak connectivity, and that are likely to be available, despite all the constraints of transiting through public spaces, for example, applications that enable the communication of key information although the batteries are almost flat, or that ensure urgent notifications receive attention, although the recipient is busy travelling to the rendezvous point. Alas, it is not until the so-called Fourth Generation of wireless network that the seamless integration of personal, local and wide area networks from a user's point of view, and "multi-mode operation", is envisaged.

5.2 Position-Aware Messaging

Messaging is quite good at seeking instant communication, despite weak connectivity. It is an asynchronous channel of communication that operates upon a principle of 'store and forward' - the sender sends a message, when his or her device has a connection, and then the message is forwarded to the recipient when the recipient's device has a connection. The contrast is with traditional voice telephony – a synchronous channel of communication, which requires both mobile devices to establish a connection simultaneously. The latter is, in practice, less likely to be established instantly particularly when both parties on en route to the rendezvous point.

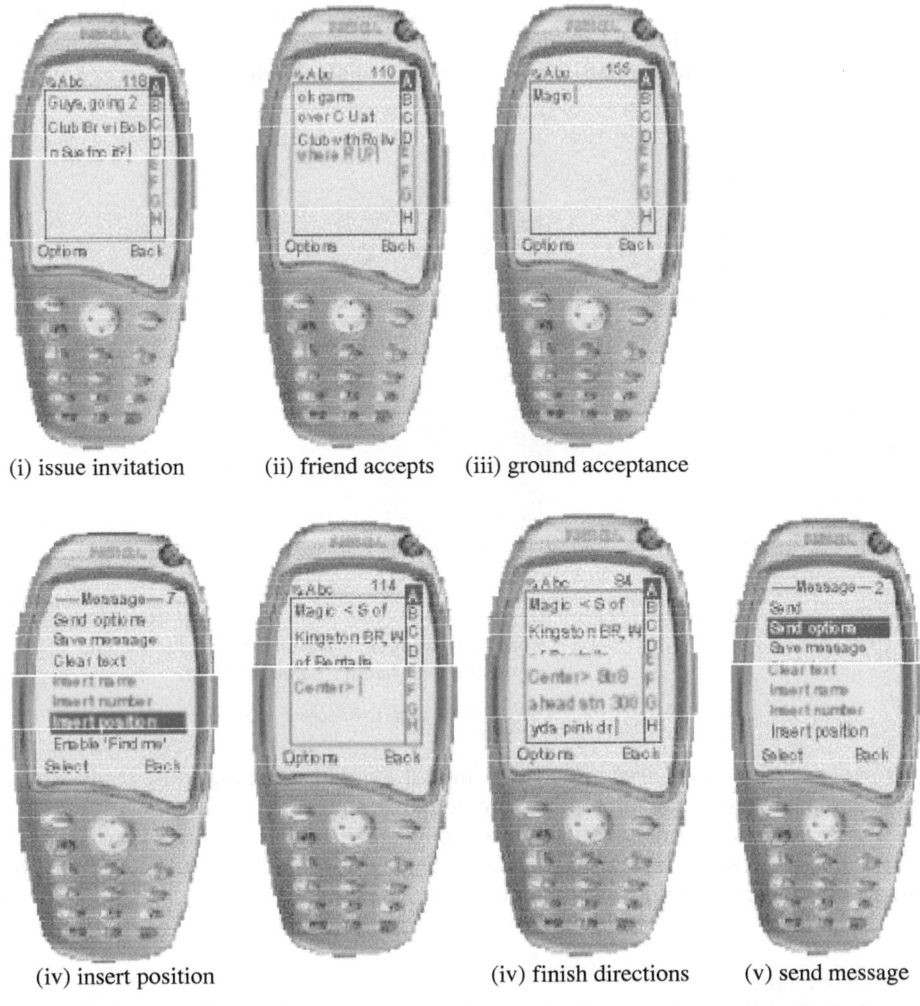

(i) issue invitation (ii) friend accepts (iii) ground acceptance

(iv) insert position (iv) finish directions (v) send message

Fig.9. A possible position-aware messaging service for second scenario in Table 1

In the design for position-aware messaging service below, basic messaging is augmented by giving users the option to include in text messages gazetteer entries/prepared directions to their approximate current position, and edit them before they are sent (see Fig. 9). The design is similar to the way users currently include names from their phone's address book. A messaging service may also be integrated with the monitoring service mentioned in Section 4, perhaps, by including in the message a permission for remote monitoring for a limited period of time. The aim of the design is to make text messaging a more appealing alternative to rendezvousers, who may have chosen to telephone to indicate their position but, for some reason, telephony is not available, or connectivity is lacking.

A messaging service that mimics group Chat session, rather than one-to-one messaging may make rendezvousing in large groups less stressful, because, although everybody may not have everybody else's telephone number, everybody could be invited to the Chat when the rendezvous is arranged (see Fig 10). (Recall larger groups include more individuals whose personal contact details are not shared by the whole group.)

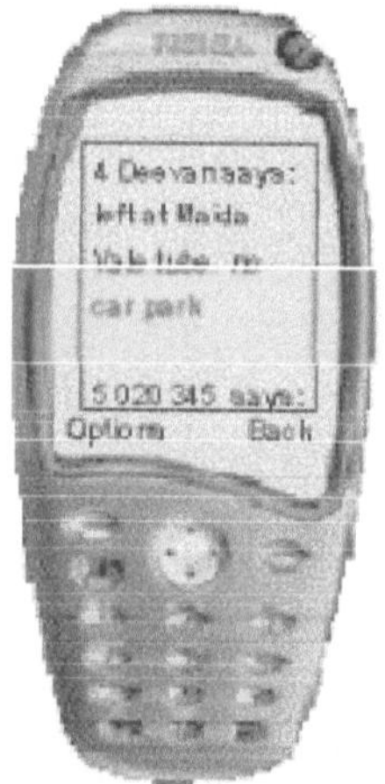

(i) log into chat session

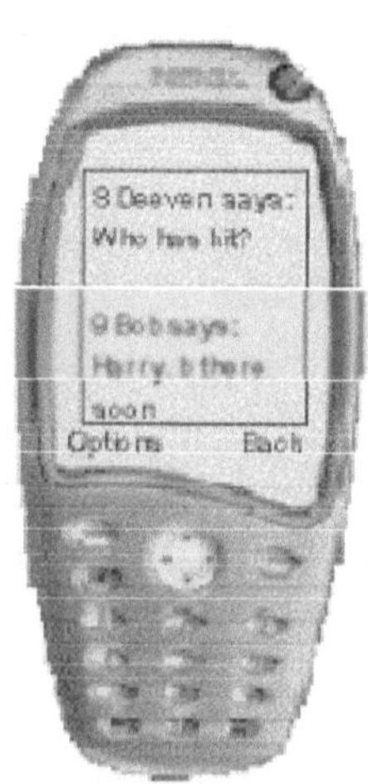

(ii) too busy to enter text himself, our user monitors the session when he gets the chance. As luck would have it, another driver had the same problem.

Fig. 10. A possible position-aware chat service for second scenario in Table 2

This paper has presented some interim results of Phase 1 of an on-going study whose implications for service development were not those sought.

Future work will analyse the data obtained in the full diary study more systematically. Many performance deficits may, or may not, exist, and many parameters and relationships may help to explain them.

Acknowledgement

The author would like to thank all those who helped out and participated at Kingston, especially Sue Watts and the undergraduate class 2001/2.

References

1. Colbert, M. A Diary Study of Rendezvousing: Implications for Position-Aware Computing and Communications for the General Public. Proceedings of ACM GROUP'01, Boulder, COL, USA, Sept 30 – Oct 3 2001. ACM, New York (2001) 15-23
2. Preece, J.: Online Communities: Designing Usability, Supporting Sociability. Wiley, New York (2001)
3. Grinter, R. E, Eldridge, M. A.: Y Do Tngrs Luv 2 Txt Msg?, Proceedings of ECSCW 2001, September 16-20, Bonn (2001)
4. Whittaker, T.: Talking to Strangers: An Evaluation of the Factors Affecting Electronic Collaboration. ACM CSCW'96, Cambridge, MA, USA. ACM, New York (1996) 409-418
5. Karasek, R, Theorell T.: Healthy Work. Basic Books, New York (1990)

Automated Multimedia Diaries of Mobile Device Users Need Summarization

M. Gelgon and K. Tilhou

IRIN/Ecole polytechnique de l'université de Nantes,
rue C. Pauc, La Chantrerie, 44306 Nantes cedex 3, France,
Tel: (33) 2 40 68 30 14 Fax: (33) 2 40 68 30 66,
marc.gelgon@polytech.univ-nantes.fr

Abstract. This paper addresses a still original issue and a solution that, while emerging from the pattern recognition point of view, certainly shares common goals with mobile HCI research goals. The contribution is at the crossroads of multimedia data analysis for content-based retrieval, and wearable computing. As users are acquiring multimedia content personal mobile devices, they are getting also undergoing information overflow. The problem of structuring the content into time-oriented meaningful episodes is addressed, and we argue that geographical location processing is crucial, as a complement to processing audiovisual material. A technique for model-based temporal structuring of one's trajectory during a day is presented, based on a Bayesian/MAP approach, that generates one or several summaries. Experimental results illustrate the applicative interest of the problem addressed and validates the proposed solution

1 Context and Objective

The work exposed in the present paper addresses needs related to information retrieval in personal mobile devices, especially aiming at those which purpose is to be continuously carried by their user, such as mobile phones and PDAs. Indeed, such small wireless terminals are undergoing considerable progress in their ability to capture, store, process and transmit data, notably of the multimedia form. Several models that incorporate image and audio acquisition, as well as geolocation of the users and that are targeting the general public, are currently being released, or are soon to be on the market.

The viewpoint of the paper is from the pattern recognition community rather than human-computer communication, but the goal of the work certainly coincides with HCI preoccupations and its purpose is cross-fertilization of ideas. As multimedia content is gathered, it progressively builds up a valuable memory of ones's life, which can be later searched, whether for practical, emotional, or much-sought fun pass-time purposes. A means of access that could prevail is the familiar time management software available on PDAs, with calendar-like views on which multimedia content could be "annotated". Yet, for the owner to quickly, reliably and comfortably retrieve a well-defined piece of information in

F. Paternò (Ed.): Mobile HCI 2002, LNCS 2411, pp. 36–44, 2002.
© Springer-Verlag Berlin Heidelberg 2002

a large collection, or merely browse in it to get an overall idea, the content ought to be organized, at least partly automatically. More precisely, bearing in mind the the stringent input and display constraints of mobile devices, we believe a crux is the ability to generate compact summaries of one's activities during a given period, that suit visualization and browsing needs, including the above-mentioned need with the straying-type of browsing. Our focus in on proposing techniques for structuring the collection along the time-axis, via analysis of the multimedia data.

Although the more general problem of content-based retrieval of audiovisual material has largely being addressed for several years (e.g. TV summaries[20,9]), understanding of needs and working out of solutions dedicated to the the particular context tackled here remains rather open. Early exploration of the automated diary concept was proposed in [13,18], yet leaving out multimedia. Recently, proposals have been made towards organizing one's personal image collection, based on image features [14]. Towards organization of video collected from a wearable computer, a summarization technique was presented in [2], which rates the interest of video shots via measurements on brain waves. The structures that are extracted from the personal data collection (e.g. through statistical learning) are utilized here for navigating within the collection, but they may also benefit other purposes: context-awareness [1] and prediction [6]. For instance in [15], the authors propose automatic discovery of the user's frequented places and assignement of location-sensitive reminders. Alternatively, a technique to learn and infer location from image sequences acquired from a worn camera is presented in [21], so as to avoid the difficulties of accurate indoor localization. Interestingly, a paper is being published in parallel to the present [3], that addresses very similar goals (automatic multiscale determination of one's pertinent locations), although aiming at prediction rather than browsing the past.

In [8], we proposed to determine, from wearable video, periods when the user had met people, by coupling probabilistic face detection with HMM, leading to visual time-oriented summaries. In the present paper, we examine the unsupervised generation of summaries from geographical position sequence recordings. The results section shows how these contributions may be combined. We advocate the use of location because it appears realistic and useful. Indeed, it is not restricted to wearable computing platforms, but is also to be available on less intrusive, lower-price, ordinary mobile phones, whether with audio-visual capabilities of not. Further, it is relatively reliable, compared to most information that may be extracted from wearable images and audio. With time and identity of people the user met, it is likely to be the major search/browse criterion for searching one's data collection, and it is indeed quite easy to express for the user (e.g. compared to image-based queries). Also, from our experiments, location accuracy fades more slowly in one's memory than alternative search/browse criteria, especially in the long run. Finally, the use of positioning only places modest storage and processing requirements.

Positioning technologies, surveyed in [7], may be GPS (possibly GSM network-assisted GPS), GSM basestation triangulation-based techniques such as E-OTD or its WCDMA successor, or cell identifier; there is likely to be time-switching or fusion among these sensors, depending on local availability, reliability or subscription. Besides, indoor performance is under encouraging R&D [10] and post-processing is likely to be performed (e.g. in the form of enhanced Kalman filtering). Due to this improving and unstable situation, the present work is not dedicated to a particular technology or noise characteristics, but remains on the white Gaussian noise assumption (whereas actual corruption on measurements is likely to be time-correlated and its variance would be time-varying). Still, the statistical framework employed is flexible enough to introduce more elaborate noise models. Let us finally mention that in our application, the sampling rate would typically be a few seconds.

In the remainder of the paper, we formalize the problem and outline its solution (section 2), providing developments in sections (2.1) and (2.2). Section 3 presents experimental results, while section 4 summarizes the contribution and suggests further work.

2 Statistical Trajectory Segmentation

The general problem may be stated as summarizing the observed spatio-temporal trajectory $o = \{o_t = (x_t, y_t)\}, t = 1, .., T$ of the user. In fact, the sub-goal tackled in this paper consists in partitioning the time-series into time segments, represented by the sequence $s = \{s_t\}, t = 1, .., T$ of hidden state labels, in which s_t indicates to which of the M models o_t is assigned. The process should be as data-driven (unsupervised) as possible, regarding M the number of segments. With [16], and in contrast to [12], we consider that the degree of homogeneity within a segment should also be as much as possible estimated from the data, rather than be set a priori. Besides, we conjecture that segmentation of the trajectory according to a piece-wise parametric model provides a reasonnably meaningful account of the episodes that compose the period to be processed, at least for short periods (e.g. a day). Let $\Theta = \{\Theta_k, k \in \{1, \ldots M\}\}$ denote the set of parameters vectors. Linear models are assumed in the remainder of the paper. We consider batch processing (that would occur e.g. every evening), but the technique proposed can be made incremental. Given these requirements, the problem comprises the two following aspects:

- the classical interwoven issues of unsupervised data clustering are gathered: estimating the model parameters, associating the data to the models, and determining the adequate number of models.
- the sequentiality of measurements should be introduced, so as to guarantee time-connexity of data assigned to a model, and limit spurious models due to very noisy measurements.

Denoting by S be the set of possible partitions, the chosen optimality criterion is the maximum a posteriori (MAP) label configuration, defined by (1).

$$\widehat{s} = \arg\max_{s \in S} p_\Theta(s|o) = \arg\max_{s \in S} p_\Theta(o|s)p(s) \tag{1}$$

The MAP criterion is a Bayesian estimator that transforms the search into an optimization problem. In our case, the search is conducted as follows: the search space is partitionned into subspaces $\{S_k\}, 1 \leq k \leq T$, where S_k contains all segmentations composed of k segments. Search within each subspace is conducted independently, as detailled in section 2.1. The solutions obtained for these various segmentation complexities are compared (section 2.2) and one or several global solutions are finally selected. For our application, we might not be only interested in a single, but in extracting several pertinent segmentations of the trajectory, especially if they can all summarize well the data, with various degrees of granularity, so as to supply the user with several alternative visual representations at multiple temporal scales, i accordion summarization [5].

2.1 Search for an Optimal Segmentation within a Subspace S_k

The search for a maximum likelihood estimate of Θ is conducted by means of the Expectation-Maximization (EM) algorithm [17]. It is a well-known iterative scheme which, provided with some initialization for Θ, replaces the (ignored) data-to-model assignments by their statistical expectation *(E step)*, given the current parameter values. Therefrom, new parameter values may be computed *(M-step)*, until (garanteed, but possibly slow) convergence. In our case, still, the data-to-model assignments are more restricted as in common clustering situations, as data assigned to a model should be connected in time. In terms of our model, the probabilities of assigning a data element to a model are not independent among the data set. This structural constraint on the search space is dealt with by setting a hidden Markov model structure on the label sequence that constraints, via the transition matrix, connectivity of identical labels. Conditionally to models parameters Θ and segmentation complexity k, the MAP sequence is straightforwardly estimated with the Viterbi algorithm (hereunder 'MAP step'), which replaces our E step in the EM algorithm. As these models parameters Θ are unknown, iterations between MAP and M steps, which is known to convergence fast in practice, could be a solution but, because of the "hard" data-to-model assignements it carries out (in constrast with EM), it is known to lead to poorer parameter estimates than EM [4], which we confirmed experimentally. To combine EM and MAP-M so that their respective shortcomings are addressed, we employ the following three-phase scheme *(algorithm 1)*, in which results of each phase initialize the next one.

2.2 Model Comparison and Selection

The attempt to fit (e.g. linear) models to the data is to be viewed as the identification of the major trends and changepoints in the trajectory, that serve the purpose of building a summary. A difficulty is that the more complex (flexible) the overall model proposed, i.e. the more segments we allow, the better it can fit to the data (e.g. in terms of likelihood), eventually undesirably fitting small local deviations due to noise. There exist principled approaches to defining the trade-off between goodness-to-fit vs. segmentation complexity, a problem well-known

---

$\widehat{s^k_{init}} \leftarrow$ `segmentation in` k `segments of equal length`
$\Rightarrow$ a reasonnable initial guess
Do `MAP-M phase (until convergence)`

- `M-step: computer a maximum likelihood estimate of model parameters` Θ `(linear regression)`
- `MAP-step: estimate a MAP segmentation`

Loop
$\Rightarrow$ fast computation of rough parameter estimates, that supply the EM phase just below with a good initialization
Do `EM phase (10 iterations)`

- `M-step: estimate a maximum likelihood of model parameters` Θ `(weighted linear regression)`
- `E-step: compute the expectation of data-to-model assignements`

Loop
$\Rightarrow$ provides a better initialisation for the the MAP-M phase, thus leading to a more reliable segmentation
Do `MAP-M phase (until convergence)`

- `M-step: maximum likelihood estimation of model parameters` Θ `(linear regression)`
- `MAP-step: estimate a MAP segmentation`

Loop
$\Rightarrow$ final segmentation $\widehat{s^k}$ with k segments and likelihood $p(o|\widehat{s^k})$.

Algorithm 1.

---

as Ockham's razor. The statistical view chosen is a well-founded framework for handling this unsupervised segmentation issue. A closely related, alternative, framework to handle this same issue is that of the information-theoretic viewpoint [11,19]. Definition (1) in fact naturally exhibits this property. Let S^k be the set of all segmentations composed of k segments. All segmentation within S^k being*a priori* equal , the prior probability distribution $p(s^k)$ is flat and $p(s^k) = 1/card(S^k)$. This can be interpreted as the general Bayesian self-penalizing property of model complexity. Stirling's approximation conveniently re-expresses this result as $ln\, p(s^k) \simeq Tln(t) - (T-k)ln(T-k)$.

The search for the optimal segmentation according to (1) can thus be carried out as defined in *Algorithm 2*. The outcome of algorithm 1 is a set of candidate segmentations (one per segmentation complexity) with their *a posteriori* probability.

An advantage retained from the Bayesian inference viewpoint is that, once provided with a posterior distribution, one is free to take decisions that suits one's dedicated needs, including extracting several pertinent solutions. The tech-

```
For all possible segmentation complexities k ∈ [k_min, k_max]
Do
search for the optimal segmentation ŝ^k using algorithm 1, this provides
ln p(o|ŝ^k).
compute the complexity penalty ln p(ŝ^k)
compute the a posteriori probability ln p(ŝ^k|o) = ln p(o|ŝ^k) + ln p(ŝ^k) + const.
Loop
```

Algorithm 2.

nique proposed for extracting *a set of segmentations* is to retain, among the solutions found in each of the subspaces, those which are local maxima, along the k-axis, within this set of solution. That is, if the trajectory inherently exhibits multiple levels of granularity (e.g. one moves between cities, but also within these cities), these levels can automatically be determined. Besides, the number of pertinent segmentations that can well explain the data is also automatically determined by the data, and is at least one. Let us finally underline that parametrization of the technique is low.

3 Experimental Results

We report experimental results from GPS measurements, on which white Gaussian noise is added to simulate alternative sensors. The period reported corresponds to activities during one day (going from home to two stores within a shopping center, then to work, temporarily to some university closeby, then to a library and finally back home. Fig 1 shows the corresponding trajectory. The two segmentations found to be relevant by the proposed scheme are overlaid (with some offset for readability), and both correspond to desirable outputs. The (minus) likelihood (penalized by segmentation complexity) for the optimal solution in each subspace S_k are plotted vs. k. Two local minima are identified. k_{max} is limited to 15, as finest summaries would anyway be unsuitable for the small GUI targeted. Fig. 2 shows a calendar-view displaying in parallel summaries from the present technique (left) and the ones based on face-detection (right). Images help summarize the face-based segments [8]. Interestingly, besides provding a good overview of the day, as the user has partial memory of the day, she can often identify the location via the face of the person that was met there, or vice-versa.

4 Conclusion

This paper has presented a contribution in a still rather open applicative field: personal multimedia data analysis for automatic annotation of electronic diaries. Arguing in favour of the use of unsupervised structuring of one's trajectory towards compact summaries to be displayed on a PDA screen, a statistical technique is proposed that, thanks to its Bayesian/MAP principle, provides an

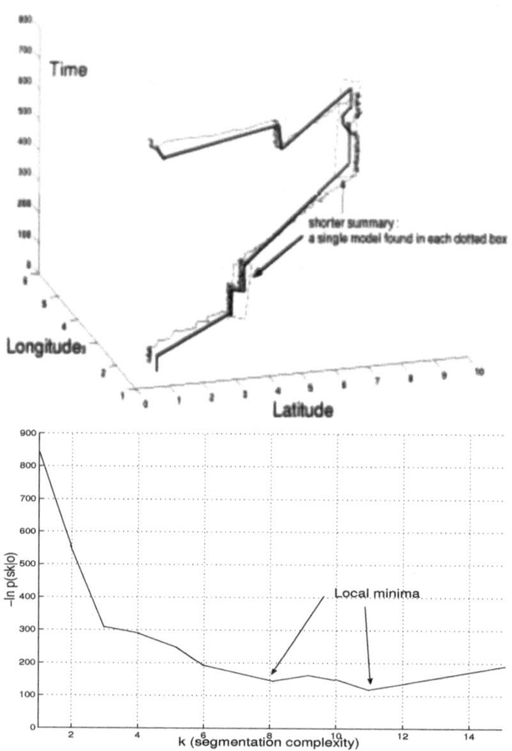

Fig. 1. Above:: spatio-temporal trajectory of the user, with the two-level summaries determined (the finest scale in bold lines; the coarsest resembles the finest, safe two grouping represented by dashed boxes).**Below:** penalized likelihood of candidate optimal segmentation with various complexities, and indication of the selected segmentations.

automatically determined number of time-oriented segmentations. This formalism would also easily enable GUI-experts, by rating the desirable complexity of summaries, to define and integrate a more elaborate prior probability distribution on the segmentation complexity. Appropriate naming of segments could partly come from a Geographical Information System, yet opening catchy issues. Finally, for longer durations, while the system is well able to separate "in town" and "out of town" remote trips, it has limitations in its "compression ability", due to the current summarization criterion. We are currently examining complementary criteria, based on detection and representation of periodicity, of usual vs. unusual.

Acknowledgements

The authors are grateful to J.Jomppanen, A.Myka and J.Yrjänäinen from Nokia Research Center, Nokia corp., Finland, for discussions related to this work.

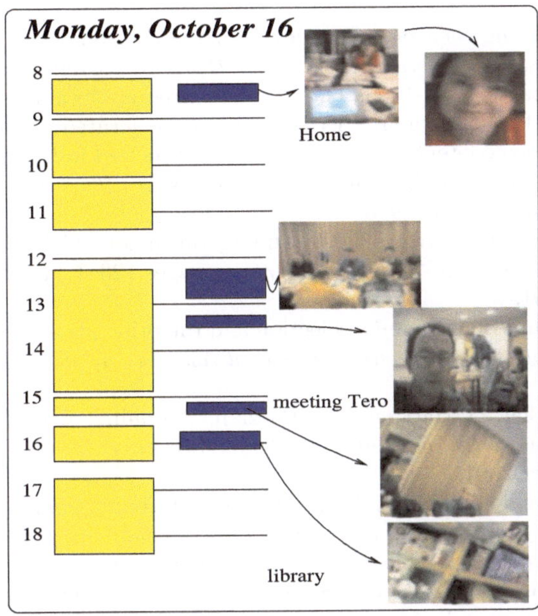

Fig. 2. Calendar-type of view constructed from the summaries built (the fine-scale location-based summary is chosen, left, and "meeting"-based summary (using face detection), right). Vertically is time of the day. Upon selection of a meeting, the face of the person met constitute an iconic summary. Images can similarly help indicate the location.

References

1. G.D. Abowd and E.D. Mynatt. Charting past, present and future research in ubiquitous computing. *ACM Trans. on Computer-Human Interaction*, 7(1):29–58, March 2000.
2. K. Aizawa, K. Ishijima, and M. Shiina. Summarizing wearable video. In *IEEE. Int. Conf. on Image Processing (ICIP'2001)*, pages 453–457, Thessaloniki, Greece, september 2001.
3. D. Ashbrook and T. Starner. Learning significant locations and predicting user movement with gps. In *To appear in IEEE Int. Symp. on Wearable Computer*, oct 2002.
4. C. Biernacki, G. Celeux, and G. Govaert. Strategies for getting the largest likelihood in mixture models. In *ASA Joint Statistical Meeting JSM 2000, invited paper*, Indianapolis, USA, 2000.
5. O Buyukkokten, H. Garcia-Molina, and A. Paepcke. Accordion summarization for end-game browsing on pdas and cellular phones. In *Proc. of ACM Computer Human Interaction (CHI'2001)*, Seattle, Washington, USA, 2001.
6. B. Clarkson and A. Pentland. Predicting daily behavior via wearable sensors. Technical Report Vismod TR 451, MIT, July 2001.
7. G.M. Djuknic and R.E. Richton. Geolocation and assisted GPS. *IEEE Computer*, pages 123–125, February 2001.

8. M. Gelgon. Using face detection for browsing personal slow video in a small terminal and worn camera context. In *IEEE. Int. Conf. on Image Processing (ICIP'2001)*, pages 1062–1065, Thessaloniki, Greece, september 2001.
9. M. Gelgon and P. Bouthemy. Determining a structured spatio-temporal representation of video content for efficient visualisation and indexing. In *5th European Conference on Computer Vision (ECCV'98), LNCS 1406-1407*, pages 595–609 (II), Freiburg, Germany, June 1998.
10. T. Haddrell and T. Pratt. Understanding the indoor GPS signal. In *Institute Of Navigation ION - GPS 2001 Conference*, pages 123–129, Salt Lake City, USA, September 2001.
11. M.H. Hansen and B. Yu. Model selection and the principle of minimum description length. *Journal of the American Statistical Association (JASA)*, 96(454):746–774, 2001.
12. E. Keogh, S. Chu, D. Hart, and M. Pazzani. An online algorithm for segmenting time-series. In *IEEE International Conference on Data Mining*, Silicon Valley, USA, December 2001.
13. M. Lamming and M. Flynn. Forget-me-not : intimate computing in support of human memory. In *Procs. of Friend21: Int. Symp. on next generation human interface*, pages 124–128, Meguro Gajoen, Japan, 1994.
14. J. Luo, A.E. Savakis, and A. Etz, S. Singhal. On the application of Bayes networks to semantic understanding of consumer photographs. In *IEEE. Int. Conf. on Image Processing (ICIP'2000)*, pages 802–807, Vancouver, Canada, september 2000.
15. N. Marmasse and C. Schmandt. Location-aware information delivery. In *IEEE Int. Symposium on handheld and ubiquitous computing*, pages 157–171, Bristol, U.K., sep 2000.
16. J. Oliver and C. Forbes. Bayesian approaches to segmenting a simple time series. In *Proc. of Proceedings of the Econometric Society Australasian Meeting, C. L. Skeels (ed), Canberry*, 1998.
17. R.A. Redner and H.F Walker. Mixture densities, maximum likelihood and the EM algorithm. *Society for Industrial and Applied Mathematics - SIAM Review*, 26(2):195–239, 1984.
18. B. Rhodes. The wearable rememberance agent : a system for augmented memory. *Personal Technologies Journal - Special issue on wearable computing*, 1(4):218–224, 1997.
19. S.J. Roberts, R. Everson, and I. Rezek. Minimum entropy data partitioning. *Proc. International Conference on Artificial Neural Networks*, 2:844–849, 1999.
20. M.A Smith and T. Kanade. Video skimming and characterization through the combination of image and language understanding techniques. In *Proc. of IEEE Conf. on Computer Vision and Pattern Recognition*, pages 775–781, Puerto-Rico, juin 1997.
21. T. Starner, B. Schiele, and A. Pentland. Visual contextual awareness in wearable computing. In *IEEE Int. Symp. on Wearable Computing*, 1998.

Personal Location Agent for Communicating Entities (PLACE)

Justin Lin[1], Robert Laddaga[1], and Hirohisa Naito[2]

[1] MIT Artificial Intelligence Laboratory,
{JL79, rladdaga}@ai.mit.edu
[2] Fujitsu Laboratories Ltd,
{naitou}@jp.fujitsu.com

Abstract. Traditionally, location systems have been built bottom-up beginning with low-level sensors and adding layers up to high-level context. Consequently, they have focused on a single location-detection technology. With sharing of user location in mind, we created Personal Location Agent for Communicating Entities (PLACE), an infrastructure that incorporates multiple location technologies for the purpose of establishing user location with better coverage, at varying granularities, and with better accuracy. PLACE supports sensor fusion and access control using a common versatile language to describe user locations in a common universe. Its design provides an alternative approach towards location systems and insight into the general problem of sharing user location information.

1 Introduction

The pervasive computing movement calls for a wealth of context information in order to perform the appropriate tasks for each person in each of their possible situations [1,2]. Of this context, location has become one of the most popular topics among research groups. This increase in location technologies has produced a large number of mobile (location) applications that fall under two categories: (1) users performing context-based tasks with his/her own location and (2) users performing context-based tasks with the locations of other users. The first category includes reminder services [3,4], navigational services [5,6], location-aware information delivery [4], environment customizations [7], location-based tour guides [8,9], and general context engines [10]. However, the second category remains largely unexplored. Some research groups have ventured into this domain (e.g. ActiveMap [11]), but possibly due to the daunting privacy concerns, there has been relatively little research in this area. In this paper, we demonstrate the potential behind sharing one's location with others, given a location system infrastructure designed specifically for this purpose.

In addition to location applications, the exploding interest in location context has resulted in a multitude of location detection technologies developed by many different groups, each of which seems to contribute something unique and interesting to the field [12]. Individually, these systems are useful, yet each faces the

F. Paternò (Ed.): Mobile HCI 2002, LNCS 2411, pp. 45–59, 2002.
© Springer-Verlag Berlin Heidelberg 2002

limitations of their methods. Working together, they can obtain location information with better coverage, at varying granularities, and with better accuracy – three qualities particularly useful in sharing user locations. Consider how people generally solve such problems. When we need to locate some object or place, we refer to several sources of location information, by asking multiple people, referring to multiple maps, using GPS receivers, and other methods. In doing so we transparently interpret multiple coordinate systems in terms of each other, resolve disparate symbolic references, and gauge the probabilities associated with each bit of evidence. These things that we do with little thought or effort are incredibly difficult for our computer systems to do. In light of this complexity, we can see why there has been little effort to bring the unique and interesting features of various location systems together into one collective system.

This paper aims to provide insight into an alternative approach toward user location systems, namely building an infrastructure with location-sharing in mind from its conception. We discuss the Personal Location Agent for Communicating Entities (PLACE), an infrastructure that utilizes (1) a semantic representation of location as a means to attain fusion of sensors at varying granularities for better coverage and accuracy, and (2) an intermediary software agent between location devices and location services that performs sensor fusion and access control on behalf of the user. We begin with a brief generalization of current location systems and the details and motivations behind our general design. Then, we move to a discussion on how to achieve a common universe of locations to allow for clear communication of location information between a world of entities. Finally, we provide brief overviews of applications that capitalize on our design, after which we give some concluding thoughts.

2 Related Work

There exists a multitude of location systems today. Each system typically chooses a location representation that suits the technology used in determining one's location. For example, the Global Positioning System (GPS) uses multiple satellites that provide specially coded signals to a GPS receiver, which then can use these signals to compute its location in latitude-longitude coordinates. Further, the Active Badges location system places a base per room that detects via infrared communication when badges enter the room, thereby providing a symbolic location representation, namely which room a user is in [14].

Each location system generally has both unique features and limitations in terms of technology, accuracy, scalability, and cost. Because these systems are usually developed in separate research facilities with different views on the ideal location system, there appears to be little effort to have two or more location systems coexist and collaborate with each other. Consequently, applications built to utilize location technologies have generally limited themselves to one location system, as shown in Fig. 1. Thus, most applications that exist today face the limitations of their chosen location systems.

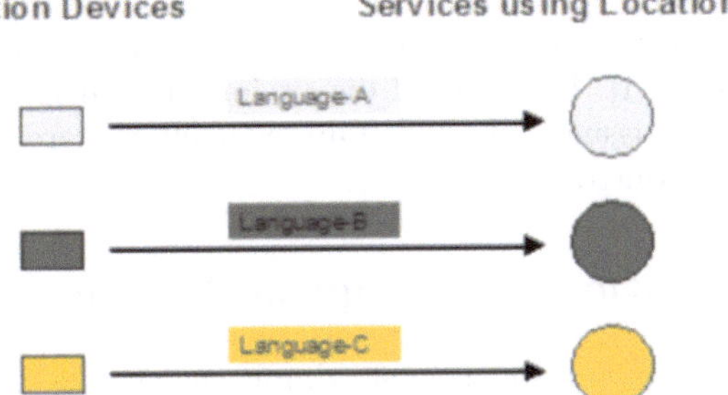

Fig. 1. Conventional Method

Korkea-aho and Tang [15] have made a similar observation that each location application seems to create its own representation. They aimed to "create a simple lowest common denominator data set that as many location information sources and applications in the Internet as possible could use." The result was called the "Common Spatial Location Data Set," consisting of geodetic latitude, longitude and altitude, accuracy and time of measurement, speed, direction, course, and orientation. This is a sensible solution for location systems in general; however, in the context of pervasive computing and describing locations of users, this geographic location representation would not be ideal. People generally tend to reason about locations as semantic places rather than as coordinates.

Furthermore, various specifications have been proposed. First, as specifications for describing geographical objects, maps, etc. G-XML [16] and GML [17] should be noted. They create a standard for encoding geospatial data into XML documents. Likewise, POIX [18] and NVML [5] are specifications for describing positions of moving and static objects and encoding this information into XML. Both sets of specifications are based on common location expressions and add additional vocabulary depending upon their objective. There are some ongoing activities attempting to standardize coordinate representations for location expressions. However, they have not attempted to standardize symbolic representation for expressions of location information.

SignalSoft Corporation [19] has also attempted to utilize multiple location sources in their product suite. The Location Manager receives data from multiple location sources and delivers the data to location applications. They have successfully created a commercial product but restrict themselves to cell-phone-based location detection technologies. Furthermore, these cell-phone-based technologies use a spatial location representation.

Multisensor data fusion is a field that has a huge amount of attention and numerous published results [20]. However, most work in the field is based on a signal processing, bottom up approach. We favor an approach that proceeds top down from a relevant decision problem, as an approach to filter away much of the irrelevant ambiguous or conflicting data.

Access control is another highly developed and explored area, with a great deal of literature. For specification of access control for purposes of maintaining military secrets, see [21]. For general discussion of security, confidentiality and information integrity issues, see [22]. Our own efforts are more influenced by Role Based Access Control, [23].

3 Sharing Location – Conceptual Requirements

In the context of this paper, location sharing is the sharing of information about locations of persons for the purpose of performing some task effectively, such as joining them for a meeting, delivering a package, predicting their near future behavior, and determining location dependent privileges. Sharing location information can be categorized into: (1) systems sharing location information with other systems, (2) systems sharing location information with humans, and (3) humans sharing location information with other humans. Wherever machines (systems) are involved, we need to be concerned about interoperable data representations. Wherever humans are involved, we need to be concerned about issues of privacy, and the ability to present information in an understandable manner. Note that humans may well have privacy concerns about data residing physically (persistently) on certain machines. Further, even in machine to machine interactions, humans may need to diagnose and debug failures, and hence need to understand data involved in the interactions.

3.1 Common Language

Ideally, all location devices and services would use a common, universal location representation and language. Unfortunately, this is not the reality. The vast contrast between location representations that exist today is not merely a consequence of philosophical differences, but rather of the inherent unique features that are specific to each location technology. GPS provides latitude-longitude coordinates while Active Badges provide symbolic spaces, but both technologies prove to be useful in different scenarios.

Our goal is to allow for bidirectional translation to and from a common language that expresses location in a representation chosen and designed specifically for user-location services. This language serves as a standard, if you will, between communicating entities, but bidirectional translation provides versatility when required. Because a universal language does not exist today, communication of location information is restricted to be from the location device to the location service (again, designed specific to the location device). With a common user-location language, location information can be freely exchanged between all communicating entities, device to service and service to service.

3.2 Common Map of Locations

With a common language, entities can communicate location information, but cannot necessarily fully understand each other without having the same map of locations. Upon finding out that a user is in "John Hancock Building," in order to be able to realize and use this location (e.g. locate on a map, get directions, build higher-level context), one must have at least a similar understanding of "John Hancock Building" as the provider of the information. For the understandings to be similar enough, we require that the processes by which each user binds an object or location to the symbol, result in roughly the "same" object or location.

3.3 Sensor Fusion

Deducing a user's location is an attempt to capture some information about the world. The multitude of location technologies that exist today provides many paths one can take to obtain this information. However, each location technology has its limitations; this motivates us to combine technologies and to "take all that we can get." With this approach comes the problem of sensor fusion.

Perhaps an even more important purpose of sensor fusion and multiple location technologies is making location systems more robust to environmental conditions. Any given location system can fail for any number of reasons, and it will often be the case that backup systems require multiple sources to recover the coverage or accuracy of the primary system. Multiple systems for providing location information, and a flexible architecture that allows for arbitrary combinations of information from disparate sensor sources and other information, is central to a more robust and self-adaptive architecture [24,25].

3.4 Access Control

The presence of an intermediary between location devices and location services provides us with a convenient place to perform access control. Traditionally, many location applications have not implemented access control because they do not involve sharing location information with other users. Those that do involve sharing of location information are often based on the user carrying a location device that the user can detach or disable, and thus address the access control issue (e.g. [11]). This sidestepping of access control issues does not adequately deal with common situations in which the user wants some people to have access to location information under certain circumstances. It also fails to deal adequately with non-personal location devices such as cameras, and with location information needed for emergency reasons. A less binary solution, and one that is less device dependent, is called for. Just as P3P [13] seeks to offer users complete control over their personal information on Web sites they visit, a location system must offer users complete control over their location information.

4 PLACE

We propose a design and implementation exploring the possibilities of a unifying location system infrastructure that allows user-location to be commonly understood among communicating entities. Our design goal is to create a user-location infrastructure supporting *sensor fusion* and *access control* using a *common versatile language* to describe locations in a *common universe*.

The ultimate goal behind our unifying location system is to allow for all communicating entities to fully understand each other, with respect to location information. In order to accomplish this, these entities must all share a common universe of locations. Sensor fusion cannot operate unless sensors are able to collaborate. Users cannot perform access control based on location context without precisely matching their perceptions of the locations. Communicated location information carries meaning exactly to the extent that there is a common universe of locations shared by the communicating parties.

In the next few sections we describe our software environment, and our design for language, representation, access control, and a common universe of locations.

4.1 Software Platform

The ubiquitous computing environment within which our location service is designed to function is the MIT AI Lab Intelligent Room [26]. The Intelligent Room is based on a distributed agent infrastructure called Metaglue [27], which provides directory and brokering services to agents. Metaglue supports lightweight agents, and provides an architecture that organizes agents into societies, in such a fashion as to allow intersociety communication. Because the service of sensor fusion occurs on behalf of each user, a distributed, agent-based software platform is quite appropriate, even if the Metaglue architecture did not dictate such a solution.

In the Metaglue environment, agents work within a society, and each user has a society of agents. Each user's society represents the computing environment devoted to performing services for the user. As seen in Fig. 2, we place the responsibility of sensor fusion and access control (described below) on an agent called the Personal Location Agent (PLA) that acts as an intermediary between location detection devices and location services.

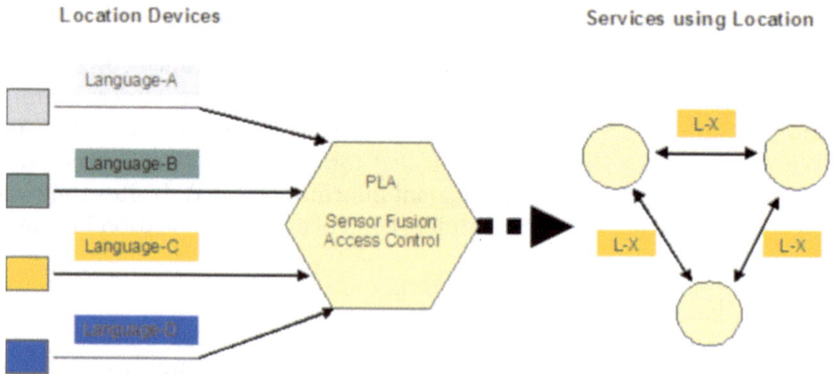

Fig. 2. General Design of PLACE

A much less obvious, yet perhaps more important motivation to using an agent-based platform for location is the vision of user-centric computation. Traditionally, location applications have simply obtained location information directly from the users' location devices, giving essentially no regard to the user as an entity, or the ultimate consumer of location information. This approach runs counter to the new trend away from application-centric computation towards user-centric computation. We propose that by performing location detection and treating location as a property of a user entity, regardless of what services actually use the information, we achieve the notion of computation acting on behalf of the user, as opposed to the application. In doing so, we separate location detection from location utilization.

Indeed, the Metaglue environment in which we have embedded the location services is itself a component of a larger ubiquitous computing environment: The MIT Oxygen prototype system, within which the Intelligent Room functions as a prototype of Oxygen's Enviro21 [28].

4.2 Language

The first step to bringing all entities into the same universe of locations is to have them understand each others' languages. There is a spectrum of possible solutions between two extreme solutions to the language problem. First, all entities can speak their own language as long as they offer translation to and from all other languages. This represents the more cumbersome (but also more flexible) of the two solutions because it places the responsibility of deciphering all incoming information on each entity. Alternatively, all entities could use the same location representation and avoid translation altogether. However, forcing one representation onto all entities would entail losing some valuable information that perhaps can only be encapsulated with a different representation.

Our current solution is to create a common default language with which all communicating entities communicate, but offers translation when necessary. The translation is carried out within the Personal Location Agent, in order to localize the translation burden. Even when using translation, it helps to reduce the problem to translating to and from a common representation in the PLA, rather than use a cross-product translation approach. Though a universal location representation for all entities and for all situations would inevitably be inadequate, a common language for all entities but for only communicating user-location seems much more reasonable. We choose to represent locations with semantic names for places and the various relations between places, simply because people seem to usually reason about locations in this way. Reasoning about location information is central to our purpose.

4.3 Relation-Map

Semantic names alone offer little with respect to a unifying location system. It is the semantic (e.g. geographical, physical) relations existing between the places that offers great value to our system. We name this knowledge representation

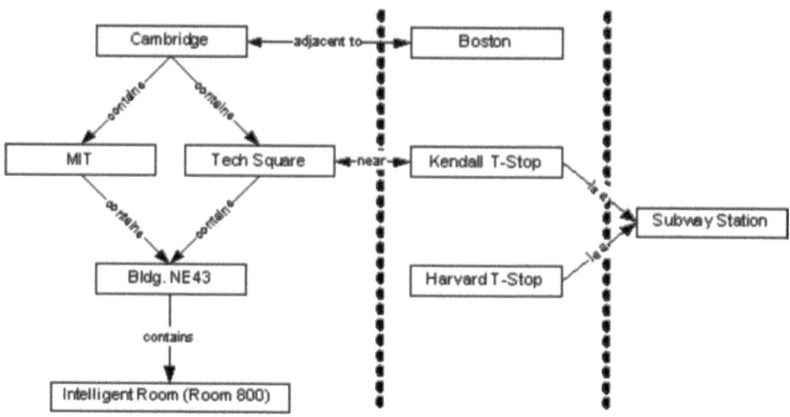

Fig. 3. Example of a Relation-Map

for locations comprised of semantic names connected by semantic relations a *relation-map.*

Collaborating Sources at Varying Granularities. Without relations, a unifying location system cannot interpret location information at varying granularities as supporting or conflicting information. Consider Fig. 3 showing a portion of a relation-map. If two location devices return "Intelligent Room" and "Bldg. NE43," the system can determine that the clues support each other only if it knows that "Bldg. NE43" contains "Intelligent Room." Otherwise, "Bldg. NE43" and "Intelligent Room" compete with each other, and the user will be concluded to be either in one or the other.

Building Context. Furthermore, PLACE uses a relation map not only to handle multiple collaborating sources at varying granularities, but also to serve as a database of information with which further inferencing can take place. Given that a user is in "Intelligent Room," we can use the relation-map shown in Fig. 3 to make several inferences, including, for example: "the user is at Cambridge," or "the user is near Kendall T-Stop." In this way, we combine sensor information with general knowledge about the world to build rich location context.

A Note on Context in a Relation-Map. In Fig. 3, the dominant relation verb is "contains." Containment is the most common relation used among the various location representations that exist today. The primary advantage of using a containment hierarchy is that it provides the most desired organization for a world of places. However, many location representations either require a strict containment hierarchy of places or make it very difficult to represent multiple parent places. Brumitt and Shafer [29] recognize that the real world cannot be neatly "partitioned into unambiguous, non-overlapping pieces that exhaustively cover the area in question." The structure of a relation-map allows for easy representation of these multiple hierarchies.

While most location representations tend to focus on containment alone, a relation-map supports other location verbs such as "near" and "adjacent to." As shown above, these verbs are very useful in establishing further context by allowing for proximity and adjacency representations along with containment. Furthermore, a relation-map admits the possibility of annotation with additional entities and relations (such as "is a"). Such annotation would typically be done by and for applications that need the additional relational information. This could very easily extend to other context analysis such as behavior or activity deduction (i.e. context building not relevant to location) by simply adding relational verbs. If, in order to provide one of these features, we need to add a particular verb, the appropriate application will add it, as an annotation.

4.4 Access Control

Within the Intelligent Room and E21 projects, access control is treated as a separate and separable service, orthogonal to services like location awareness. Consequently, our design does not solve any access control problems per se, but simply uses access control mechanisms supplied as a service by the intelligent environment. However, there still remains significant design work on the location side, because we must specify the expressive power that we require of the access control system in order to allow the flexible access control that we need.

Location tracking is a delicate issue. The ability to find other people at any time is a very powerful utility; however, the ability for other people to find you at any time is a rather uncomfortable notion. People do not want to be stalked by strangers, suffocated by friends and family, or always locatable by business associates. Therefore, when using a PLACE service, it is important for users to have the functionality to precisely fine-tune the access control to their location information. Though widely used as a standard access control mechanism, Access Control Lists (ACLs) alone do not suffice because they do not take advantage of contextual information that users may find pertinent in their distribution of location information [26].

PLACE approaches this problem with access control based on who, why, when, how, what, and where. The following highlights potential uses for each type of control.

- Who is asking?
  - no strangers – keep undesired people from tracking you.
  - family only – parents feel more secure knowing the locations of their children.
  - travel groups – prevents individual members from getting lost.
- Why does someone want to know?
  - emergencies – be locatable by anyone during emergency situations that are relevant to you.
  - technical support – IT employees only want to be found if there is a legitimate technical support issue.

- When does someone want to know?
  - during scheduled meetings times – let other members of meetings know how far you are from the meeting.
  - during business hours (9am–5pm) only – the ability to quickly find resources is conducive to productivity and efficiency.
- How are they asking, or how is the location information provided?
  - give abstract location information, but not output of a camera.
  - give location information if they are asking over a secure network path.
- What am I doing?
  - sleeping/taking a vacation – regardless of where, do not disclose location.
  - giving lecture – allow all who are interested in your lecture to find out where you are speaking.
- Where am I?
  - not in library – prevent people from distracting you while studying.
  - at home – know when friends get home from school.
- Combination
  - no business contacts outside work hours – provide increased productivity at work while not allowing work to travel with you after work hours.
  - friends during designated social hours – allow friends to socialize with you only when desired.

The information of who, why, etc. simply contributes to user context, and PLACE utilizes user context to deduce permissions on user information, such as location. If additional context is available, users can add further precision to the access control of their information. Thus, as shown in the medical database domain [30], users can enhance their access control by incorporating context as additional variables and constraints.

With our software platform having an agent society per user, we can allow users to constantly obtain location information but control distribution of this information to services and other users. More specifically, the Personal Location Agent (PLA) working in a user's society can keep track of the user's location at all times, but control distribution of this information to those outside of the user's society. The primary benefit to this design is that the user can allow services working on his/her behalf to access his/her location at all times, while restricting access to foreign services.

Semantic Expressions. The ability to establish location at varying granularities with a relation-map enables a user to perform such access control with greater ease. Using the convenient organization of places portrayed by a relation map, users can create more intuitive distribution rules. For example, instead of creating a rule stating, "If a user is in room-1, room-2, ..., or room-n, then tell those in room-1, room-2, ..., or room-n that the user is at work," we can instead create a rule stating, "If a user is in his work-building, then tell those in his work-building that the user is at work."

Furthermore, by customizing a relation map and inserting one's own perception of places and relations between places, a user can create rules that more

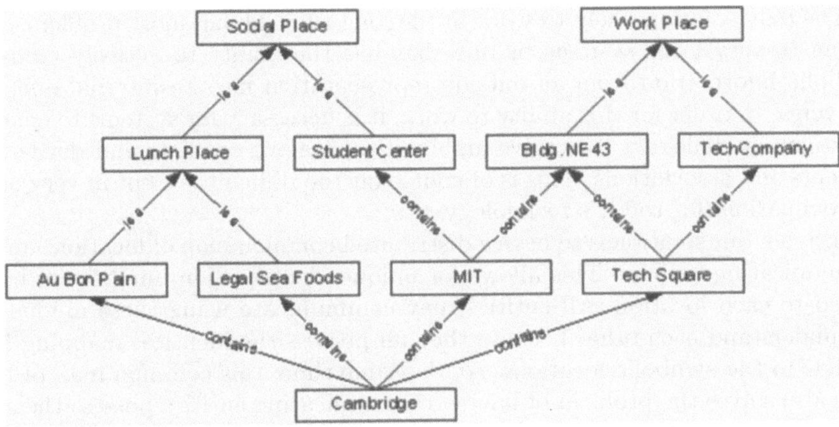

Fig. 4. Relation-Map with User Customizations

adequately portrays his/her personal life. For example, with the customized relation map shown in Fig. 4, a user can create access control rules using semantic terms like "social place," "work place," "lunch place," etc.

Distribution at Varying Granularities. Users may also use the additional context available in a relation map to distribute location information at varying granularities. For example, a user may wish to have the following distribution:

- Let family members know the user's exact location.
- Let friends know the user's location at building granularity.
- Let acquaintances know the user's location at city granularity.
- Let strangers know the user's location at planet granularity.

4.5 Uniquely Identifiable Places

In the above examples, the semantic names for the places are casually assumed to be unique. Uniqueness is essential when communicating location because entities must not confuse two places with identical names. For example, more than one building can have a room called "Room 800." More than one institution can have a building called "Bldg. NE43." In the relation-map shown in Fig. 3, we do not explicitly address the potential ambiguities.

One solution would be to include a specification in each name, such as "Cambridge.MIT.Bldg-NE43.Room150." However, this approach has two problems. First, the specification must address all granularities in the universe of locations in order to ensure that there is not a duplicate place specification. That is, we must ensure that the universe of locations does not include more than one room with the name "Cambridge.MIT.Bldg-NE43.Room800." Second, the specifications imply a strict hierarchy of containment, an unrealistic perception of the world.

The only solution likely to work in the long run, solving most problems, and scaling to very large systems, is one that has the ability to actively translate symbolic information from an outside representation into an internal one, and vice versa. In order for this ability to work, it is necessary for systems to conduct investigatory dialogues to resolve ambiguities, remove conflicts and derive new concepts and associations. This is of course far too difficult, except in very crude approximation, for today's technology.

Instead, our solution is to have a distributed common map of locations among communicating entities. This allows for unique identification numbers to be attached to each location. All entities may communicate using these unique IDs and understand each other because they all possess the identical mapping from the IDs to the symbolic locations. At the same time, this common map of locations also solves the problem of having communicating entities possess the same vocabulary of locations.

We want to be clear that this solution is only a stopgap. It won't scale well, and carries with it enormous problems about generating, distributing, and coordinating information. We are only interested in solving those problems as much as is required to support our stopgap. Furthermore, it simply sweeps under the rug the fact that existing and new applications that depend on other representations for object and locations will still need to be semantically coordinated with the unique identifiers.

5 Applications

Our general design was inspired by the issues that we found inherent in the problem of sharing user locations. The following describes a series of applications that demonstrate the motivations behind the features of our design.

5.1 Patient Tracker

Patient Tracker uses location information of patients to track a patient in a medical facility and to provide doctors with appropriate and relevant information when approaching a patient. Whereas doctors in the ER might want to immediately know the patient's vital signs (including temperature, blood pressure, heart rate, respiratory rate, etc.), those in the recovery department might want to immediately know the patient's laboratory values (cell blood count, electrolytes, etc.) or study results (x-rays, CAT scans, etc.). In this way, Patient Tracker not only utilizes symbolic representation of physical location but also a relation-map semantic relation that categorizes the rooms in the hospital (see Fig. 5). In any case, whether given in certain departments or all departments, patient information given to doctors before or upon arrival in preparation for treatment can save doctors time and energy. Furthermore, in terms of privacy issues, patients do not want to show their location to persons other than doctors and nurses responsible for their care, and appropriate visitors.

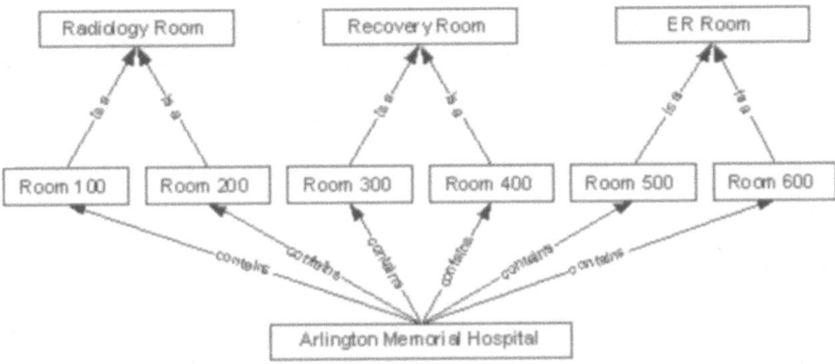

Fig. 5. Relation-Map used in Patient Tracker

5.2 Expert Finder

Expert Finder uses location information to find people declared as experts of a specified topic. In some cases, it is much easier to ask someone for help rather than search through manuals and documents. For example, one might want to find the nearest IT employee for immediate on-site technical support. Unlike the Patient Tracker, Expert Finder potentially deals with people outside the scope of a building. Upon medical emergency, one might want to find the nearest doctor in the area. Thus, in addition to semantic representation and relation maps, the ability to interpret multiple location technologies and to perform sensor fusion is necessary to ensure that one is able to understand the doctor's location.

Even experts do not want to show their location while they are in a private location. However, in a significant enough emergency, that privacy concern may need to be over-ridden. Flexible, Role Based Access Control can solve this type of problem. In this case, when an expert comes close to a user in public space, the system will tell that to the user.

6 Conclusion

Driven by varying motivations, research groups have developed numerous location technologies and services that exist today. Traditionally, applications are rapidly built in response to the new and exciting location technologies available. As a consequence, applications are usually specific to one location technology.

We present a general design of a location system aimed to capitalize on multiple location technologies for the purpose of selectively sharing location information. Our design does not hope to serve as the general location infrastructure for all location services, but for services specific to user-location. Using a location representation consisting of semantic names and relations, we permit collaboration between location sources with different, yet related perspectives of the world. Similarly, we allow for further context inferencing that enables PLACE to distribute location at varying granularities and with intuitive access control. Finally, we demonstrate the utility of such features with a series of applications.

We hope the issues addressed in this paper will stimulate discussion about sharing location information, and will eventually enable us to communicate about locations seamlessly. We also think this idea can apply to other categories of context [31,32].

Acknowledgements

The authors are grateful for support and advice from Howie Shrobe, for advice and suggestions from Krzysztof Gajos, and for support from Rod Brooks and the MIT Oxygen Alliance Partners: Acer, Delta, Hewlett-Packard, Nokia, Philips and NTT.

References

1. Dertouzos, M.: The Future of Computing. Scientific American, July 1999.
2. Ark, S. , Selker, T.: A Look at Human Interaction with Pervasive Computers. IBM Systems Journal, Vol.38, No.4, 1999.
3. Dey, A., Abowd, G.: CyberMinder: A Context-Aware System for Supporting Reminders. Proceedings of International Symposium on Handheld and Ubiquitous Computing (HUC 99), 1999.
4. Marmasse, N., Schmandt, C.: Location-Aware Information Delivery with ComMotion. In Proceedings of International Symposium on Handheld and Ubiquitous Computing (HUC 99), 1999.
5. Sekiguchi, M., Takayama K., Naito, H., Maeda Y., Horai H., Toriumi M.: NaVigation Markup Language (NVML). World Wide Web Consortium (W3C) Note, 6 Aug. 1999. http://www.w3.org/TR/1999/NOTE-NVML-19990806.
6. Long, S., Aust, D., Abowd, G., Atkeson, C.: Cyberguide: Prototyping context-aware mobile applications Proceedings of ACM CHI'96 Project Note, 1996.
7. Snoeren, A., Balakrishnan, H.: An End-to-End Approach to Host Mobility. Proceedings of the 6th Annual International Conference on Mobile Computing and Networking (MOBICOM 2000), Boston, MA, Aug. 2000.
8. Jose, R., Davies, R.: Scalable and Flexible Location-Based Services for Ubiquitous Information Access. Proceedings of International Symposium on Handheld and Ubiquitous Computing (HUC 99), 1999.
9. Oppermann, R., Specht, M.: A Context-Sensitive Nomadic Exhibition Guide. Proceedings of International Symposium on Handheld and Ubiquitous Computing (HUC 99), 1999.
10. Schmidt, A., et al. Advanced Interaction in Context. In Proceedings of International Symposium on Handheld and Ubiquitous Computing (HUC 99), 1999.
11. McCarthy, J., Meidel, E.: ActiveMap: A Visualization Tool for Location Awareness to Support Informal Interactions. Proceedings of International Symposium on Handheld and Ubiquitous Computing (HUC 99), 1999.
12. Hightower, J., Borriello, G.: Location Systems for Ubiquitous Computing. Computer, Aug. 2001, pp.57-66.
13. World Wide Web Consortium: Platform for Privacy Preferences Project. http://www.w3.org/P3P/.
14. Want, R., et al.: The Active Badge Location System. ACM Trans. Information Systems, Jan. 1992, pp. 91-102.

15. Korkea-aho, M., Tang, H.: Experiences of Expressing Location Information for Applications in the Internet. Workshop Proceedings of Ubicomp 2001: Location Modeling for Ubiquitous Computing, Sept. 2001.
16. Database Promotion Center: G-XML (Geospatial-eXtensible Markup Language) Protocol 2.0, March 2001. http://gisclh.dpc.or.jp/gxml/contents/index.htm
17. Open GIS Consortium: Geography Markup Language (GML) 1.0, Dec. 1999. http://www.opengis.org/techno/rfc11info.htm
18. MOSTEC: MObile Information Standard TEchnical Committee: POIX: Point Of Interest eXchange language, version 2.0 Document Revision 1, July 1999. http://www.w3.org/TR/poix/.
19. SignalSoft Corp. Wireless Location Services. http://www.signalsoftcorp.com.
20. Hall, D., Llinas, J.: Handbook of Multisensor Data Fusion, CRC Press, Boca Raton, June 2001.
21. Trusted Computer Security Evaluation Criteria, DOD 5200.28-STD. Department of Defense, 1985.
22. Krause, M., Tipton, H. F.: Handbook of Information Security Management. CRC Press LLC, 1998.
23. Baldwin, R.W. Naming and Grouping Privileges to Simplify Security Management in Large Databases. In IEEE Symposium on Computer Security and Privacy, 1990.
24. Robertson, P., Laddaga, R., and Shrobe, H.: Proceedings of the First International Workshop on Self-Adaptive Software Lecture Notes in Computer Science 1936, Springer-Verlag, 2001.
25. Laddaga, R.: Creating Robust Software Through Self-Adaptation. IEEE Intelligent Systems, Vol 14, May/June 1999.
26. Hanssens, N., Kulkarni, A., Tuchinda, R., Horton, T.: Building Agent-Based Intelligent Workspaces. In submission.
27. Coen, M., Phillips, B. Warshawsky, N., Weisman, L., Peters, S., Finin, P. Meeting the Computational Needs of Intelligent Environments: The Metaglue System. Proceedings of MANSE'99, Dublin, Ireland, 1999.
28. Gajos, K. and Kulkrani, A.: FIRE: An Information Retrieval Interface for Intelligent Environments. Proceedings of the International Workshop on Information Presnetations and Natural Multimodal Dialogue (IPNMD 2001) Verona, Italy, 2001.
29. Brumitt, B., Shafer, S.: Topological World Modeling Using Semantic Spaces. Workshop Proceedings of Ubicomp 2001: Location Modeling for Ubiquitous Computing, Sept. 2001.
30. Tzelepi, S., Koukopoulos, D., Pangalos, G.: A Flexible Content and Context-Based Access Control Model for Multimedia Medical Image Database Systems. Proceedings of ACM MMSig'01, 2001.
31. Naito, H., Takayama, K., Maeda, Y.: Situated Information. In the Proceedings of the International Workshop on AI in mobile Systems (AIMS2001), IJCAI 2001, pp.43-48, 2001.
32. Shafer, S.A.N., Brumitt, B., Cadiz, J.: Interaction Issues in Context-Aware Interactive Environments. Journal of Interactions, to appear.

Distributing Event Information by Simulating Word-of-Mouth Exchanges

Elaine M. Huang, Michael Terry, Elizabeth Mynatt, Kent Lyons, and Alan Chen

College of Computing, Georgia Institute of Technology
Atlanta, GA 30332, USA
+1 404 394-0673
{elaine,mterry,mynatt,kent,smile}@cc.gatech.edu

Abstract. Word-of-mouth is a persuasive but error-prone and unreliable mode of communicating personally relevant event information in a university environment. In this paper we present a design, early prototype, and the results of preliminary usability tests for Augmented Word-of-mouth Exchange (AWE), a portable system that models and enhances word-of-mouth communications. AWE simulates word-of-mouth exchanges by automatically transmitting accurate and persistent information about community events between physically proximate devices, and by visualizing the popularity of each event. The system uses physical proximity between mobile devices to help users filter incoming information and determine its relevance.

1 Introduction

As the number of communication channels we use and amount of information we receive through them increases, so does the quantity of irrelevant and redundant information that people receive. Filtering what information is of interest or importance becomes a difficult task that requires time and attention. In determining what transmissions are most important, people take into account several factors, such as the source of the information, the channel through which it was transmitted, and the party to which it was communicated. As they become increasingly bombarded by various electronic and traditional methods of communication, the utility of lightweight interfaces to help them determine the relevance of information and facilitate filtering becomes apparent.

Predicting the potential relevance of information is a challenging task for which there are many possible approaches. We chose to take advantage of the fact that a person is likely to find something relevant if people with similar interests, or in a similar context, found it relevant. Furthermore, people who share space or interact frequently are likely to share some common ground and have similar interests. Using increasingly popular personal digital assistants (PDAs) with wireless networking capabilities, we can take advantage of encounters to transfer information between people who are likely to share interests. Our goal was to design a system that would

F. Paternò (Ed.): Mobile HCI 2002, LNCS 2411, pp. 60–68, 2002.
© Springer-Verlag Berlin Heidelberg 2002

help users filter information by using the relevance of proximity that handheld devices afford.

The Graphics, Visualization, and Usability Center at Georgia Tech is an interdisciplinary organization that holds over 40 events a year. To broadcast event information to its members, the GVU Center relies on email, posters, web pages, and word-of-mouth. We researched how GVU members receive information about events and found that while mass email, posters, and web pages present event information accurately and completely, and are persistent artifacts, they tend to broadcast information to a broad group of recipients, some of whom may find the information irrelevant or not of interest. We found word-of-mouth to be a highly influential but error-prone and unreliable mode of communication. The close, personal context typical of word-of-mouth heightens the influence and relevance of this mode of communication. There is no guarantee, however, that the information communicated is accurate or complete. In addition, the messages communicated are transient, thus requiring the recipient to commit the information to memory, or actively record it. For example, when a student is informed of an event by her colleague, she is more likely to pay attention to that communication than to a mass email advertising the same event, though the information may be inaccurate or forgotten later.

In this paper, we describe Augmented Word-of-mouth Exchange (AWE), an event information dissemination system designed to overcome the shortcomings of word-of-mouth communication while preserving the advantages it offers.

2 Design Motivation

Our design is motivated largely by the observation that people who are frequently in close physical proximity often share interests or activities, and information relevant to one person in this context is likely to be relevant to others. In the environment that we observed, individuals who are located near each other, or who encounter each other frequently, are likely to share some form of context. For example, individuals who work in the same lab will have colleagues in common, share space, be engaged in similar types of research, or be affiliated with the same organizations. Likewise, people who work in labs on the same floor of the same building may be performing research in similar areas and will come into contact each other on a regular basis. These people are likely to chat while working or mention events when they meet, and it is these exchanges motivated by proximity and collocation that we wished to mimic with our system.

In our interviews with students, we found that one of the factors that influences participation or attendance at an event is the knowledge of labmates' or colleagues' interest in it. We aimed to take advantage of the social factors that influence people to attend an event by providing users with information about the attendance plans of the groups of people with whom they share context or interests.

Taking advantage of potential shared interest may help to simulate the personal nature and likelihood of relevance in word-of-mouth communications, but we were also interested in preserving the fidelity and persistence of information that electronic means provide. We did not set out to create a system that would replace either word-of-mouth or more formal means of transmitting event information. Rather, we wanted to incorporate the benefits of email or posters into a word-of-mouth style communication, to give users both the potential relevance of face-to-face communication and complete and accurate information.

3 Design

We first illustrate our design with this scenario:

> *The GVU Center plans a talk on mobile computing and a barbecue for students. In addition to advertising using mass emails, they also beam the event information using RF beacons located in the building. Matt, an architecture student who is running AWE on his PDA, walks by one of these beacons in the morning and receives notification of these two events. He decides that he is not interested in the talk and specifies that he does not plan to attend it, but chooses to attend the cookout. Later he passes by Liza, a computer science student who has not yet received information about these events on her handheld. The event information is transmitted to her device, as are Matt's attendance plans. Liza immediately decides to attend the barbecue, but is unsure whether the talk will be important to her. The fact that Matt is not planning on attending it is shown to her as a low planned attendance for the general public at this event. As her indicated interest groups are the Everyday Computing Lab and the Future Computing Environments groups, Matt's plan to not attend the talk does not affect the group popularity metrics on her device. She then goes to her algorithms class, and finds that the planned attendance rate of the public for the talk continues to be low. However, when she enters her lab and information from her labmates devices is transmitted to hers, she finds that a high percentage of her interest groups are planning on attending it, so she decides it must be relevant and chooses to attend.*

The AWE system is designed to allow users to automatically transmit event information to each other, and provide users with a view of outside interest in the event, a dynamic indicator of its potential relevance to the user.

When the GVU Center advertises an upcoming event, necessary information such as the event's title, description, time, and location are beamed from beacons located near the GVU office to devices running AWE that pass within close range. These devices continually transmit the information to other devices within a range of a few meters, thus propagating the event information through proximity.

When users receive notification of a new event, they can indicate tentatively whether or not they plan to attend the event (see Fig. 1). AWE keeps track of the percentage of encountered users who plan on attending the event, and updates this metric when it passes new devices. This propagation of dynamic interest information allows users to see the interest level of the general public in an event, information she may use to determine its relevance to her.

Fig. 1. The listing for the event through which the user can schedule his planned attendance and view the event's current popularity levels

This model of information transfer, however, does not guarantee that all information gathered through proximity is relevant to the recipient. Some degree of filtering is necessary to prevent users of the system from being bombarded by irrelevant event or attendance information as they will likely pass by other users with whom they share little or no context, and in whose attendance they are not interested.

To reduce the effects of this irrelevant information, we utilize interest groups. Interest groups help the user to judge whether the event is relevant to him by indicating if other people who share his interests are planning on attending. Our system allows the user to select groups with whom he shares common interests, or in whose event attendance he is interested. When the user comes into contact with members of these groups, metrics that track the popularity of events among the user's interest groups are updated, reflecting the planned attendance of the people encountered. This information is visualized alongside indicators of the overall popularity of the event. By allowing the user to view the popularity of upcoming events among specific groups as well as the general popularity of the event, the user is better informed of what events will be most relevant to him (see Figs. 1 & 2).

3.1 System Details

AWE requires an application installed on a Personal Digital Assistant (PDA); radio frequency (RF), peer-to-peer communication services available on the same PDA; a desktop application to install both the AWE application and the user's group preferences on the PDA; RF-enabled beacons to broadcast new events to users; and an application to enter new events to be broadcast by the beacons. The AWE application is the heart of the system for end-users and consists of a main screen listing all events received by the application. The name, date, and public and group rankings for each event are listed in the main screen. Users can sort on any of these dimensions to facilitate navigation of the events. A checkbox next to each event indicates whether the user has scheduled this event in her calendar (see Fig. 3).

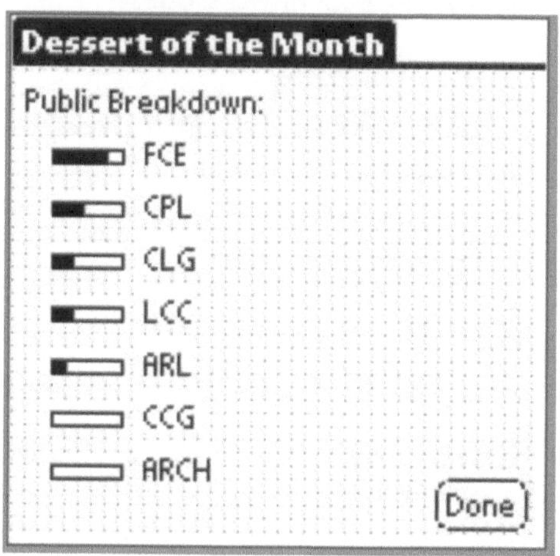

Fig. 2. The display showing the event's planned attendance among groups of users

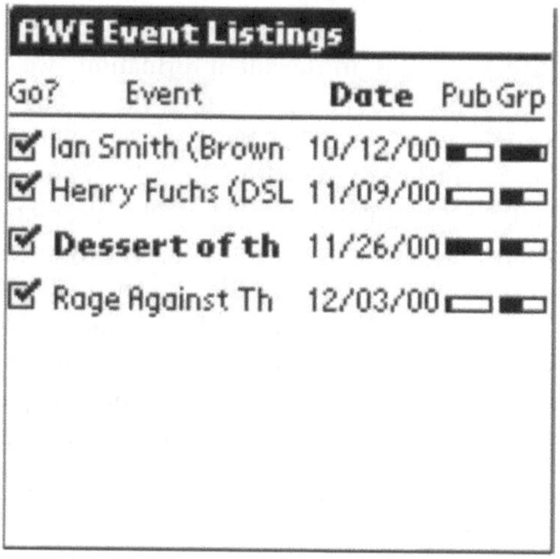

AWE Event Listings

Go?	Event	Date	Pub	Grp
☑	Ian Smith (Brown	10/12/00		
☑	Henry Fuchs (DSL	11/09/00		
☑	**Dessert of th**	11/26/00		
☑	Rage Against Th	12/03/00		

Fig. 3. The main AWE screen listing all events received by the user

Selecting an event brings up a detailed description of the event. From this screen, users can automatically add the event to the PDA's calendaring program, delete it, view their schedule on the day of the event, or see a finer-grained view of the popularity of the event for individual groups.

When users add the event to their calendar through AWE, the application records this information, then later exchanges this information with other AWE-enabled devices to generate popularity rankings of events. That is, AWE uses a person's intent on attending an event as the method of determining the relative popularity of an event. To maintain privacy, a user's scheduling information is not made available to others; it is only used to generate the event rankings.

To implement the communications model promoted by AWE, information must be exchanged automatically without requiring intervention from the user. For this reason, we require a PDA enhanced with RF, peer-to-peer communications, for example, those offered by Bluetooth[11].

Event rankings show the popularity of the event with the general public and with user-selected groups. AWE therefore needs to know in what groups the user is interested, as well as of what groups the user is a member. The web-based desktop program for AWE provides the facilities to collect this information and to install both the application and user's group information on the PDA. The creation and subscription to user groups in our system is currently left to administrators as opposed to the individual users, much in the way that administered email groups are often handled. This feature also reduces opportunities for users to monitor the behavior of specific individuals.

We place "beacons," or stationary, RF-enabled broadcasting devices, in the environment to transmit new events to users who pass by. The location of the beacons influences the propagation of the message; thus they are placed in locations with meaningful context, for example, outside the GVU Center's office. A form-based application enables individuals to create new events to broadcast via the beacons. In our design, we have focused on the case where one organization generates new events, and do not consider the possibility of users generating their own events.

4 Relevant Research and Related Work

MemeTags [1] are digitally enhanced name tags that store and exchange "memes", or text-based sayings, with other name tags in line of sight of each other. When a user finds a meme they like on another's device, they can copy it to their own name tag. The most popular memes are presented on a "community mirror" for viewing by all. AWE shares with MemeTags a similar communication scheme and a similar "ranking" of the most popular information, but differs in its application space and intent.

Recommender systems [5, 10] promote the concept of peer review to create meaningful and relevant reviews of everything from books to websites. In particular, AWE promotes a form of *social navigation* [2] by implicitly capturing and sharing information regarding planned attendance. AWE also differs from most recommender systems by being completely decentralized in nature.

Previous work in wearable communities [5, 8] has addressed similar needs for peer review and filtering. Much of this prior work requires explicit action on the user's part to maintain relevant filters, such as rating the quality of an interaction. AWE gleans this information from an action that directly benefits the user, in particular, adding an event to the calendar.

Horvitz's work on mixed-initiative user interfaces outlines design principles for intelligent systems that attempt to infer user goals. The current AWE prototype uses a simple metric of proximity for filtering, but any future work incorporating more sophisticated intelligence will be influenced by these principles.

5 Status and Results

At this time, we have built a prototype of the AWE application which runs on a Palm III. Using this prototype, the user can read about an event, indicate whether she plans to attend it, and view the group and public planned attendance ratings. Using Wizard of Oz[1] data, the user also receives notification of new events, and changes in group and public attendance information. We have also created prototypes of the web-based forms through which new events are created, but this information is not automatically propagated to the devices.

We performed some initial usability tests in which we gave the device with the application to ten GVU members and students and asked them to evaluate the interface. Many of the evaluators were skilled in HCI techniques; to these participants we administered either an hour-long cognitive walkthrough exercise or heuristic evaluation exercise [4, 6]. The cognitive walkthrough exercise allowed participants to evaluate the usability of the interface by following a prescribed set of steps including finding and reading about a particular event, examining and assessing the group popularity of the event, viewing one's own calendar, and scheduling the event. The heuristic evaluation entailed that the participants explore the prototype, evaluating it specifically in regards to a set of system-specific heuristics, including visibility of system status, consistence of interface design, recognition over recall, and user control. With the few participants unfamiliar with HCI techniques, we performed more informal evaluations, allowing them to try the interface and give freeform comments.

Based on user feedback, we made several modifications to our design. In the earlier designs of this system, group selection was a feature of the application on the device, allowing users to add or edit their interest groups directly on the PDA. We found, however, that having setup and configuration features directly on the device greatly increased the complexity of the interface, necessitating more interaction screens, and making the application, as a whole, less lightweight. In addition, the

[1] We simulated portions of the system in order to evaluate the initial design concept and user interaction.

small screen did not lend itself well to the potential for having long scrolling lists of interest groups, which were better viewed on a standard monitor.

While our tests were concerned primarily with the usability of the interface, we also wanted to gauge the extent to which people found the attendance indicators to be useful and desirable features. We found that most test participants appreciated having this data available and would use it to help them decide what events to attend, but were also concerned about others tracking their attendance as they passed.

6 Future Work

Our next step is to fully implement AWE using real RF sensing and real GVU event information so that we can deploy the system and test its utility as a communication tool that provides awareness to aid its users. In addition to completing the system as we designed it, there are also several changes in the design that we would like to try.

One area of exploration is the creation and maintenance of interest groups. As users' interests change, or as they find themselves becoming increasingly involved with other groups, how can we allow them to easily edit their group membership in a way that does not require too much attention? In addition, could the system automatically classify users into groups based on their attendance patterns in relation to others?

Finally, we need to consider the implications of using physical proximity as a means of inferring common interest or context. Deployment of our device will help us to determine to what extent shared space is an accurate and reliable metric. We will also need to experiment with various thresholding functions to determine what kind of physical proximity is nontrivial for assuming shared context; we will need to consider the duration and frequency of encounters to decide whether they are relevant. Understanding the relation between common interest and physical proximity will be especially necessary if we are to scale up the number of events, broaden the user base, or consider a physical space larger than the few buildings for which our prototype was built. [9]

7 Conclusion

We have presented the design of AWE, a decentralized system designed to mimic and enhance the word-of-mouth communications model. By using an augmented word-of-mouth model of communication, the system gives users complete and accurate information while simulating the personal relevance of face-to-face communication. To do so, the system takes advantage of the fact that one can gather data through proximity using handheld devices. By paying attention to the event attendance of people nearby, who are likely to share context, the system enables the user to filter event information based on others' planned attendance. It gathers and provides data that help users determine the relevance of upcoming events.

References

1. Borovoy, R., Martin, F., Vemuri, S., Resnick, M.: Meme Tags and Community Mirrors: Moving from Conferences to Collaboration. In Proceedings of the ACM 1998 Conference on Computer Supported Cooperative Work, CSCW '98, Seattle, WA, 1998, pp. 159-168.
2. Dourish, P., Chalmers, M.: Running Out of Space: Models of Information Navigation. In Proceedings of HCI '94, Glasgow, 1994.
3. Horvitz, E.: Principles of Mixed-Initiative User Interfaces. In Proceedings of CHI 1999, pp 159-166.
4. John, B., Packer, H.: Learning and Using the Cognitive Walkthrough Method. In Proceedings of CHI 1995, pp. 429-436.
5. Kortuem, G., Schneider, J., Suruda, J., Fickas, S., Segall, Z.: When Cyborgs Meet: Building Communities of Cooperating Wearable Agents. In Third International Symposium on Wearable Computers (ISWC'99), San Francisco, CA, 1999.
6. Nielsen, J., Molich, R.: Heuristic Evaluation of User Interfaces. In Proceedings of CHI 1990, pp. 249-256.
7. Resnick, P., Varian, H.: Recommender Systems. Introduction to special section of Communications of the ACM, vol. 40, 3. March, 1997, pp. 56-58.
8. Schneider, J., Kortuem, G., Jager, J., Fickas, S., Segall, Z.: Disseminating Trust Information in Wearable Communities. In Proceedings of 2nd International Symposium on Handheld and Ubiquitous Computing (HUC2K), Bristol, England, 2000.
9. Terry, M., Mynatt, E., Ryall, K., Leigh, D.: Social Net: Using Patterns of Physical Proximity Over Time to Infer Shared Interests. In CHI 2002 Extended Abstracts, Minneapolis, MN.
10. Terveen, L., Hill, W., Amento, B., McDonald, D., Creter, J.: Building Task Specific Interfaces to High Volume Conversational Data. In Proceedings of CHI 1997, pp. 226-233.
11. Bluetooth Specificatons: http://www.bluetooth.com/developer/specification/specification.asp

A New Transcoding Technique for PDA Browsers, Based on Content Hierarchy*

F.J. González-Castaño, L. Anido-Rifón, and E. Costa-Montenegro

Departamento de Ingeniería Telemática, Universidad de Vigo, ETSI Telecomunicación, Campus, 36200 Vigo, Spain,
{javier,lanido,kike,jvales,rasorey}@det.uvigo.es

Abstract. This paper presents a new transcoding technique for WWW navigation on small display devices: Hierarchical Atomic Navigation. Unlike previous techniques, Hierarchical Atomic Navigation keeps all original information in a readable way, without imposing the use of a specific browser. To achieve this goal, a *navigator page* is used to represent original contents in a symbolic way. A set of representative icons replaces unreadable elements. These icons are linked to actual individual contents, as a set of atomic pages. Hierarchical Atomic Navigation can be used on any PDA, regardless of OS and browser choice, since both navigator and atomic pages use widely supported standard formats (e.g. XML, HTML).

1 Introduction

In the future, many WWW services will be accessed from cell phones and wireless Personal Digital Assistants (PDA) [1,2,3,4]. Although GPRS and 3G systems may provide enough mobile bandwidth, there is an intrinsic problem to be solved: mobile terminals are small and, therefore, have small displays. Most WWW contents are not designed for them [5,6]. Searching information through a page is a complex task, users are prone to incorrect selections and scroll up and down most of their time. It has been reported [6] that users with small screens are 50% less effective than users with large screens. A trivial (and obviously undesirable) solution is limiting mobile WWW services to specialized WAP-like ones, like stock quotes, weather forecasts and sports scores.

This paper presents a new philosophy to improve WWW navigation on small displays. Our system, described in section 3, is based on the Hierarchical Atomic Navigation concept (HANd): some or all elements embedded in a WWW page are identified by means of a reduced page preview, the *navigator page*, which is generated automatically. Section 2 discusses related work. Section 4 evaluates usability. Finally, section 5 concludes.

* Additional authors: J. Vales-Alonso, R. Asorey-Cacheda, J. García-Reinoso, D. Conde-Lagoa, J. M. Pousada-Carballo and P. S. Rodríguez-Hernández. This research has been supported by European Commission FEDER grant 1FD97-1485-C02-02 TIC.

F. Paternò (Ed.): Mobile HCI 2002, LNCS 2411, pp. 69–80, 2002.
© Springer-Verlag Berlin Heidelberg 2002

2 Related Work

2.1 Web Clipping

Web clipping is based on delivering a "clipped" version of the original WWW page to a small display device. First, a page fragmentation process is carried out. Then, a human or an automated process specifies an importance value for every page fragment. Low importance fragments are ignored when display space is limited. For example, elements that do not carry significant information in relation with their size.

Hori et al. designed a page fragmentation system based on external annotation files [7]. A transcoding process uses annotations to decide which elements will be delivered to the client. Annotation files follow the RDF data model [8] to specify the relative priority of page elements (punctuation). The system is basically composed of a page splitting module and a generator, which creates the client page. The system includes an authoring tool, to create annotation files with page fragment priorities.

Although clipped versions of WWW pages are adequate for small displays, there is information loss according to criteria that are ignored by the end-user. Also, page fragments must be punctuated with their priority (this process may be accelerated by means of advanced authoring tools [7]).

2.2 Handy-Fit-to-Screen

Microsoft Pocket Internet Explorer [9] implements the Handy-Fit-to-Screen feature [10], which resizes WWW pages to fit into small displays. Obviously, many page elements are not properly displayed. To cope with this drawback, Pocket Internet Explorer includes a zoom menu option that displays text in several sizes.

Handy-Fit supports typical commercial standards, including HTML and XML [11], and keeps all information. Although it is a particular feature of Microsoft Pocket Internet Explorer, similar tools could be developed for other browsers. Nevertheless, it does not work properly on large text pages that use small fonts.

2.3 WML

WML WAP devices [12] use microbrowsers to access WAP-supporting WWW sites. WML is designed to optimize text delivery over limited-bandwidth wireless networks. It supports WMLScript, which is similar to EcmaScript, but imposes minimal system resource demands. It is unlikely that WML will handle color, audio and video in the near future.

Typical WAP sites present a series of options for visitor access. Whilst clearly limited, this operation mode is satisfactory for the delivery of news, stock quotes, sports results, travel inquiries and similar applications. However, it cannot support the capabilities of most WWW sites. A WML version of a WWW page is, in a sense, a "clipped" version for the WAP-enabled cell phone world.

2.4 i-MODE

In February 1999, the wireless branch of Nippon Telegraph & Telephone, NTT DoCoMo, launched the i-MODE network [13]. By May 2000, the service had more than seven million subscribers in Japan and the demand was so great that NTT stopped advertising for a while, in order to install extra service capacity. i-MODE terminals access WWW contents over the PDC wireless network, using a proprietary DoCoMo protocol.

A typical i-MODE terminal has a 95×65-pixel screen, which is sharper than most current cell phone resolutions. In order to display WWW contents on such a small screen, i-MODE employs subsets of HTML 2.0, 3.2 and 4.0. These subsets are called c-HTML [14], which stands for compatible-HTML. This language is the only requirement for porting a WWW site to the i-MODE system. Since c-HTML relies on ordinary HTML coding and uses common HTML tags, c-HTML pages can also be displayed on PCs. i-MODE terminals can display GIF images and half-width Kana characters, but do not support Java or script languages.

Unquestionably, i-MODE is a successful experience in the Japanese market. Its main advantage is the lack of punctuation processing, since clipping is an implicit consequence of using c-HTML. Consequently, an i-MODE WWW site is just a "clipped" version of the original site. All elements outside the subset are no longer included in the i-MODE page and, therefore, no longer viewable by i-MODE terminals. Again, we are facing the problem of information loss, according, in this case, to fixed criteria over a proprietary network.

From this analysis, it can be concluded that most approaches rely on some form of clipping, which yields high readability but has information loss. Microsoft Handy-Fit avoids that disadvantage, but it is not adequate for large text pages. In the next section, we present a new philosophy that keeps all original information in a readable way, and does not impose a browser choice.

3 HANd

3.1 Philosophy

The HANd philosophy is based on the fragmentation of the original page into zones. These zones are sets of multimedia elements included in the original page (e.g. images, text paragraphs, headers, hyperlinks, Java applets, forms, etc.). The user may select any zone, but only one can be displayed at original scale at a given time.

The main page, or navigator page, is a reduced overview of the original page. The navigator page is always displayed on a side frame or pop-up window. It facilitates access to different atomic readable elements (i.e. pieces of the original page), using links to auxiliary or *atomic pages* (on a second side frame or pop-up window, according to user preferences and display capabilities). There are different navigator page representations for a given element in the original WWW page:

– A reduced version of the element, if still readable.
– A representative icon, if it is decided that the reduced element is unreadable.

A key issue is element dependence. All original elements are ordered in a tree hierarchy, and any element is considered unreadable if all its descendants are unreadable. In the navigator page, the representative icon of an ancestor hides all its descendants, but, if that icon is selected, all descendants are shown on the atomic page.

Navigator page representatives are placed in the same relative positions the corresponding elements occupied on the original page. The goal is to provide the user with an *idea* of the original appearance of the page, in a reduced space.

Note that HANd can be considered a sort of visual clipping (see section 2). However, unlike web clipping techniques, every single information item is kept and the end user is free to decide which elements are shown.

Also, HANd can be used on any PDA, regardless of OS and browser choice, since both navigator and atomic pages use standard formats. This is a major difference with Microsoft Handy-Fit, a technique that is implemented by specific browsers.

3.2 HANd Page Examples

Figures 2 to 5 show the navigator/atomic page set after processing the page in figure 1. All figures are captures taken on an iPAQ H3630 PDA. The width ratio between figures 2 to 5 and figure 1 on a real screen is approximately 20%. The left screen in figures 2 to 5 shows some possible user selections on the navigator page. The right screen shows the corresponding atomic pages (the "atom" icon is used to return to the navigator page, in an implementation without frames).

Figure 6 shows a selection of representative icons used in navigator pages. On a real PDA screen, the icons are blue if the corresponding atomic pages contain hyperlinks.

3.3 HANd Page Generator Prototype

The previous section described a navigator/atomic HANd page set. These pages are created by a HANd page generator, which is shown in figure 7. The generator receives documents written in any language, and processes them in two stages:

– *Preprocessing stage.* Non-XML documents (HTML, TEX,...) are converted into XHTML, which is a reformulation of HTML 4.01 in XML. The conversion is performed by JTidy [15], the Java version of HTML TIDY [16]. Once a XHTML document is available, we obtain its DOM tree [17] using a DOM parser. A DOM tree is a representation of a XHTML document as a tree of nodes, each one corresponding to a element in the XHTML document.
– *Conversion stage.* This stage applies visibility rules to the DOM tree, and obtains the navigator/atomic page set from the resulting new DOM trees.

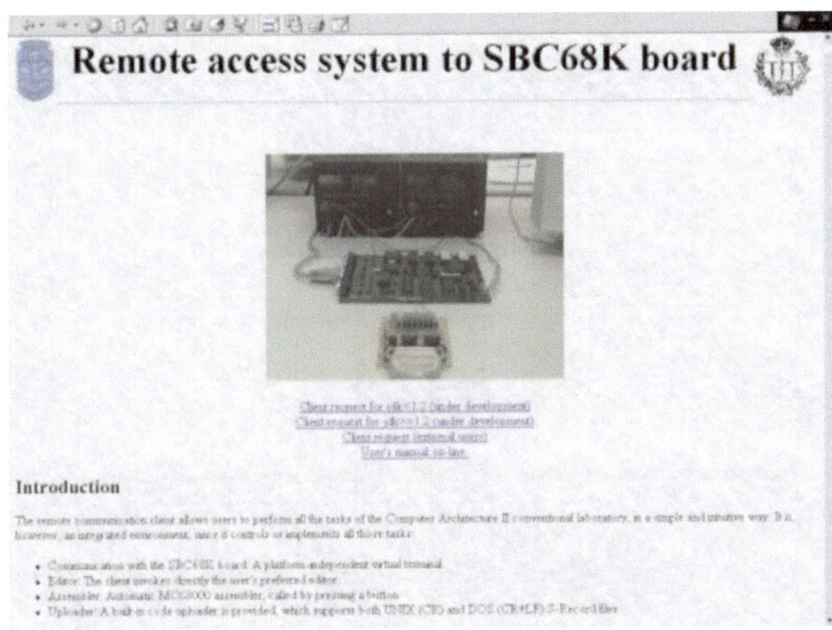

Fig. 1. Original page (it occupies a full 17" 1024×768 screen)

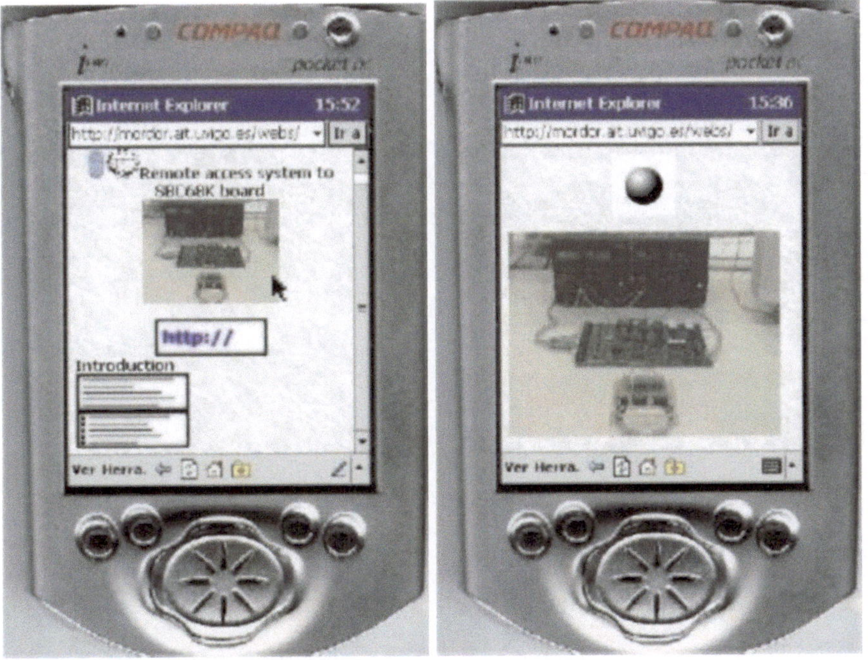

Fig. 2. Navigator/atomic page set

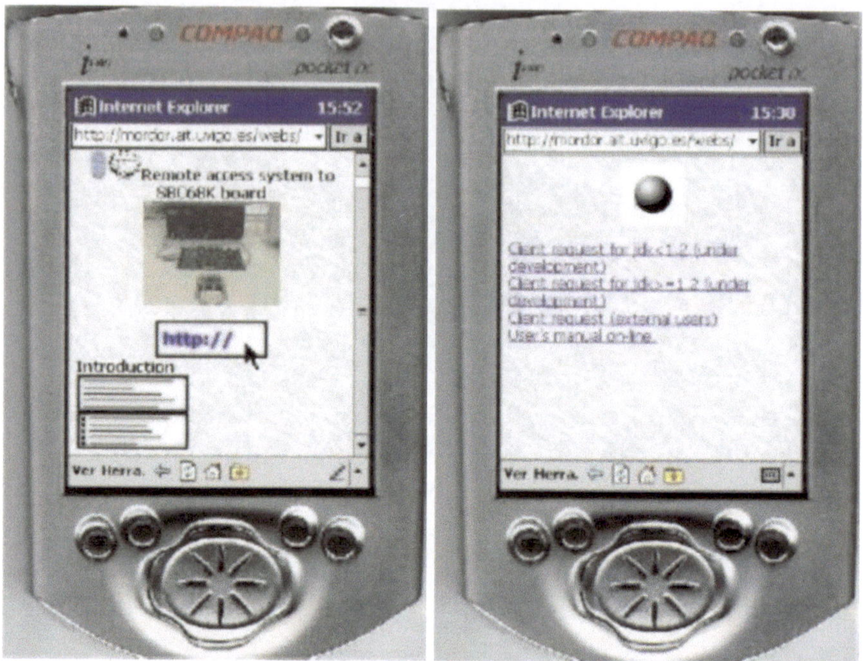

Fig. 3. Navigator/atomic page set

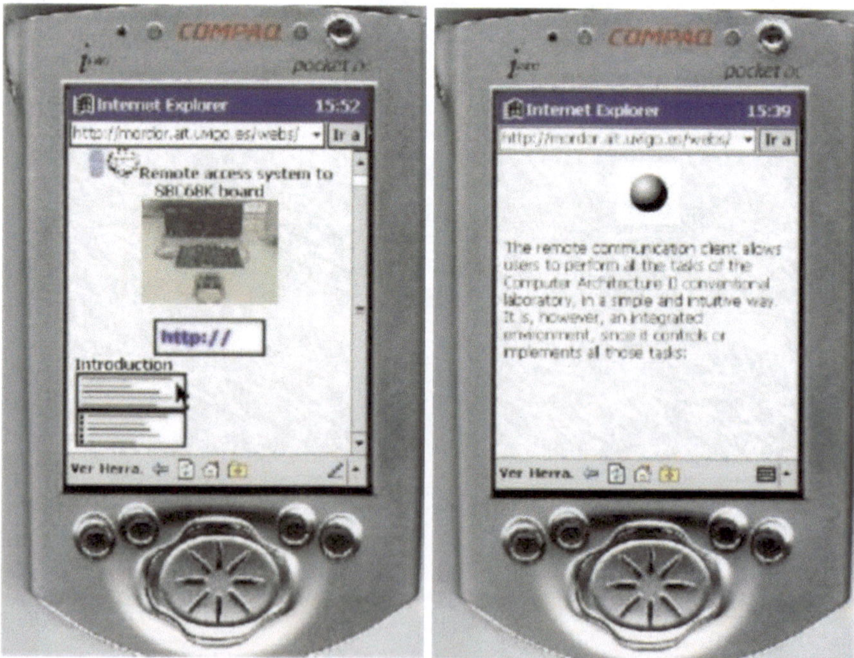

Fig. 4. Navigator/atomic page set

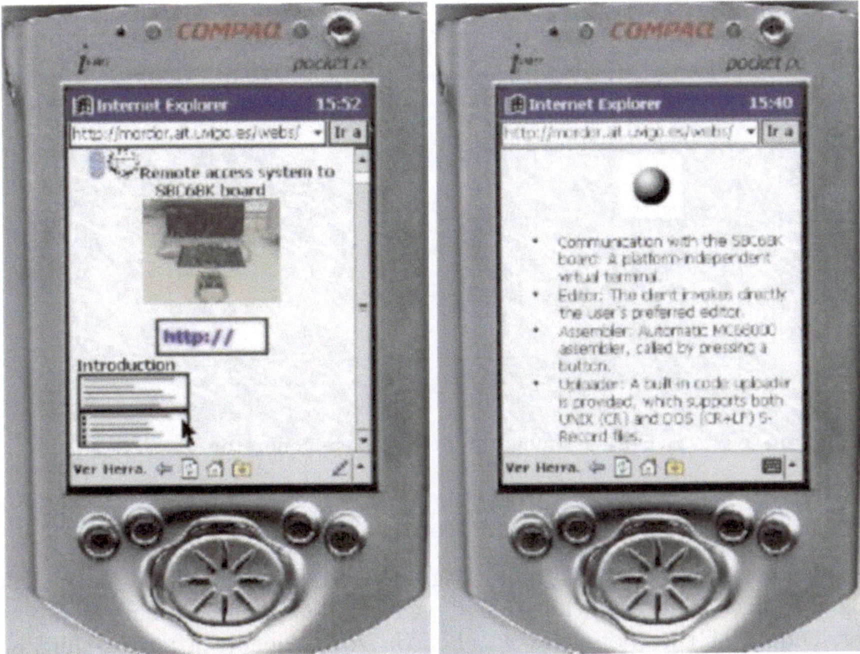

Fig. 5. Navigator/atomic page set

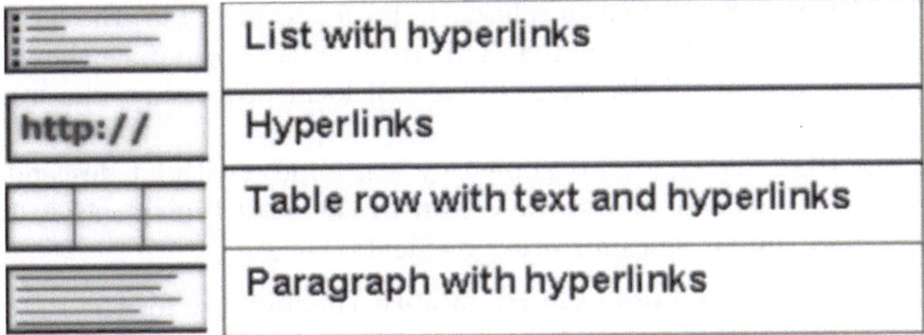

Fig. 6. Representative icons

DOM generates the tree hierarchy required in our philosophy and the new trees to obtain the navigator/page set. For non-XHTML WWW contents we need a JTidy preprocessing stage to generate the appropriate input for the conversion stage. Note that this conversion does not impose constraints on original WWW contents (this is not the case of WML and c-HTML).

A final remark: HANd implementation is based on language conversion, in a broad sense. It can be considered a particular case of transcoding strategy [7].

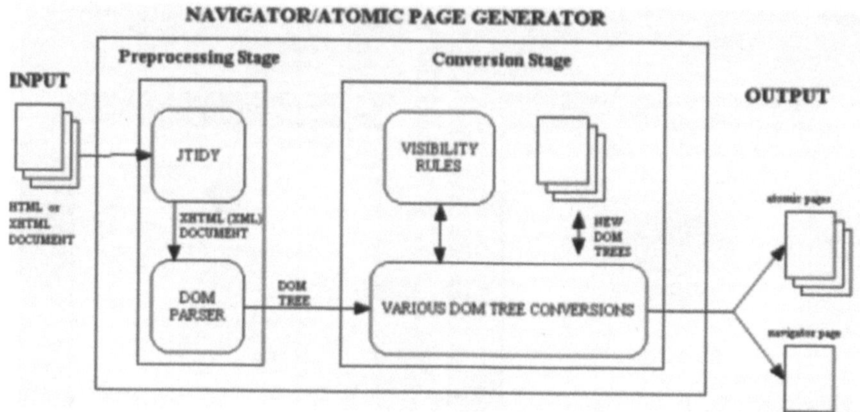

Fig. 7. Navigator/atomic page generator

3.4 Transformation Rules and Representatives

The following rules associate original page elements to their representatives in the navigator page. The tree of the original page is followed in increasing hierarchical order, ending at the root node, n_s. Let (n_i, n_{i-1}) denote a parent-child node relationship between levels i and $i-1$, $i = 2, \ldots, s$.

For a XML page that has not been explicitly created for a small display device, it is assumed that a generic large display produces a good presentation. Using a 17" 1024×768 screen as a reference, for a specific PDA, the PDA-to-PC screen area ratio $R < 1$ is calculated. This ratio is used to reduce the size of all elements in the original XML page. In case of local processing, the user could set R manually.

Then, the PDA screen area occupied by leaf nodes (without descendants) is calculated. For the sake of clarity, we comment the most typical cases, although it is easy to find similar rules for other elements:

- *Text*: Since we are interested in readability, the font size for the PDA screen (after reduction) is calculated.
- *Images*: Given the PDA screen resolution and the (reduced) image size in pixels, the occupied PDA screen area is estimated.

For any reduced element, a readability threshold is applied. If a reduced element is smaller than its threshold, it is considered unreadable, and is labeled with 1. Otherwise, it is labeled with 0. There are some exceptions: for example, HR lines are always unreadable. Let $l(n_i)$ be the label of node n_i. Then, element labels are propagated across the XML tree as follows:

For $i = 2, \ldots, s$, *do:*
For all n_i, *do:*

> *Let* I_{n_i} *be the set of all nodes* n_{i-1} *such that* (n_i, n_{i-1}) *holds. Then,* $l(n_i)$ *is the product of the labels of all nodes in* I_{n_i}

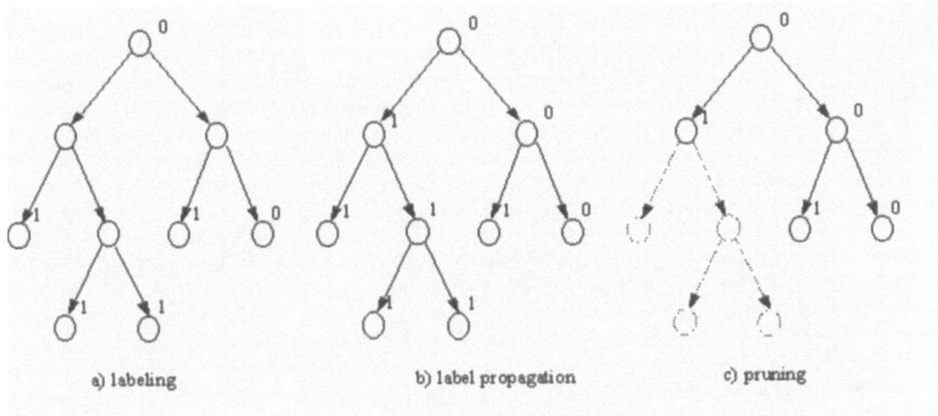

Fig. 8. Basic transformation algorithm

Finally, all branches that do not contain a 0 label are pruned. For the resulting subtree, the following rules apply:

- All nodes whose labels are 0 (readable nodes) are substituted in the navigation page by reduced representatives.
- All nodes whose labels are 1 are replaced by an icon. There is a different icon for each element (tables, lists, text...). For a given element, there are two possibilities, with a different background color: normal icon (black) and hyperlink icon (blue, if the original element is associated to a hyperlink). Specific icons are resized so that their area is approximately equal to the screen area covered by all associated descendants.
- Each icon is linked to an atomic page, containing either the associated original element or *all original elements associated to the corresponding pruned branch*, for intermediate nodes.
- *Adjacency rule*: all adjacent icons associated to elements of the same type are represented by a single icon. Elements such as HR or BR are transparent when applying this rule. Also, other information-irrelevant elements like small pictures (bullets, etc.) are considered transparent.

Figure 8 shows the three stages of the algorithm for a simple tree example: (a) labeling, (b) label propagation and (c) pruning.

Figure 9 shows the hierarchy tree of the page in figure 1. Note that several elements are pruned and replaced by the corresponding representative icon. The adjacency rule applies to the object group in the dotted box.

4 Usability Tests

In order to test the effectiveness of the new transcoding technique, we performed usability tests based on the WWW site that starts with the page in figure 1. The

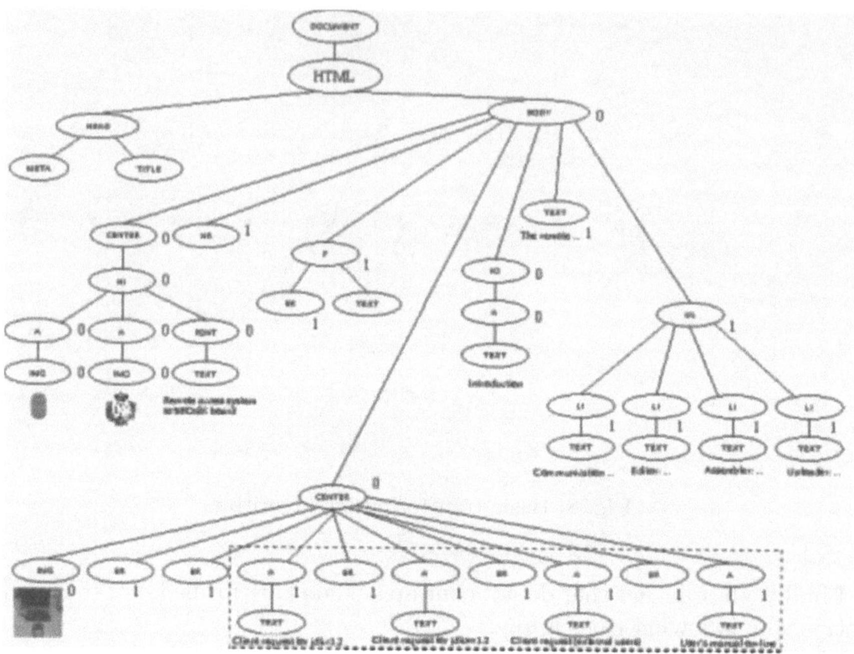

Fig. 9. Hierarchy tree of the page in figure 1

site contains a user manual that explains how to use the remote access system to a third-year laboratory course, *Arquitectura de Ordenadores II.*

We divided a group of students into three testing groups. The first one accessed the WWW site from the desktops in the laboratory room. The second one accessed the WWW site from an iPAQ H3630 PDA. The third one accessed the transcoded version from an iPAQ H3630 PDA. They had to solve a quiz on laboratory usage.

We tracked two quality measures per group: *average success* (percentage of right answers) and *average access time* (time to finish the quiz).

In order to calculate access time, we served all pages (original and transcoded) with Apache. Apache traces follow the CLF standard. Each line displays the IP address of the accessing computer, GMT access date with seconds-resolution, submitted command (GET, POST ...) and server answer (OK, 404 ...). Individual user times can be easily extracted with a filter.

As we expected, it was hard for group 2 to visualize normal pages on a PDA display. This made users prone to commit more mistakes. The first time the users in group 3 accessed the transcoded pages, their access time was slightly longer than the access time in group 1, but success was similar (and clearly better than success in group 2). In subsequent tests, the performance of users in groups 1 and 3 was similar, which suggests that users got used to the transcoding technique easily. Nevertheless, there was no improvement in group 2, neither in success nor in access time.

It should be understood the results cannot be generalized, since the users were skilled students. However, we think that they demonstrate the potential of the approach for small-to-medium traditional pages.

5 Conclusions and Future Research

In this paper, we have presented a new transcoding technique for small display devices: Hierarchical Atomic Navigation (HANd). Like the alternative transcoding techniques in section 2, it may need manual adjustment when processing visually complex pages like Amazon.com or www.elpais.es. Nevertheless, it has the following advantages over them:

- No information loss. All original elements can be shown on demand, as atomic pages.
- The navigator page is a readable overview of the original page, using a reduced space. Actual contents are accessible via their representatives (atomic navigation).
- Atomic navigation can be adapted to any WWW document, including pages created by accessing ASP or PHP pages. There is no need for a mark-up language that introduces content constraints, or a specialized WWW browser (both navigator and atomic pages are supported by standard mark-up languages).

Current work is oriented towards completing the following objectives:

- Integration of script languages in the framework. A language processor is required to generate a hierarchy of explicit objects. This could be achieved by means of tools like Rhino [18].
- Further enhancements of the HANd philosophy. For example, design of rules to generate hierarchies of navigator pages (or, in other words, a representative icon could be linked to a new navigator page in a lower hierarchical level).

References

1. "Market Risks: Security: The Downside of .com". Research briefs. http://www.infoworld.com/articles/hn/xml/99/07/19/990719hnrbcom.xml
2. PDAs at PC technology guide. http://www.pctechguide.com/25mob3.htm
3. Personal mobile communicators. http://www.pdastreet.com/
4. Lewis, T. "UbiNet: the ubiquitous Internet will be wireless". IEEE Computer 32(1999), pp. 126-128.
5. Gessler, S. and Kotulla, A. "PDAs as mobile WWW browsers". Proc. of the 2nd Int. WWW Conference. Chicago, USA. October 1994.
6. Jones, M., Mardesen, G., Mohd-Nasir, N., Boone, K. and Buchanan, G. "Improving Web interaction on small displays". Proc. of the 8th Int. WWW Conference. Toronto, Canada. May 1999.

7. Hori, M., Kondoh, G., Ono, K., Hirose, S. and Singhai S. "Annotation-based web content transcoding". Proc. of the 9th Int. WWW Conference. Amsterdam, The Netherlands. May 2000.
8. Resource Description Framework (RDF). http://www.w3.org/TR/REC-rdf-syntax/
9. Microsoft Pocket Internet Explorer. http://www.microsoft.com/mobile/pocketpc/features/pie.asp
10. Handy-Fit-to-Screen. http://www.microsoft.com/mobile/pocketpc/features/articles/web.asp
11. XML at W3C site. http://www.w3.org/
12. WAP. http://www.wapforum.com/
13. NTT DoCoMo. http://www.nttdocomo.com/
14. c-HTML. http://www.nttdocomo.com/source/tag/index.html
15. JTidy. http://lempinen.net/sami/jtidy/
16. HTML TIDY. http://www.w3.org/People/Raggett/tidy/
17. DOM at W3C site. http://www.w3.org/DOM/
18. Rhino at Mozilla site. http://www.mozilla.org/rhino/

Sorting Out Searching on Small Screen Devices

Matt Jones[1], George Buchanan[2], and Harold Thimbleby[3]

1 Dept. of Computer Science, University of Waikato, New Zealand,
always@acm.org
[2] Interaction Design Centre Middlesex University, London, UK,
george10@mdx.ac.uk
[3] UCLIC, University College London Gower St, London UK,
H.Thimbleby@cs.ucl.ac.uk

Abstract. Small handheld devices – mobile phones, PDAs etc – are increasingly being used to access the Web. Search engines are the most used Web services and are an important user support. Recently, Google™ (and other search engine providers) have started to offer their services on the small screen. This paper presents a detailed evaluation of the how easy to use such services are in these new contexts. An experiment was carried out to compare users' abilities to complete realistic tourist orientated search tasks using a WAP, PDA-sized and conventional, desktop interface to the full Google™ index. With all three interfaces, when users succeed in completing a task, they do so quickly (within 2 to 3 minutes) and using few interactions with the search engine. When they fail, though, they fail badly. The paper examines the causes of failures in small screen searching and proposes guidelines for improving these interfaces.

1 Introduction

Research into usable, useful and effective approaches for users to search the Web is vital. Unless effective user-centered approaches are developed and applied, the promise of wide access to the information resources will be lost, with users left frustrated and overwhelmed [16]. There is much work on the social impact of small screen devices [e.g., 13], which more-or-less take the technology and the user interface design in particular as given; this paper, in contrast, reviews and discusses experiments on the usability impacts of software and structural aspects of user interface design on small screens for supporting web applications.

Recently, Google™, one of the most comprehensive Web search engine, has offered its service to WAP phone users and via PDA type handheld computers . We evaluated the usability of these services to explore the impact of screen size. We looked at the performance of users under realistic search task situations on the different platforms. We were interested in differing patterns in user behaviour with the three services studied.

On a typical desktop screen, the user has many different ways to interact, often with varied interaction styles (menus, direct manipulation, text, etc). The rich desktop environment is in contrast to the "impoverished" interfaces of mobile, handheld devices. While a range of interesting search result visualization and manipulation schemes have been proposed for large screen devices (e.g., [6]), these schemes, on the

F. Paternò (Ed.): Mobile HCI 2002, LNCS 2411, pp. 81–94, 2002.
© Springer-Verlag Berlin Heidelberg 2002

whole, are not appropriate to handheld devices. Our work presents some first guidelines for those involved in designing approaches that are more appropriate for the small screen contexts.

2 Background and Review

2.1 Interfaces for Mobile Web Browsing

Most Web pages are designed for conventional large screen viewing. In earlier work we assessed the impact on user interaction of using such pages with the small display areas found on handheld computers. The study [11] suggested that users did not want to use the conventional page-to-page navigation as it was interactively very costly on the small screen. Rather, a much more direct, systematic approach requiring less scrolling was seen as appropriate.

WebTwig [12] was developed to demonstrate a direct approach to handheld browsing that takes account of the limited display. The tool presents an hierarchical outline view of any site and users can manipulate this view as they attempt to identify useful areas of the site. User evaluations of our system suggest benefits of this approach. Recent work by others amplifies our findings [2].

2.2 Interfaces for Searching

Most search systems simply present the results of a user query as a (long) ranked list, often broken into pages, of matching documents. With such an interface users have to scroll and page through the often-overwhelming list, examining documents in detail as they proceed to make relevance judgements. Such approaches mean that even on conventional large displays search interfaces are not highly usable. As Shneiderman *et al* [16] put it, "…the result is confusion and frustration."

Hearst [9] has identified two key types of interaction search interfaces should support. First, users should be able to scan search results quickly getting a feel for the effectiveness of their query and the sorts of information available. Second, interfaces must also facilitate the flexible, dynamic way users search. Users rarely view the information retrieval task as one of successively narrowing down a set of retrieved documents until a perfect match is found for some original information goal. Instead, goals change as the search proceeds: results can refine the original goal and trigger off tangential searches.

Search Visualisation Using Large Screen Devices
Information visualisation is a well-established research area [7]. Much work has been put into the use of highly graphically sophisticated approaches to help the user make sense of large sets of information. Such graphical schemes have been applied to the fields of information retrieval and exploration in an attempt to overcome search problems on conventional displays. For instance, the Information Visualiser [6] allows users to manipulate an animated 3-D categorical view of search results.

These types of visualisation scheme may not be appropriate for small screen devices. Even if the display technology can deliver the high resolution required, the available screen space is not necessarily adequate for meaningful presentations and manipulation by the user. Adaptations of certain of approaches may, though, be possible [8].

Visualisation schemes that are not graphically highly intensive have also been proposed for large screen devices. An example is the Scatter/Gather approach. Similar documents are automatically clustered together and key term summaries can be displayed for each cluster. By scanning the cluster descriptions, users are able to gain an understanding of the topics available. The approach has been applied to search result output and small studies indicate it may improve users' effectiveness [15]. Schemes like the Scatter/Gather system may bring gains in the small screen context as a significant amount of information about query results can be displayed in a small space.

Search Interfaces for Small Screen Devices

We have developed a new search interface that uses the WebTwig tree outlining technique; a screenshot is shown in Figure 1. The rationale behind the scheme is that the outline view not only limits the amount of scrolling required to make sense of the search results but provides context information which should help users make decisions about which alternatives to pursue.

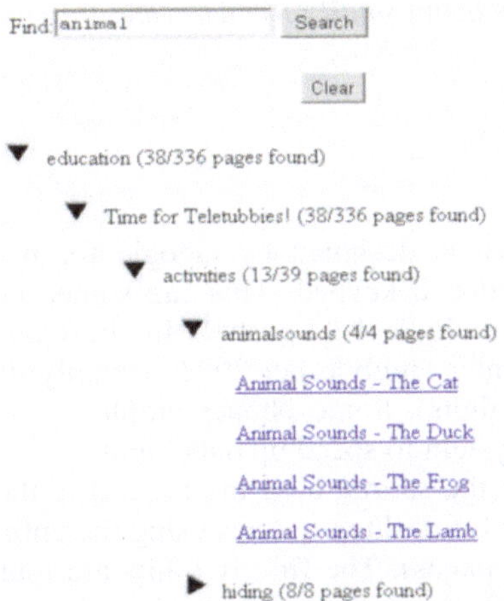

Fig. 1. WebTwig prototype search interface for handheld browser. User entered search term "animal". User was then presented with list of top-level nodes that contained hits (e.g., 'education'). By expanding nodes that contain hits, the user can progressively gain information on the context of hit. The prototype also allows the full path to a hit to be shown with one click

The PowerBrowser uses a similar approach to WebTwig for browsing, but a very different one for cross-site searching [3]. With each new search keyword, the user is shown the number of pages in the Web site that contain the term(s). Individual page details are only shown when the user feels the number of pages in the retrieval set is small enough to deal with on the small space of the screen. The danger, of course, is that relevant and important pages may be overlooked while the user focuses on reducing the number of pages retrieved. Unfortunately, the published papers on their novel schema do not evaluate the impact of this mode of interaction, so its impact, whether beneficial or negative, is not known.

3 Experiment: Evaluation of Small Screen Searching

Given the many unanswered questions about search on the small screen given above, we decided to embark upon an evaluation that permitted some of the outstanding questions to be investigated further. As already mentioned, we wanted to compare the performance of users using a traditional web search engine on different sized screen displays.

Google is one the Web's most comprehensive search engines with over 2 billion web pages. Although most people experience Google using a desktop computer, the company has recently started to introduce services for small screen devices.

Google on the large screen is viewed as useful and usable. Our aim here is to compare that success with user experiences when the search engine is used in both the WAP and PDA type contexts. It is easily accessible by other workers should others wish to replicate our work.

The sorts of questions we wanted to answer included: *does the small screen environment reduce users effectiveness and, if so, how?* ; and, *do users alter their searching behaviour when using small screen devices?*

3.1 The Three Interfaces

WAP Interface

Figure 2 shows the interface designed by Google for mobile phones. Users enter search terms using the reduced keypad. This can sometimes be a lengthy process – most text items require multiple key pushes. In the example shown here, 56 key presses were needed to enter "mobile hci 2002" (mainly due to the large number of key pushes required for digits). Some phones employ predictive text entry schemes such as the Tegic T9™[1] system to speed up data input.

When the user presses the *search* key, the request is then handled by the Google server in the same way as for desktop queries using the entire Google index (not some subset restricted to WAP pages). The first five hits are returned to the mobile phone (see Fig. 2b) and to view further hits the user can scroll past the last result and select a *Next 5…* link.

Unlike the standard Google interface, because of the very limited screen real estate, only the title and URL is displayed for each result. WAP's horizontal scroll feature is used to display this information: for example, Fig 2b shows the first part of the first search result's title ("MobileHCI 2002 (gio"); this is replaced with the remaining portion of the URL after a short delay (see Fig 2c).

---

[1] http://www.t9.com/

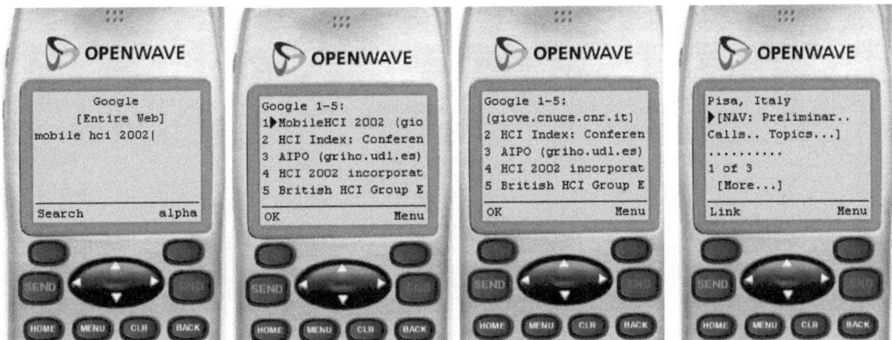

Fig. 2. (a,b,c,d : left to right): Google on WAP type device

When a user selects a search result link, the relevant page is delivered to the device. Less than 1% of the Web's content is marked up specifically for WAP devices using WML (WAP Markup Language)[2]. Standard HTML pages cannot be displayed, so Google pre-processes each page returned. As well as basic reformatting and coding, large HTML pages are broken down into several smaller linked WML pages to fit the micro-browsers "page" (deck) size limit (around 1400 bytes). Figure 2d shows the last few lines of the first page of the "MobileHCI 2002" site; users can access further pages by selecting the *More...* link.

Previous research suggests ways of improving the WAP interface described above. First, work on news headline search result display [1] indicated that users completed search result selection tasks quicker where the full details were shown as wrapped text instead of being horizontally scrolled. Second, the basic splitting of HTML pages could be enhanced using summarisation and outlining techniques [4][5].

The PDA Type Interface

The PDA type Google interface is designed for small pocket/PDA sized computers. Figure 3 shows an example interaction. First (Fig. 3a), the user enters search terms using the device's input mechanism that might involve stylus-based handwriting or an on-screen full keyboard, for instance. Then (Fig. 3b), five search results are displayed at a time. Here, more information is provided than with the WAP scheme: the document summary and a link (*sp*) to locate similar pages. When a page is selected (Fig. 3c) the full Web page is accessible, with the user having to scroll possibly both horizontally and vertically to view information. Google does no pre-processing of the selected documents.

Earlier work [11] showed that accessing pages designed for large screen devices on small display areas could lead to significant user problems. The search service, then, might be greatly improved if Google carried out some pre-processing on returned pages. Again, the summarisation and outlining techniques discussed in [4][5] are of interest.

---

[2] http://www.google.com/wireless/link_wap.html

Fig. 3. (a, b,c left to right) PDA-size Interface for Google

Conventional Interface

The standard desktop size Google interface is well known and is widely considered usable and useful. By default double (ten) the number of results is displayed on each page than in the WAP or PDA case. Additional metadata (description and category) and facilities (e.g. to view cached copy) are also presented.

3.2 Experimental Evaluations

A controlled experiment was carried out using three interfaces, a set of volunteer users and a range of information retrieval tasks. The aim of the experiment was to gauge the effectiveness of users using three different screen sizes (micro/WAP, small/PDA and large/conventional), and through this identify the impact on screen size on performance.

Apparatus

All 3 interfaces were presented using a conventional desktop computer and appropriate emulators/ screen sizes. All data entry was done using the standard desktop keyboard and mouse.

Use of the desktop platform allowed us to focus on the display-based interface differences. This was important for two reasons. First, it removed effects that might arise from the differing physical form factors, data entry and network conditions of the three platforms. Second, whereas the input approaches vary widely in actual devices (e.g. PDA users might use an on-screen keyboard, stylised text entry or cursive script), the display characteristics are more standardised. Search service designers, then will more easily gain benefits by targeting screen size issues. To reduce bias against the WAP interface, users were also not able to use the advanced search engine features available via the PDA and full screen interfaces.

For the WAP interface, the Openwave™ emulator[3] was used (see Figure 2, above). For the PDA (Figure 3) sized and conventional interfaces we built a browser that

[3] Downloadable from http://www.openwave.com/products/developer_products/sdk/index.html

allowed us to fix the screen dimensions appropriately. The Microsoft IE WebBrowser2 library was used. This technology is also used in the PocketPC versions of Internet Explorer. For the PDA display version, the sizes of interactive widgets were reduced to a size consistent with those on PDA-sized devices. The usual basic range of navigational tools (forward, back, home, etc.) was provided. However, the ability of the user to use many windows, and certain other functions, were removed to avoid accidental use of multiple displays and other anomalous features in the test environment. Furthermore, the browser was extended with its own, client-side, logging system to track the progress and interaction of the user accurately.

Subjects and Tasks

We recruited 12 volunteer subjects for our experiments from within the staff and student population at a university. Before taking part, volunteers were asked to provide some information so we could assess their prior use of the Web, search engines and mobile phone and PDA technology. Most of the subjects were not computer scientists, however most saw themselves as Web 'experts'. All had used a variety of search engines (including Yahoo!, Alta Vista and Lycos) and only 3 had never used Google before. Almost all (10/12) owned a mobile phone (with the average time of ownership over 28 months), 7/11 had previously used SMS services and 5 owned a PDA

Three realistic information retrieval scenarios were used in the experiment. Each scenario involved the user being in a city (London, San Francisco and Venice) and required them to complete 3 tourist type tasks. For example, in the London scenario the users had to find the opening hours of the National Gallery, a train schedule for London to Cambridge trains and the weather forecast. For each scenario the tasks were chosen to as similarly challenging as possible.

Experimental Method

Each user attempted to complete all 3 scenarios (that is 9 tasks). The scenarios were always presented in the order: London, San Francisco and Venice. The interface presentation order was varied for each user: 4 users were given the WAP, then the PDA and finally the conventional interface; 4 saw the conventional first, the PDA second and the WAP last; and 4 saw the PDA first, the conventional second and WAP last. Thus, each scenario was used with four users on each interface. This balanced ordering meant that all interfaces were used equally with all task scenarios and any learning or task-interface biases were reduced. In order to reduce any performance influence due to familiarity and experience, users were given a training session with each of the interface schemes immediately before attempting to carry out the tasks.

An observer sat next to the user during the trials and all sessions recorded on videotape. As the user carried out the tasks, they were free to verbalise their thoughts and the observer noted comments.

For each task, the time taken to complete a task and the number of tasks successfully completed was recorded. The number of search queries entered for each task, the number of search results that the user selected to look at and the number of Google search result pages examined were also recorded. After using each interface, users were asked to fill in a questionnaire to give their opinions on its usability and usefulness.

3.3 Results

The quantitative data we gathered related to users' ability to complete tasks with each interface and the interactions required for their performance. Performance was measured in time to complete the tasks and the number of tasks actually completed successfully. A task was completed either by a user giving an answer or when the user decided to stop looking for the information. A task was viewed as successfully completed if the user found information on a Web page that answered the task question. The interaction data relates to the logs of search engine user actions. Tables 1 and 2 present the data for task performance and search engine interactions for the three interfaces:

Table 1: Task performance using the different interfaces. Performance measured in time to complete (successfully or unsuccessfully) and number of tasks completed with correct answer

Interface	Time to Complete Task		Success (of 36 tasks)
	Mean (secs.)	Std. Dev.	
WAP	318	217	13 (36%)
PDA	207	157	29 (81%)
Desktop	165	160	34 (94%)

Table 2: mean search engine interactions per task. *Search attempts* is number of individual queries used in task, *search results selected* gives number of Web pages viewed as result of searches and *Google results pages viewed* indicates number of search results scanned by users (there were 5 results per page on WAP and PDA interface and 10 on conventional).

Interface	Search attempts	Search results selected	# Google results pages viewed
WAP	1.8	2.1	2.9
PDA	1.6	1.5	2.0
Conventional	1.8	1.7	1.9

Table 3: mean rating by users of Google search result information for the three interfaces

Interface	Mean rating of information *quantity*	Mean rating of information *quality*
WAP	3.4	3.75
PDA	4.4	4.8
Conventional	4.5	4.9

After using each interface, users were asked to rate search result information presented by Google in terms of the quantity and quality of information. Quantity was rated on a scale of 1 (too little) to 7 (too much); rating 4 on the questionnaire was specified as "good". Quality was rated on a scale of 1 (poor) to 7 (good). Table 3 shows the results of these ratings.

Users were also asked to order four pre-specified factors in terms of how helpful they were in assisting them to decide whether to select a search result for viewing. Table 4 shows the modal ordering given by users for each of the interfaces. Finally, the subjects were asked to order six factors in terms of their negative impact the use of their search engine. The modal ordering for each interface is shown in Table 5.

Table 4: modal ranking of factor most helpful to users when interacting with search results

WAP interface	PDA interface	Conventional interface
1. First few words of search result (the title) 2. URL of search result 3. Summary text of search result 4. Position of search result in list	1. First few words of search result (the title) 2. URL 3. Summary text of search result 4. Position of search result in list	1. Summary text of search result 2. First few words of search result (the title) 3. URL 4. Position of search result in list

Table 5: modal ranking of factors affecting users most adversely when interacting with search results

WAP	PDA	Conventional
1. Screen size 2. Navigation facilities 3. Search result description 4. Text/data entry facility 5. Responsiveness 6. Colours used in display	1. Screen size 2. Navigation facilities 3. =Search result description 3. =Text/data entry facility 4. Responsiveness 5. Colours used in the display	1. =Navigation facilities 1. =Search result descriptions 2. Responsiveness 3. Text/data entry facilities 4. Screen size 5. Colours used in display

3.4 Discussion

A striking result is the very poor performance of users when they used the WAP interface. Users took almost twice as long on average to succeed or give-up than when the conventional large-screen interface was used. They were also and were almost 60% less successful in completing tasks than the conventional case. The mean time to complete was compared with the PDA and conventional interfaces using the analysis of variance test (ANOVA) and found to be statistically significant at the 5% level ($p=0.001$). One possible explanation would be that users were obstructed by their unfamiliarity with the WAP interface. However, as stated earlier, users had been trained with each interface prior to the test, and examination of the video and transcript data gives little evidence to support such a hypothesis; there were few instances where users could be observed or expressed having problems with recalling the interaction, and these were not particularly present within this one interface.

The PDA interface users failed to complete 14% less tasks than large screen users. This is encouraging: a large proportion of tasks were completed by users with the small screen interface. In our earlier reported work [11], the performance difference was much bigger with small screen users failing 50% more often. In that work, though, users mainly browsed rather than directly accessed through search. We concluded in that study that direct systematic search for small screen contexts would lead to improved usability. The results here help to validate this observation.

In all three interfaces, the numbers of search engine interface actions is small. This is consistent with other studies that show that show for instance that users usually only make one search query and rarely go beyond the second results page [10]. Interestingly, even though both the PDA and WAP interfaces display only half (five) the number of results on the first (and subsequent pages) than the large screen display, the number of results pages viewed in all cases is within the 2 to 3 range. So on average, users base their search result selection on 20 possible choices on the conventional interface and only 10 on the PDA/WAP interfaces.

It should be emphasised that for the WAP and PDA cases, the performance we noted suggests upper limits of performance. In real-life use, WAP and PDA users will have to enter search terms using a much impoverished input device (numeric keypad and handwriting for instance) and will have to navigate with a less sophisticated tool than the conventional PC mouse.
Although there is a trend of improved performance (both time and successful completion rate) as the screen size increases (WAP-to-PDA-to-conventional), on testing the mean time to complete tasks for PDA versus conventional screens we found no statistical significance. The reason for this lack of significance is the very high variability in completion times. We investigated these large variances further by looking at two groups of data for each interface: performance when users successfully completed a task versus that when they failed to complete. The results for each interface are shown in Table 6 (over).

For both the WAP and PDA cases, users spend over twice as long on a failed task and then give up than when they succeed. The differences in mean time of success and failure cases in WAP and PDA contexts are statistically significant. The distinction is even greater in the conventional interface case but there are only 2 failure cases and 34 successes.

With all three interfaces, then, when users succeed in completing a task, they do so quickly (within 2 to 3 minutes) and with few interactions with the search engine. When they fail, though, they fail badly.

We reviewed the logs we made during the studies and found some explanations for these two distinct patterns of use – quick successes and prolonged failures. First, we noted that problem cases were not "outlier" effects: 8 users with 9 separate tasks types accounted for the WAP failures, 5 users (and 4 different tasks) for the PDA and 2 users (and 2 different tasks) for the conventional interface.

Table 6: Task performance on WAP, PDA and conventional (Conv.) interfaces for successfully completed tasks (answer provided) and failed cases (user gave up). Levels of search engine user interactions (search attempts etc) also are shown

Interface	Outcome	Mean time to complete	Std. Dev. of completion time	Number of search attempts	Number of results selected	Number of Google pages viewed
WAP	Success	192	128	1.4	1.8	2.0
	Failure	430	222	2.2	2.4	3.7
PDA	Success	165	135	1.3	1.4	1.7
	Failure	381	117	2.6	2.3	3.1
Conv.	Success	137	113	1.7	1.5	1.8
	Failure	627	122	3.5	4.0	2.5

Exploring a search result on the WAP and PDA interfaces can involve a very high user cost in terms of time and effort. As Figures 1 and 2 illustrate, finding information within a conventional HTML page, which is being redisplayed on the smaller interface, can be a tedious, time consuming and frustrating task.

When users failed using the WAP and PDA interfaces, the main reason for failure, and the associated large task timings, was the great difficulties they had in navigating the site selected from the search result. Most of their wasted time and effort was spent in becoming increasingly lost within the small window. As Table 6 indicates, failing users also carried out more search engine interactions: they carried out a greater number of search attempts, browsed more of the search result pages and selected more of the search results.

The impression when observing these cases was of users 'thrashing' to try and solve the problem. They would carry out an initial search attempt, spend more time scanning the search result outputs, explore a search result and become lost and frustrated, then return to the search engine for another fruitless attempt.

In the unsuccessful cases, often it seems that users were very uncertain about whether a search result they were about to explore was going to be of any use. They then made blind leaps of faith into a usually disappointing unknown.

The successful cases for the WAP and PDA contexts were where the search engine results contained "obviously" good candidates. These results were the ones where even the limited information about the page (title and URL for WAP and title, URL and limited summary for PDA) was enough to suggest the page was worth exploring.

In real world use (using the physical devices rather than emulators) we might expect to see more of the unsuccessful cases – as search term entry is expensive on the impoverished interface, less expressive queries might be entered, leading to poorer search results. Furthermore, users might be less inclined to review search lists due to the navigation costs.

User Subjective Ratings

As well as differences in the measurable performance, as the screen size increases, so do users' satisfaction with the quantity and quality of information provided by Google (see Table 3). However, the range in ratings, around one point from lowest to highest, is not as wide as might be expected. A possible explanation for this is that Google have produced a very simple, uncluttered interface for all three devices.

Interestingly, Table 4 indicates users see the more descriptive elements of search results as most important. Titles are favoured over URLs and where detailed summary text is available (the conventional interface case) this is rated the most important.

The difference in users' views about factors that adversely affected their behaviour (see Table 5), show that screen size was seen as biggest limitation for WAP and PDA (it is rated fourth out of fifth in the conventional case). Leaving aside screen space, for all three interfaces, the search engines navigation facilities (manipulating the result sets) and search result information were rated as negatively affecting user behaviour.

4 Design Guidelines for Small Screen Search Engine Interfaces

Clearly screen size has a major impact on user performance. Success rates drop and even the time to complete successful searches increases. From our evaluations and observations, we propose several ways that the Google WAP and PDA interfaces might be improved. These guidelines will also be of interest to others developing search interfaces for small screen contexts.

1. Reduce the amount of page-to-page navigation needed to view search results. Users do not look at many search result pages and also prefer not to shuffle with groups of pages to view information. As we observed in [11], page-to-page navigation is very costly when browsing in general, and in our current observations we have seen similar behaviour when users are browsing search results. Although increasing the number of results on a WAP card or PDA screen will lead to increased vertical scrolling this additional user effort affects performance to a lesser extent than the page-to-page navigation.

2. Provide more rather than less information for each search result. Users value good quality information about search results. As we have seen, selecting a search result particularly for WAP is a very "risky" action. Users were clearly observed seeking information to guide their next step as they browsed the search result list, and expressed uncertainty when given what they felt was inadequate information. Better quality information should support user confidence, and if appropriate should also enhance performance. For the WAP interface especially, more information should be provided and should be presented using the wrapped round text rather than the automatic horizontal scroll method [1]. Clearly, for WAP given the limited deck size, there needs to be a technical trade-off between this guideline and the first.

3. Provide a quick way for users to know whether a search result points to a conventional HTML page or a small screen optimised page. If search results are not optimised for WAP pages, there is very little point in WAP users selecting them: users will simply become lost as they struggle through the many WAP cards needed to represent the HTML page. We observed that frames-based sites can be particularly damaging, even with sophisticated conversion. Although the larger display area on PDA type computers reduces the problem, pages adapted for these devices will be easier to use. The search result list could use a small icon or text device to let users scan and find small screen suitable information. It may not be possible, for any one of a host of reasons, to provide a small-screen optimised version of a site. Where an optimised form is available, users will generally perform better, through reduced scrolling, so assist them in making an informed choice before committing (and then often failing on poorly converted pages).

4. Pre-process conventional pages for better usability in small screen contexts. Google already pre-processes non-WAP pages so they can be displayed on WAP devices. More sophisticated adaptations for both WAP and PDA sized screens are possible (see Section 3.1). This could lead to much increased user effectiveness.

5. Adapt for vertical scrolling – in our first evaluation [18] and our observations in this evaluation, users tend to scroll *vertically* rather than *horizontally* – design with this bias in mind; information which requires significant sideways scrolling will often never be seen.

5 Conclusions and Future Work

More and more people will soon be using small, handheld devices to search the Web. Anticipating this, Google (along with other search providers) have begun to introduce services for these new platforms. Although the technology has been 'optimised' for the *device* capabilities (bandwidth, memory sizes etc), our work suggests that optimising for *user* capabilities would improve such services greatly. User-based experiments are important, particularly as mobile Internet devices are developing quickly without a transparent process for careful user-centred design.

As screen size is reduced, from full screen to PDA-sized and yet further to mobile phone dimensions, user performance drops. The main reason for this is that on a smaller screen it becomes increasingly more difficult for a user to make good quality judgements about the usefulness of any particular search result. Poor search result choices can be disastrous in human-computer interaction terms: some of our users became completely lost, spending 10 minutes trying to find information on a WAP screen that took 10 seconds to locate on a conventional desktop computer.

Using the search engine via a WAP phone was very ineffective. However, the performance on the PDA-sized screen was encouraging. Our previous studies suggested that for this sort of screen area, user performance would be good if direct, search-based access was provided to Web resources; the results here add weight to this claim.

This study focussed on the screen-size issues. Now that we have "upper-limit" indications on performance, we are extending our evaluations to look in more detail at interaction problems when the search services are used on the physical devices. We expect the patterns seen in this work to be repeated but are interested in measuring, for example, search query length used given different text entry mechanisms.

For improvements in both WAP and PDA-sized devices, search engine designers need to develop interaction schemes that allow users to better assess search results. Users should be able to make good choices quickly. Further, when a conventional Web page is re-displayed on the smaller devices pre-processing is needed to help users navigate within the information. We are using the results of this study to further develop and evaluate the WebTwig [12] and alternative approaches for small screen search.

Acknowledgements

Thanks to Craig Neville-Manning of the Research Group at Google Inc. who provided technical information and advice about the WAP and PDA search index and pre-processing.

References

1. Buchanan, G., Farrant S, Jones, M., Thimbleby, H., Marsden, G. & Pazzani M. (2001). Improving mobile Internet usability. Proceedings of the Web 10 conference on World Wide Web, 673 – 680. Hong Kong, April 2001

2. Buyukkokten, O., Garcia-Molina, H., Paepcke, A. & Winograd, T. (2000a). Power browser: efficient Web browsing for PDAs. Proceedings of the CHI 2000 conference on Human factors in computing systems, pp 430 – 437. Amsterdam.
3. Buyukkokten, O., Garcia-Molina, H. & Paepcke, A. (2000b). Focused web searching with PDAs. Proceedings of Web 9 conference, Amsterdam.
4. Buyukkokten, O., Garcia-Molina, H. & Paepcke, A..(2001a) Accordion Summarization for End-Game Browsing on PDAs and Cellular Phones. Human-Computer Interaction Conference 2001 (CHI 2001). Seattle, Washington - 31 March-5 April, 2001
5. Buyukkokten, O., Garcia-Molina, H. & Paepcke, A.. (2001b) Seeing the Whole in Parts: Text Summarization for Web Browsing on Handheld Devices. The 10th International WWW Conference (WWW10). Hong Kong, China - May 1-5, 2001.
6. Card S.K, Robertson, G.G. & Mackinlay, J. D. (1991). The information visualizer, an information workspace. Proceedings of Human factors in computing systems (CHI 91), pp181–186
7. Card, S. K., Mackinlay, J. D. & Schneiderman, B. (1999). Readings in Information visualisation – using vision to think. Morgan Kaufmann Publishers, Inc.
8. Dunlop, M. D. & Davidson, N. (2000). Visual information seeking on PDAtop devices. Volume II Proceedings of BCS HCI 2000, pp19-20, Sunderland, September 2000.
9. Hearst, M. (1999). User interfaces and visualisation. In Modern Information Retrieval, R. Baeza-Yates & B. Ribeiro-Neto (Eds). ACM Press/Addison-Wesley.
10. Jansen, B J, Spink, A, Saracevic, T. (2000). Real life, Real Users and Real Needs: A Study and Analysis of User Queries on the Web, Information Processing and Management, 36(2), 207-227
11. Jones, M, Marsden, G., Mohd-Nasir, N, Boone, K, & Buchanan, G. (1999a) Improving web interaction in small screen displays. Proceedings of Web 8 conference, pp51–59, Toronto 1999
12. Jones M, Buchanan, G & Mohd-Nasir, N (1999c). Evaluation of WebTwig — a site outliner for handheld Web access. International Symposium on Handheld and Ubiquitous Computing, Karlsrhue, Germany. Gellerson, H-W (Ed.), Lecture Notes in Computer Science 1707:343–345, Springer-Verlag.
13. Ling R., editor, *Personal and Ubiquitous Computing*, Special Issue on Mobile Communication, **5**(2), 2001.
14. Marsden, G., Gillary, P., Thimbleby, H. & Jones. M. The Use of Algorithms in Interface Design. International Journal of Personal and Ubiquitous Technologies. Springer-Verlag Volume 6 issue 1, January (2002)
15. Pirolli, P., Schank, P., Hearst, M. & Diehl, C. (1996). Scatter/ Gather browsing communicates the topic structure of a very large text collection. Proceedings of Human factors in computing systems conference (CHI'96):213–220.
16. Shneiderman, B., Byrd, D. & Croft, B (1998). Sorting out searching. Communications of the ACM, 41(4):95–98.

Adapting Applications in Mobile Terminals Using Fuzzy Context Information

Jani Mäntyjärvi[1] and Tapio Seppänen[2]

[1] Nokia Research Center, P.O. Box 407,
FIN-00045 NOKIA GROUP, Helsinki, Finland
Jani.Mantyjarvi@nokia.com
[2] University of Oulu, P.O. Box 4500,
FIN-90014 University of Oulu, Oulu, Finland
Tapio.Seppanen@oulu.fi

Abstract. Context-aware appliances are able to take advantage of fusing sensory and application specific information to provide proper information for situation, more flexible services, and adaptive user interfaces. Characteristic for mobile devices and their users is that they are continuously moving in several simultaneous fuzzy contexts. We present an approach for controlling context aware applications in the case of multiple fuzzy contexts. The design of controllers and experiments with real user data are presented. Experimental results show that the proposed approach enhances the capability of adapting information representation in a mobile terminal.

1 Introduction

Context aware computing enhances the usability of mobile terminals by enabling the adaptation of various applications and user interfaces according to the usage situation of a mobile device user. The implementation of context awareness into a mobile terminal requires sensing of a physical environment via on-board sensors, exploiting of context information sources located in the Internet, as well as, acquiring terminal specific knowledge of applications and operating networks.

There are several studies concerning implementation of sensing systems for mobile terminals. Brown proposes common triggering mechanisms for executing context aware applications [1]. The work of Schmidt et al. presents the implementation of several low cost sensors into a mobile phone and into a personal digital assistant (PDA) to recognize the context of a mobile device user [14]. Hinckley et al. studied sensor integration into a PDA for developing novel applications, and for examining usability issues [7]. They presented several new user interface applications for mobile terminals enabled by a sensor fusion. The study also included tests with real users.

F. Paternò (Ed.): Mobile HCI 2002, LNCS 2411, pp. 95–107, 2002.
© Springer-Verlag Berlin Heidelberg 2002

A number of research efforts concerning the development of context recognition and extraction systems have been carried out. Context recognition methods for wearable computers have been studied with the means of a wearable camera, environmental audio signal processing, and Hidden Markov Models [3]. The context recognition from multidimensional sensor signals can be carried out for example by using algorithms based on neural networks [14]. Signals from multiple sensors can be processed into multidimensional context vectors, from which relevant context information can be extracted and compressed by using statistical methods [5]. To extract higher-level descriptions of contexts, the time series segmentation and clustering methods can be utilized [6, 12].

Exploiting context awareness in mobile terminals sets challenges for software architectures. Salber et al. introduce an idea of context toolkit for supporting the development of context aware applications [13]. The architecture introduces context widgets mediating between the environment and application. Widgets handle the sensing of the environment and they provide context information to applications. The study made by McCarthy et al. presents a tool for supporting peripheral awareness in workplaces [10]. The tool allows users to be notified when a predefined event occurs. Mandato et al. present a concept of a modular mobile Internet portal enhanced with context-aware features [9]. An architecture of an IP based context-aware mobile portal provides a framework for hosting services, a middleware for authenticating users, managing users preferences, access functions to the context, an adaptation process e.g. for personalization.

In most of the studies concerning the development of software architectures for context-aware applications the format of context information is assumed to be crisp, 1 or 0. However, mobile devices and their users are continuously moving in several simultaneous fuzzy contexts, which increases the risk of launching applications in unsuitable situations. Context awareness for dynamic environments sets special requirements related to usability issues and social acceptance of novel functionalities of applications. Context aware applications must be able to operate according to multiple fuzzy context information with a minimum misinterpretation of current contexts.

In this paper, we present an approach for adapting applications in mobile terminals according to fuzzy context information. The application adaptation, explained in Section 2, is carried out by directly controlling different features of user interface (UI) applications. The adaptation exploits fuzzy logic controllers (FLCs) in handling multiple fuzzy contexts. In Section 3 the design of FLCs is presented and the performance of the application adaptation is examined with concrete examples with real user data. Section 4 concludes the work.

The main motivation for utilizing fuzzy sets and rules is that they provide continuous control signals for applications unlike Boolean rules used with threshold levels that produce discontinuous control signals. Another motivation is that context information can be represented in a uniform format.

2 Fuzzy Control of Applications

This section explains methods for converting raw signals from context information sources into a uniform format for representing context information. A framework for fuzzy control of context aware applications is also explained.

2.1 Fuzzy Sets

The theory of fuzzy sets defines the relations and operations between variables and fuzzy sets [4, 17]. Operations of fuzzy sets can be utilized effectively in decision-making systems [4, 17]. Fuzzy logic control has been widely used for control purposes in many fields of research [2, 8, 15]. A variable can be considered as an input x, which is related to a fuzzy set A by a membership function μ. The relation is expressed as $\mu_A(x)$. The value of membership function is between 0 and 1, $\mu_A(x) \in [0,1]$, 0 denoting a null membership and 1 denoting full membership. An example of fuzzy sets with membership functions related to a variable x is presented in Fig. 2.

Basic operations for fuzzy sets are complement, intersection and union. The complement of a fuzzy set A is

$$\mu_{\neg A}(x) = 1 - \mu_A(x) \quad (1).$$

The intersection between two fuzzy sets A and B for aggregating two membership functions can be expressed as

$$\mu_{A \cap B}(x) = \min(\mu_A(x), \mu_B(x)) \quad (2).$$

The union between two fuzzy sets A and B is

$$\mu_{A \cup B}(x) = \max(\mu_A(x), \mu_B(x)) \quad (3).$$

Fuzzy sets form basics for developing the format for context representation. Operations of fuzzy sets are used in designing rule bases for FLCs.

2.2 Context Representation

The first task in context extraction is to abstract raw sensor signals and compress information by using different signal processing methods. The purpose is to advance to processing perceptions instead of raw measurements. The features to be extracted are chosen according to how well they describe some parts of the real world context. The illustration of the process for increasing the abstraction level of raw context information is presented in Fig. 1. After sampling raw signals are preprocessed; for instance, in the case of sensor measurements, they are filtered or converted to the frequency domain. Several types of preprocessed signals are used as inputs to various feature extraction methods producing features to describe low-level context information. For example, root mean square (RMS) value of an audio signal sequence describes the loudness of audio in the proximate environment.

We have chosen to apply fuzzy quantization resulting in a more expressive representation of context information. Fuzzy quantization can be viewed as a granulation of information [16]. Two different methods can be used to quantize features to represent context information: i) Set crisp limits for features. The result is logic style true false labeling for fuzzy membership function $\mu_L(x)$. For example, in the recognition of loudness of environment audio, the processed feature is divided into three quantities labeled Silent, Modest, and Loud, corresponding to three membership functions as presented in Fig. 2. If one of these is true, others are false. ii) Apply a fuzzy set for features resulting in a more expressive fuzzy labeling. For example: $\mu_L(x)$ = 0.7 / Silent + 0.3 / Modest + 0 / Loud.

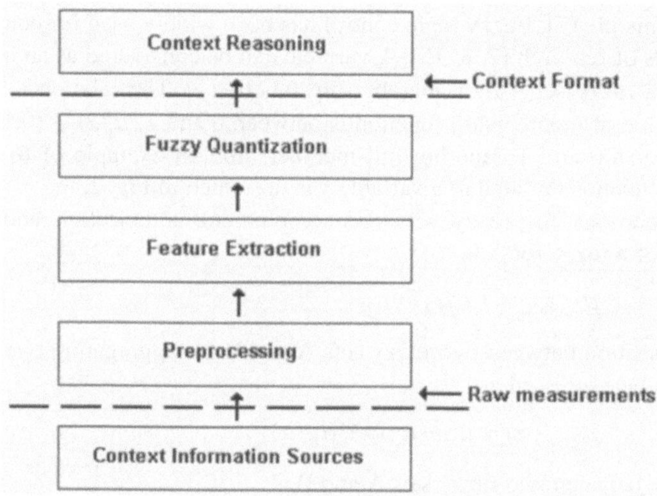

Fig. 1. Illustration of context information processing procedure. Context reasoning produces higher-level context information presented in fuzzified format

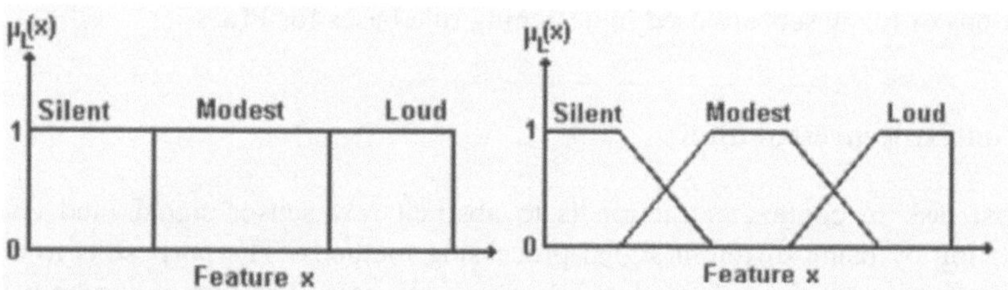

Fig. 2. Examples of crisp (on the left) and fuzzy (on the right) quantizations

The context information processing procedure combines quantized features into a context atom vector $\boldsymbol{x(n)} = [x_1, x_2, \ldots, x_k]^T$ at each quantization time step *n*. The values of individual context atoms are between 0 and 1. The context format presented here has many advantages: i) Elements of the context atom vector $\boldsymbol{x(n)}$ can be utilized directly for application purposes. ii) Context atom vector $\boldsymbol{x(n)}$ is normalized to the range [0, 1], enabling the utilization of e.g. explorative data analysis methods for examining higher-level contexts [5, 6, 12]. The format for context representation presented here can also be utilized for higher-level contexts. iii) It is straightforward to add new context atoms into a vector $\boldsymbol{x(n)}$.

2.3 Application Control

The implementation of context awareness into mobile terminals requires novel approach for developing architectures for applications. Several promising architectures have been presented [9, 10, 13]. We propose an additional block for handling context aware applications utilizing low and higher-level context information. The approach differs from others because the block consists of fuzzy logic controllers, which are designed to use predefined sets of contexts as inputs in controlling directly different applications.

As in most architecture models, an application control is located between an application and information source, for instance, a server. The purpose of the application control is to process information from the information source into the functionality of application. Context information recognized by a terminal may affect on the control of applications. In the case of the user interface (UI) application of a mobile terminal, the application can be controlled according to user context by adjusting various features, such as, font size, spatial resolution, size of images, colors and contrast of images showed on a display, and the volume of operating sounds. The information content adaptation and adaptive compression of information presented on the display improve the capabilities of presenting only the task relevant information to a user.

We propose an additional block called 'Context based application control' to be inserted into the architecture. The overview of the proposed architecture is presented in Fig. 3. The control scheme is guided by the context information on many levels of abstraction, the requirements set by the usability models of applications, and the requirements set by the service information content domain. The 'Context based application control' -block resides between UI application and the source of context information. The block consists of i pieces of fuzzy logic controllers. Each FLC uses a predefined set of contexts as inputs and a particular rule base for handling information from multiple fuzzy contexts. A predefined set of contexts of a particular controller can be defined as a group of the most important contexts from the viewpoint of the functionality of an application or a certain feature of the UI application.

The functionality of the 'Context based application control' block includes the capability to deliver control information for the 'application control' block and directly for the features of UI applications. This ensures that it is possible to launch applications according to context information, and at the same time, to control the functionality of the features of UI applications.

Service adaptation covers the handling of physical, semantic and functional entities of media content of the service and it provides functional scaling for the content and representation [11]. The service adaptation can be exploited in representing only task relevant highlighted information according to context. Information presented by a

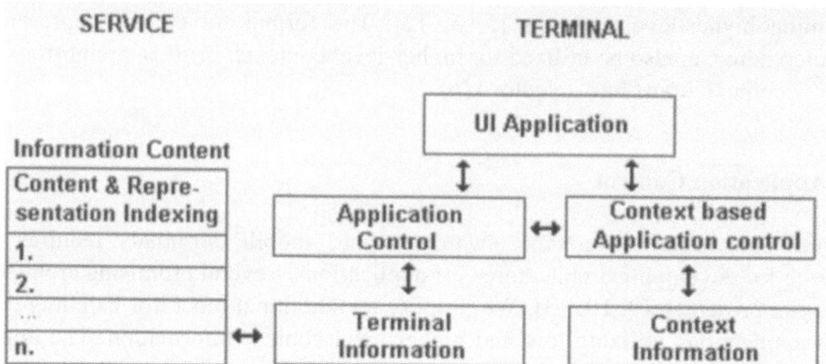

Fig. 3. Overview of architecture for controlling applications based on fuzzy context information. Service provides information to a terminal with indexed content. Context information-block consists of context information available in a terminal, e.g. low and higher level context information

terminal, e.g. a bus timetable, includes information content of a service with representation indexing. The indexing can be used to control the content and the format of service information according to csontexts. The adaptation can be realized by matching indexing information with appropriate control signals from the FLCs. Adaptation on application level takes place in the 'application control' block, while the adaptation on the application feature level takes place in the 'UI application' block.

3 Experiments and Results

This section explains the design of FLCs for adjusting applications according to contexts. Experiments with real user data demonstrate the control of content and information representation. Controlled application features include: Volume of operating sounds, font size of text presented on the display, display illumination, and the compression of information content, which is presented as text on the display. Data from a user scenario is logged with a sampling frequency of 256 Hz by using a portable laptop a user carries. Analog signals from accelerometers and illumination sensor are A/D-converted and sampled with National Instruments DaqCard 1200 measurement board, which is connected to a PC-CARD slot in a laptop PC. Audio data is sampled at 16 kHz by using the standard audio measurement board of the laptop. Data is stored in a file by using a measurement program made in LabView® graphical programming environment. Conversion from raw sensor signals to context atoms is performed by using self-programmed context recognition software, which utilizes fuzzy sets to produce context atom vectors $\boldsymbol{x(n)} = [x_1, x_2, \ldots, x_k]^T$, for describing three context types: *'User Activity'*, *'Loudness of Environment Audio'* and *'Environment Illumination'*. Each context type contains three context atoms resulting to a total number of nine context atom vectors. Context atom values are calculated every one second and a sliding window with the length of four seconds is used in calculating averaged context atoms to prevent transient interference in control signals. Fuzzy logic controllers are designed in a Matlab® Fuzzy Logic Toolbox. The experiments with a user scenario are performed using a Matlab® Simulink® simulating environment.

3.1 Design of Fuzzy Logic Controllers

Operating Sound Controller

One FLC is designed for controlling operating sounds. Inputs to the controller are context types *'User Activity'* and *'Loudness of Environment Audio'*. Feature type *'User Activity'* contains context atoms *Movements*, *Walking*, *Running* while context type *'Loudness of Environment Audio'* includes context atoms *Silent*, *Modest*, *Loud*. The context types are chosen based on tests with real sensor measurements. These types turned out to be the most important ones affecting to the audibility of operating sounds in dynamic environments. Membership functions of fuzzy sets i.e. context atoms used in the design of the controller are presented in Fig. 4. The output of the controller guides one specific feature of a user interface application: The volume of operating sounds including ringing tones, message alerts, and operating tones.

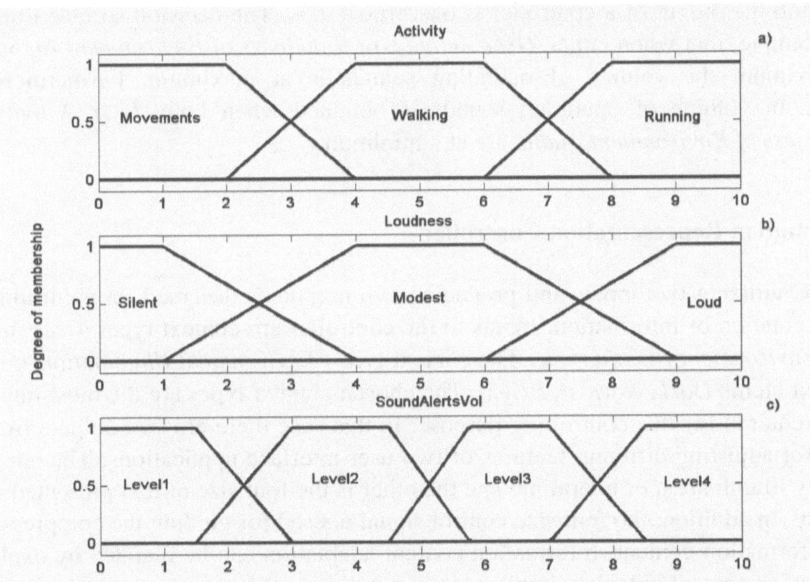

Fig. 4. Trapezoidal membership functions of context atoms used in designing the FLC. a) User activity, b) Loudness of environment c) Volume of operating sounds. Values of input and output features are on the range of [0, 10], and values of membership functions are on the range of [0, 1]

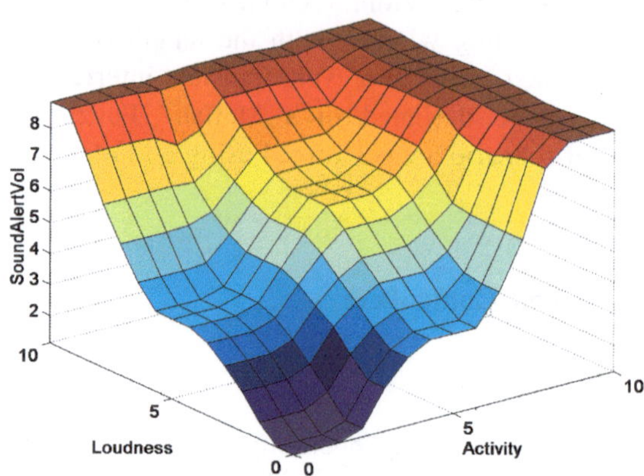

Figure 5. The decision surface of fuzzy sound alert controller. Input features are on horizontal axes and output of a controller is on vertical axis

Input and output are used in generating manually the fuzzy rule base for the controller. The rule base describes how a controller maps inputs into outputs according to the predefined relations, which are set by using basic operations of fuzzy sets described in equations (1), (2), and (3). The fuzzy rule base can be presented as a 3D-decision surface, which is presented in Fig. 5. Inputs to a controller are on horizontal axes and the output of a controller is on vertical axis. The decision surface illustrates for example, that when either *'User Activity'* or *'Loudness of Environment Audio'* is at a maximum the volume of operating sounds is at maximum. Furthermore, the minimum volume of operating sounds is obtained when both *'User Activity'* and *'Loudness of Environment Audio'* are at a minimum.

Information Representation Controller

A FLC utilizing two inputs and producing two outputs is designed for controlling the representation of information. Inputs to the controller are context types *'User Activity'* and *'Environment Illumination'*. The context type *'Environment Illumination'* contains context atoms *Dark*, *Normal*, *Bright*. The chosen context types are the most important ones required for this controlling purpose. In this case there are two outputs from the FLC for adjusting different features of two user interface applications: The one is the display illumination of a terminal and the other is the font size of text presented on the display. In addition, the font size control signal is used for guiding the compression of the information content. Information content adaptation can be realized by exploiting the service content indexing, implemented in mobile services.

Fuzzy rule bases for controlling features of information representation are illustrated as 3D-decision surfaces in Fig. 6a and Fig. 6b.

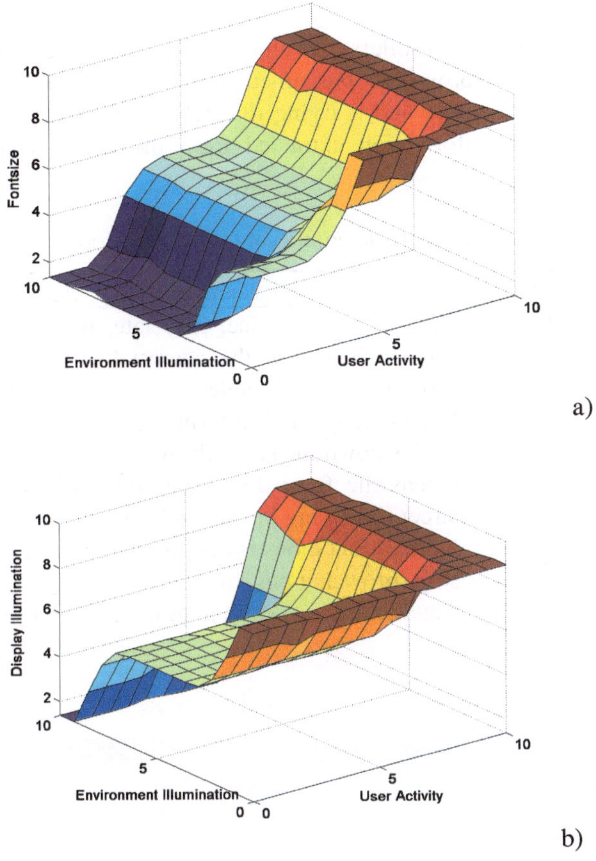

Figure 6. a) The decision surface of fontsize FLC. b) The decision surface of display illumination FLC. Input features are on horizontal axes and output of a controller is on vertical axis

The decision surface in Fig. 6a illustrates the behavior of font size control. When *'User Activity'* is at maximum the font size is large, whereas when *'User Activity'* is at minimum and *'Environment Illumination'* is quite bright the font size is small. The decision surface presented in Fig. 6b illustrates the behavior of display illumination control. The brightest display is obtained when *'Environment Illumination'* is low, and also when *'User Activity'* is at a maximum. The display is dark when *'Environment Illumination'* is high, except when *'User Activity'* is at a maximum. Both controllers are designed to maintain the maximum contrast of information presented on the display.

3.2 Experiments with User Scenario

Performance of designed controllers is evaluated with real context data recorded from a scenario where a user browses a bus timetable service by using a mobile terminal, while he is leaving the office and is trying to catch a bus. A scenario is described in Table 1. The context information extracted from the user scenario recordings is presented in Fig. 7.

Table 1. Description of the user scenario

Segment n:o	Explanation of the action
1	User is at the office, it is late, he is planning to catch a bus from the nearest bus stop, and he is browsing bus timetable.
2	Walks along a hallway (browsing bus timetable).
3	Walks down the stairs (browsing bus timetable).
4	Opens the front door, goes out and walks on the street (browsing bus timetable).
5	Notices that a bus is coming, begins to run to a bus stop (browsing bus timetable).
6	Catches the bus, steps in to it, pays, and sits down.

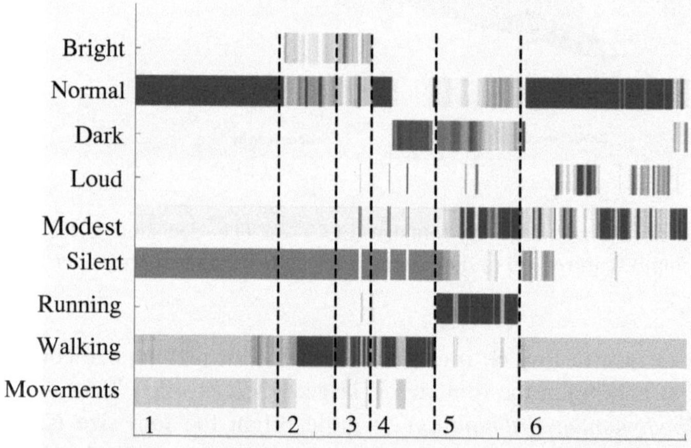

Fig. 7. Multiple fuzzy context signals from user scenario recordings. Values of contexts atoms are represented as gray level bars, black means 1 and white means 0 activity. Numbers of segments correspond to certain activities of a user

Experiments with a user scenario are performed as follows:

1) A user scenario presented in Tab. 1 is recorded and processed into context atom vectors.
2) Inputs are assigned to each FLC.

3) The output of the font size controller is used in adapting the information content of a text presented on the display. An information content indexing of a service is mapped to the quantized control signal of the font size FLC.

Fig. 8 presents outputs of FLCs as continuous lines. The numbers of various segments correspond to actions explained in Table 1. Adapted information content of a bus time table-services in different phases of the scenario is also presented.

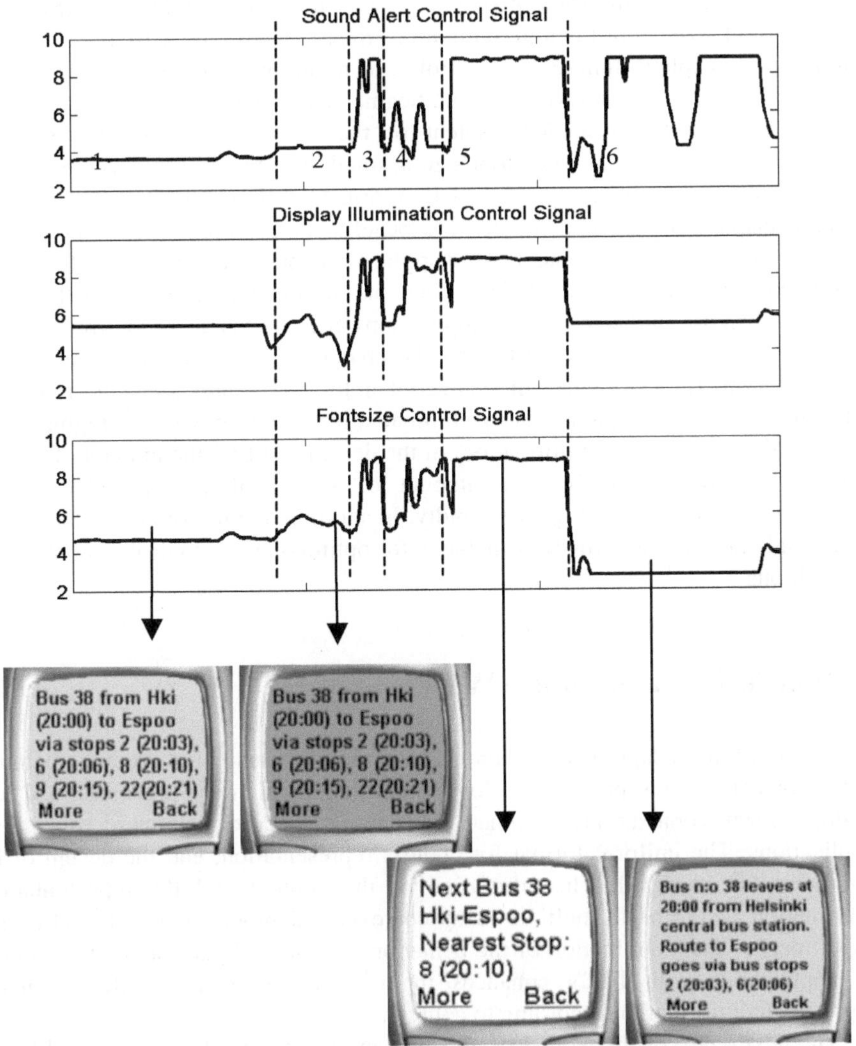

Fig. 8. a) Volume control, b) Display illumination control and c) Font size control signals from the user scenario. Numbers correspond the segments of the scenario. Outputs of FLCs are illustrated with continuous lines, while outputs of controllers based on traditional logic are presented with dashed lines. Adapted information contents of bus timetable service with illumination control in different phases of the scenario are also presented

The behavior of control signals, presented in Fig. 8, can be examined by comparing them to user actions expressed as context atoms in Fig. 7. Sound alert control signal is to a minimum level when a user is inside (Segment 1), and in some points when a bus is silent (Segment 6). The control signal starts to rise when a user begins to walk (Segment 2). The control signal rises considerably when a user walks down the stairs (Segment 3), reaching the maximum when a user is running (Segment 5), and in a noisy bus (Segment 6). Signals of font size and display illumination controllers behave in the same manner, except when a user is in a bus (Segment 6) when the font size control is to a minimum. This is because the activity of a user is to a minimum. Display illumination control is to minimum when the user is under the bright light when going outside (between Segments 2 and 3).

In the adapted information content of the bus timetable service very basic information is represented when font size is small. The bigger font size gets the more compressed is the content of presented information. It demonstrates the effect of content adaptation, which combines the benefits of the content and representation indexing of a service and context based application control in a terminal. It is required that the indexing is delivered via the service, or a terminal is able to index the content.

Control signals indicate that the applications are able to adapt to current context of a mobile device user by using FLCs and chosen context information. However, control signals tend to fluctuate when the values of individual contexts change rapidly. The fluctuation of control signals can be eliminated using e.g. moving averaging methods for smoothing the context information. In the design of FLCs the amount of important features can be larger, but the amount of fuzzy rules, which determine the functionality of a controller, grows rapidly. In addition, the use of controllers based on fuzzy logic provides a number of methods for optimizing the control of context aware applications.

4 Conclusions and Future Work

An approach for adapting applications in a mobile terminal according to multiple fuzzy contexts is presented. Fuzzy logic controllers, which use predetermined sets of multiple fuzzy contexts, are designed for adapting specific features of user interface applications. The uniform format for context representation, and the design of FLCs are presented in detail. The adaptation of the content and the representation of information according to multiple fuzzy contexts is demonstrated with real user data experiments. The examination of the behavior of control signals suggests that control of applications with FLCs enhances the capabilities of a mobile terminal for representing information according to context.

The next phase of the work is to develop and test the application control by using high-level contexts.

References

1. Brown,P.J., Triggering Information by Context, Personal Technologies, Vol.l. (1998) 18-27
2. Castro, A.P.A.; Da Silva, J.D.S.; Simoni, P.O., Image Based Autonomous Navigation with Fuzzy Logic Control, Proceedings on the International Joint Conference on Neural Networks (IJCNN), Vol. 3. (2001) 2200–2205
3. Clarkson, B., Mase, K., Pentland. A., Recognizing User Context via Wearable Sensors. Proceedings on the fourth International Symposium on Wearable Computers (ISWC), (2000) 69-75
4. Driankov, D., Hellendoorn, H., Reinfrank. M., An Introduction to Fuzzy Control, Springer-Verlag. Berlin, (1993)
5. Himberg, J., Korpiaho, K., Mannila, H., Tikanmäki, J., Toivonen, H., Time Series Segmentation for Context Recognition in Mobile Devices, International Conference on Data Mining, (2001) 203-210
6. Himberg, J., Mäntyjärvi, J., Korpipää, P., Using PCA and ICA for Exploratory Data Analysis in Situation Awareness, International Conference on Multisensor Fusion and Integration for Intelligent Systems, (2001) 127-131
7. Hinckley, K., Pierce, J., Sinclair, M., Horwitz, E., Sensing Techniques for Mobile Interaction, Symposium on User Interface Software and Technology (2000) 91-100
8. Ko; Y.C., Park; S.C., Chun; C.Y., Lee; H.W., Cho, C.H., An Adaptive QoS Provisioning Distributed Call Admission Control Using Fuzzy Logic Control, International Conference on Communications, Vol.2. (2001) 356 –360
9. Mandato, D., Kovacs, E., Hohl, F., Amir-Alikhani, H., CAMP: A Context-Aware Mobile Portal, Workshop on Service Portability and Virtual Customer Environments, (2001) 52–61
10. McCarthy, J., Anagnost, T., EventManager: Support for the Peripheral Awareness of Events, 2nd International Symposium on Handheld and Ubiquitous Computing (HUC), (2000) 227-235
11. Metso, M., Koivisto, A., Sauvola, J., Multimedia Adaptation for Dynamic Environments, 2nd Workshop on Multimedia Signal Processing, (1998) 203 -208
12. Mäntyjärvi, J., Himberg, J., Korpipää, P., Mannila, H., Extracting the Context of a Mobile Device User, 8th Symposium on Human-Machine Systems, (2001) 445-450
13. Salber, D., Dey, A.K., Abowd, G.D. The Context Toolkit: Aiding the Development of Context-Enabled Applications, Conference on Human Factors in Computing Systems (CHI), (1999) 434-441
14. Schmidt, A., Aidoo, K.A., Takaluoma, A, Tuomela, U., Van Laerhoven, K., Van de Velde. W., Advanced Interaction In Context, 2nd International Symposium on Hand Held and Ubiquitous Computing (HUC), (1999) 89-101
15. Sepehri, N.; Lawrence, P.D., Fuzzy Logic Control of a Teleoperated Log Loader Machine, International Conference on Intelligent Robots and Systems, Vol.3. (1998) 1571 -1577
16. Zadeh, L.A., Fuzzy Logic and the Calculi of Fuzzy Rules and Fuzzy Graphs: A Precis. Journal of Multiple Valued Logic, 1, (1996) 1-38
17. Zadeh, L.A., Fuzzy sets, Information and Control Vol.8(3), (1965) 338-353

Direct Combination: A New User Interaction Principle for Mobile and Ubiquitous HCI

Simon Holland, David R. Morse, and Henrik Gedenryd

Computing Department, The Open University, Walton Hall
Milton Keynes, MK7 6AA, United Kingdom
Fax +44 1908 652140
{S.Holland,D.R.Morse,H.Gedenryd}@open.ac.uk

Abstract. Direct Combination (DC) is a recently introduced user interaction principle. The principle (previously applied to desktop computing) can greatly reduce the degree of search, time, and attention required to operate user interfaces. We argue that Direct Combination applies particularly aptly to mobile computing devices, given appropriate interaction techniques, examples of which are presented here. The reduction in search afforded to users can be applied to address several issues in mobile and ubiquitous user interaction including: limited feedback bandwidth; minimal attention situations; and the need for ad-hoc spontaneous interoperation and dynamic reconfiguration of multiple devices. When Direct Combination is extended and adapted to fit the demands of mobile and ubiquitous HCI, we refer to it as *Ambient Combination (AC)*. Direct Combination allows the user to exploit objects in the environment to narrow down the range of interactions that need be considered (by system and user). When the DC technique of pairwise or n-fold combination is applicable, it can greatly lessen the demands on users for memorisation and interface navigation. Direct Combination also appears to offers a new way of applying context-aware information. In this paper, we present Direct Combination as applied ambiently through a series of interaction scenarios, using an implemented prototype system.

1 Introduction: Problems with Mobile HCI

User interfaces for mobile devices must typically deal with four general problems. Firstly, only limited screen real estate is available; more generally, taking into account non-visual forms of feedback such as auditory displays, *feedback bandwidth* is limited. Secondly, the bandwidth, precision and convenience of *input devices* are generally restricted. A third problem is that many mobile devices are typically used in *minimal attention situations* [1], where the user has only limited, intermittent attention available for the interface. In such situations, interactions with the real world are generally more important than interactions with the computer, the users hands and eyes may be busy elsewhere, and the user may be busy avoiding the normal hazards of moving around, as well as engaging with real-world tasks. Fourthly, as devices

F. Paternò (Ed.): Mobile HCI 2002, LNCS 2411, pp. 108–122, 2002.
© Springer-Verlag Berlin Heidelberg 2002

diversify and proliferate, users increasingly face the need to make two or more devices *interoperate for some ad-hoc purpose*. Even where each device has a well-designed user interface, this kind of task can be hard to arrange.

These four factors, singly and in combination, can cause difficulties in situations where the user does not know, cannot recall or cannot locate the commands needed to make the computer carry out a desired action. For mobile devices and their typical contexts of use, it is often inconvenient, impractical or too time-consuming to navigate to the appropriate screen or to otherwise search the space of available commands to effect the appropriate action.

In the case of context-aware devices, a final cluster of problems applies: there is no generally agreed uniform framework for applying contextual information. Context-aware systems tend to make incorrect guesses; and the user generally has little scope for correcting or capitalising on incorrect guesses [2]. This paper argues that Direct Combination has a part to play in addressing all of these problems, and has a useful role in any future framework for mobile and ubiquitous HCI.

2 Ambient Combination: Direct Combination Applied to Mobile HCI

The four general problem areas related to mobile user interaction noted above can be viewed in part as problems of *search*. That is to say, they all cause problems whenever a mobile device does not immediately afford the currently desired action, and users are consequently forced to navigate the interface, drill down, scroll or otherwise search the interface for the item needed to perform the intended action.

Direct Combination (DC) [3], previously applied to desktop computing, can often significantly reduce the degree of search required to operate user interfaces. In this paper, we argue that Direct Combination can be applied even more aptly to mobile computing and inter-device interactions. We will use the term *Ambient Combination* to indicate specifically the adaption, extension and application of DC principles to Mobile and Ubiquitous Computing [4]: however the most general term for the framework remains *Direct Combination.*

One convenient way to introduce Direct Combination in its Ambient Combination guise is by means of an imaginary interaction scenario featuring a magic wand. Binsted [5] has argued that imagined magic is a valuable source of inspiration when designing innovative forms of technology. We will use a hypothetical magic scenario to illustrate a crucial usability problem with magic wands (and other mobile devices). We will then illustrate how Direct Combination can address this, and other problems. Later in the paper, we will present interaction scenarios that we have investigated with a prototype implementation.

2.1 A Hypothetical Scenario: Harry and the DC Wand

Harry raised his wand towards the menacingly advancing Gator[1] *and tried to remember the spell for turning it into something harmless. It was no good; he just couldn't remember the right spell….*

Problems of this sort with magic wands are common in fiction and folklore. For example, the story of the Sorcerer's Apprentice deals with an inexperienced wizard who has memorised enough commands to start a process, but does not know, or cannot recall, the commands needed to control or stop it.

Harry suddenly remembered that this was one of the new Direct Combination Wands. He wouldn't need to recall the spell. Quickly looking around, Harry noticed a small stone on the floor. Pointing the wand at the Gator, Harry made the select *gesture and then made a second* select *gesture at the stone. A glowing list next to the wand presented the two available actions applicable to this particular* pair *of things:* propel the stone at the Gator *and turn* the Gator into stone. *Gratefully Harry activated the second command and the Gator froze into grey immobility.*

A key insight of Direct Combination is that if the user is allowed to indicate in advance *two or more interaction objects* involved in an intended action, the system can often use this information to constrain significantly the search space of possible commands. This allows the system to present to the user a space of focused relevant options to choose from, instead of the unrestricted space of commands. In an environment rich in objects of interest, users often know, or can recognize, what objects they want to use (a printer, a wallet, a car, a door, a document, etc) - but operations, apart from commonly used operations, tend to be relatively more abstract, and harder for people to recall. The contrast is between *recognition* (easy) and *recall* (hard). Given this insight, Direct Combination user interfaces give the user the freedom to specify the parts of commands in any order desired, for example *noun noun* (to be followed later by a verb), as well as the more conventional *noun verb.* At each stage, feedback is given on how this constrains the available options (examples follow later).

However, while this kind of pairwise interaction is sometimes invaluable, it is not always convenient. For example, sometimes the objects that would help to constrain the space of relevant actions are not to hand in the environment. Hence, Direct Combination requires that pairwise interaction always be made available in such as way that it does not get in the way of pre-existing interaction methods or practices. Direct Combination offers more ways of getting things done, but must never leave the user worse off than if DC was not provided - conventional methods should always be available. This can be illustrated by replaying Harry's imaginary encounter in more detail.

Harry raised his wand towards the menacingly advancing Gator and tried to remember a spell for making it harmless. Harry could see in glowing letters next to the wand the most common general spells, and category titles grouping all known spells, but he had no time to search this list. Pointing the wand at the Gator with the select

---

[1] An imaginary monster.

gesture, the list of general spells vanished, to be replaced by a list of the most common spells applicable only to Gators. The entire list of Gator-specific spells was now available - but Harry had no time to deal even with this extensive list. Remembering at last that this was a DC compliant wand, Harry quickly looked around. He couldn't see a pig, or a pig medallion, a mouse, a spider or even an ant, but he could see a stone on the floor and he made a second select gesture at the stone. The glowing list next to the wand shrunk to the two available actions applicable to this particular pair of things: propel the stone at the Gator *or turn* the Gator into stone. *Gratefully Harry activated the second command.*

The list of Gator-specific commands that appears when Harry selects the Gator alone is nothing new to DC enabled devices: many existing systems have such a feature. Even less remarkable is the fact that the DC wand would have allowed Harry to point at the Gator and then invoke the verb (spell) directly, if his memory had permitted. Although Direct Combination allows users to employ pairwise interaction *(noun noun ….)* whenever they wish, they are never forced to work in this way. A DC command processor must accept pairwise interaction, but also the more conventional ways of specifying commands such as *noun verb* and *verb arguments.* Indeed the Direct Combination principle of *Subsumption* (explained later) requires this freedom. DC also requires the system to provide appropriate feedback at all points that the user partially specifies the command, in order to show how choices so far constrain or suggest further choices. Finally, the power of Direct Combination is greatly enhanced [3] when the assignment of operations to pairs of objects need not be specified exhaustively by the designer, but can be propagated by inheritance from definitions in well-factored abstract classes, either held locally or in distributed fashion.

3 Scenarios Illustrating Ambient Combination

We have informally illustrated two principles of Ambient Combination, pairwise interaction and Subsumption. Ambient Combination is most easily understood further by considering more detailed scenarios.

3.1 Technology Required to Support Scenarios

Later in the paper we will detail the mix of technologies used in our implementation. However, for simplicity of exposition, for now we will describe the infrastructure in more general terms. Figure 1a shows the screen of a PDA. The PDA has a stylus for detailed work, but can also be operated one-handedly using buttons. Plugged into the system is a scanner for reading ID-tags indicating the *object* or *class* identity of devices and objects in the environment. The scanner allows the user to select items of interest in the environment, such as tagged documents, other PDAs, printers, book covers, telephones, merchandise, room nameplates, captioned photographs on the wall, etc., by scanning their IDs. The user may scan objects at a distance. In the context of Ambient Combination, selecting an object by scanning it is referred to as 'zapping'. The user always has full control of what is zapped and what is not zapped. Ideally, the scanner should form an integral part of the PDA, for example as a PCMCIA module. In our current prototypes we use 30-foot range barcode readers, but other technologies such as infrared scanners and RFID tags could serve equally well. The PDAs are wireless networked, so that they can access object directories, retrieve representations of objects stored elsewhere, and consult DC servers, as explained explain later. When initially switched on, the PDA gives indirect access, via applications and hierarchical menus etc, to all of the commands available in the system, just as usual - as shown in Fig 1a.

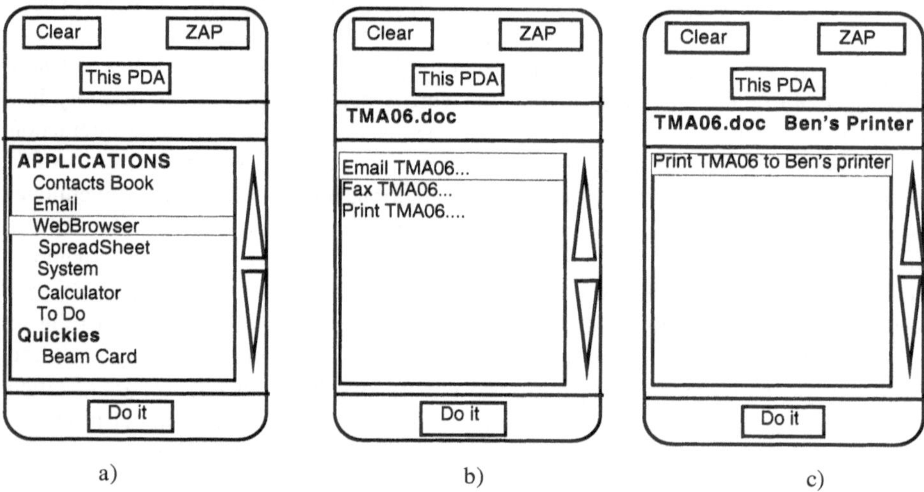

Fig. 1 a) shows no items zapped; b) shows document alone zapped; c) shows document and printer zapped, all in Scenario 1

3.2 Scope of Scenarios

The primary purpose of the scenarios reported here was to help us prototype the interaction techniques and infrastructure necessary for AC, rather than to spotlight application areas where it would be most valuable. For an a priori attempt to characterise such situations and domains, see Section 7.

Scenario 1 (Document/Printer)
Anne walks into Ben's office one evening to use his coffee machine. While Anne leaves with her coffee in one hand, Ben mentions that he is reading a paper document that might interest her. Anne wants a printed copy to read later. To save the inconvenience, time and trouble of searching for the document on a server (Ben can't recall the pathname or even what the file is called) Anne uses her PDA to 'zap' the bar code on the document and then 'zaps' the printer in Ben's office. A suggested action is presented on the PDA (figure 1c). Anne presses the 'Do it' button on her PDA to accept this action. The document is printed on the printer. The suggested action is:

Print 'TMA06' to Ben's Printer (see figure 1c)

To understand this interaction fully, it is instructive to consider the state of the interface on Anne's PDA at three successive stages, as shown by figures 1a, 1b and 1c.

Stage 1: No items zapped
Figure 1a shows the options available before any items were zapped. The PDA simply displays a choice of application programs and some immediate actions - precisely what a PDA would offer if Direct Combination were not available.

Stage 2: Document alone zapped
Figure 1b shows the options available after the paper document alone has been zapped, with a default highlighted. The options are

Email 'TMA06.doc' ……
Fax ' TMA06.doc' …….
Print ' TMA06.doc'. …c9….

These actions are those that are relevant when all that is known is that the document has been selected. Any of these options could be chosen by the user straight away and an argument filled in manually or by more zapping. There is nothing new about such a facility - its equivalent on the desktop is called a "contextual menu". In this situation, the lack of novelty is a virtue: it means that, whenever only a single item is zapped, DC does not get in the way of the conventional way of doing things. If the user decides that the wrong item was zapped, the cancel button may be pressed to deselect that object, in which case the situation reverts to figure 1a.

Stage 2: Both items zapped
Figure 1c shows the options available after both items are zapped. This illustrates a canonical pairwise interaction, as spelt out at the beginning of the scenario. Zapping both the document and printer icon greatly reduces the search space of applicable actions, avoiding the need to drill down or otherwise search on the PDA or desktop machine. Note that the options do not depend on the document or printer alone - they depend on the ordered pair of object types that have been zapped - and would in general be different for a different pair of objects. Typically, there is still more than one option presented - though just one in this case. If the presented action(s) did not seem appropriate or if something had been zapped by mistake, the user need merely press *cancel* to get back to the state where only a single item, the document, was selected. At this point, the options available would revert to those in figure 1b. Anne might choose to select some other zappable object e.g. one of the wallcons (see below) to initiate another way of achieving her goal. Or Anne might press *cancel* again to revert to the situation where nothing was selected.

Scenario 2 (Document/Person)
Anne might want to have the document emailed to her rather than printed. Anne can achieve this using DC in many ways. For example, Anne could use her PDA to zap the document and then zap her own id badge, if she is wearing one. Her PDA then offers the following options, with the first as the default.

Email 'Report42.doc' to self
Fax 'Report42.doc' to self

Scenario 3 (PDA as Proxy for Person)
If Anne had not been wearing a badge, then she could have scanned the document and pressed the "this PDA" button on her PDA instead. This is equivalent to the PDA zapping itself. Amongst other roles, PDAs often act as proxies or representatives of their owner. Hence, the same options are presented in this case as when Anne scanned her own badge.

As a minor variant of scenario 1, Ben might not have a printer in his room, but might have 'wallcons' on his wall (physical or projected captioned wall icons) representing various commonly used entities or resources, such as colleagues, departments, companies, projects, printers, locations, etc. Anne could use her PDA to zap the bar code on the document and then zap the printer wallcon on Ben's office wall. The following two suggested actions are presented on the PDA.

Print Document 'Report42.doc' to Room 308 printer
Print Document 'Report42.doc' to Meeting Room printer

4 Principles of Direct Combination

We are now in a position to state the principles of Direct Combination, extended and adapted slightly to deal with mobile and ubiquitous interaction (See box below). We will now comment on the three principles stated in the box in turn.

Principles of Direct Combination

idctlpar*Principles of visibility and pairwise (and more generally n-fold) interaction*
Subject to social, aesthetic, organisational and legal issues including privacy, ownership and security:

Every object of interest, both virtual and in the environment should be,
- visible (or perceptible),
- capable of a **range** of **usefuls**18 interactions with **any other** object of interest.

The available interactions between pairs (or n-tuples) of objects should be:
- diverse,
- tailored to each pair (or n-tuple) of object types.

Principle of Subsumption
DC interaction styles must be implemented in such a way that they are immediately available, but do not impede access to pre-existing interaction styles.

4.1 Visibility

Ambient Combination requires that as far as possible, every object of interest in a given environment (and in a user's computer) should be visible and capable of zapping (i.e. selection and identification using a pointing device). Issues of privacy, dignity, permission, safety and security must be met. The principle of *visibility* is borrowed from Direct Manipulation. However, many desktop interfaces that pay lip service to Direct Manipulation often fail to make objects of interest visible. In the area of mobile and ubiquitous computing, despite the popular notion of invisible computers, with its welcome stress on aesthetic minimalism, visibility remains vital for many kinds of interaction, especially, for example, to avoid the possibility of uninvited context-aware interactions [2].

4.2 Pairwise Interaction (More Generally, N-fold Interaction)

In most conventional systems, commands may be constructed by the user only in a restricted, unidirectional manner. The initial noun followed by the verb must be selected before any arguments can be specified. The order of assembling the command would not matter if it were not for the fact that these restrictions have significant implications for the search strategies available to the user. The user may have a clear intention, but may be unable to recall, or may be ignorant of, the appropriate verb to achieve it. Unknowns for the user may include the name of a command, how it is categorised, whether it exists, or where to find it. Even once the user has selected the initial object of interest, this does not always greatly constrain the space of applicable commands, so the user may still have a large space of verbs to search. As already noted, if the user can indicate in advance *two or more* of the interaction objects involved in an intended command this typically greatly *constrains the search space of possible interactions*. This allows the system to present a space of focused relevant commands. If the choice of focused commands offered is inappropriate, the user may deselect items or select other items of interest until the system makes suggestions that are more appropriate, or that can be easily modified to be appropriate. The requirement that **every** object of interest be capable of diverse interactions with **any other** object of interest is a heuristic to encourage designers to seek meaningful pairwise interactions (and to endow objects of interest with suitable operations) rather than an absolute universal requirement. The reference to *virtual* objects refers to objects of interest represented in a computer. When using Direct Combination, it is possible to mix both virtual and real objects in a single interaction [6].

4.3 Subsumption

Subsumption ensures that users are never worse off with a system that includes DC than with a system that excludes it. For example, in the scenarios, all of the traditional ways of using a PDA are available when nothing is selected: hence users can still make use of the familiar *verb, arguments* pattern or the *noun, verb, arguments* pattern. DC may be viewed as, and is easily implemented as, *subsuming* these patterns of interaction. The principle of Subsumption is a common sense precaution, as it is not always convenient to use pairwise interaction every time. Thus DC does not impose anything on users: it simply provides greater freedom. In many ways it is surprising that we didn't identify and rid ourselves of this imposition earlier.

5 Context-Aware Interaction

In this section, we will present some scenarios that illustrate some ways in which Direct Combination can afford context-aware interactions. These scenarios are analysed in more detail in [6]. The scenarios contrast what happens when Anne zaps someone else's PDA with her PDA in four contrasting situations. In each case, Anne includes her own PDA in the interaction by using the 'this PDA' button noted in scenario 3.

Scenario 4
Anne meets a stranger, and zaps their PDA with her PDA (including her own PDA as described in scenario 3). The following option is offered.
Beam card to strange PDA
This scenario is hardly novel, though the underlying processing behind it is slightly unconventional, as we will discover in a moment. However, it makes a useful initial benchmark for the subsequent scenarios.

Scenario 5
Anne meets a research colleague Fred in the corridor. Anne zaps Fred's PDA with her PDA. The options offered are as follows.
Email 'Research draft' to Fred
Email 'Research notes' to Fred
par Search folder 'ACDC 'for documents to email to Fred
Arrange meeting with Fred

Scenario 6
Anne meets a teaching colleague Tony in the corridor. Anne zaps Tony's PDA with her PDA. The options offered this time are as follows.
Email 'TMA04.doc' to Tony
Email 'TMA tutor notes' to ny
Search folder 'TMA04 2002' for documents to email to Tony
Arrange meeting with Tony
In each of these scenarios, different interactions are offered, although the ordered pair of object types is the same each time - two PDAs. In scenarios 5 and 6, which have been implemented, the crucial difference is the identity of the owner of the zapped PDAs, and the nature of their work relationship with Anne, as previously recorded directly or indirectly by Anne, as we will outline below.

The technique currently used to support this simple contextual interaction is association matching: in each case, the direct combination of PDAs instantiates an operation where the initiating PDA will act as a disseminator of information. The key is to determine from the context how to prioritise or order the information to be offered for dissemination. The current mechanism is simple. Where the identity of the owner of the zapped PDA is available for the DC server to look up, activities associated with that owner can be retrieved (e.g. in this case particular research or teaching activities) and compared with associations noted against files or folders stored by, or accessible to, the initiating PDA. Documents whose associated activities or groups match the activities or groups associated with the owner of the zapped PDA are then prioritised for dissemination. In the case of the stranger, no previous relationships are recorded, so only options appropriate for strangers are offered. However, if electronic business cards were to be exchanged, and if either party went on to annotate the other's card at a later point with discovered interests or roles in common (in a manner outlined below), more tailored options might be offered at a later encounter.

The next scenario, unlike the others, is not implemented in prototype. The necessary contextual inference used in this combination could be addressed by much the same technique as in the three previous scenarios, but seems better addressed by more powerful mechanisms, such as aspect-oriented programming and related techniques, as we are currently exploring. The next scenario, in contrast with the others, also involves *automatic sensing* of context (such as location) to complement the elective zapping we have considered exclusively so far (figure 2).

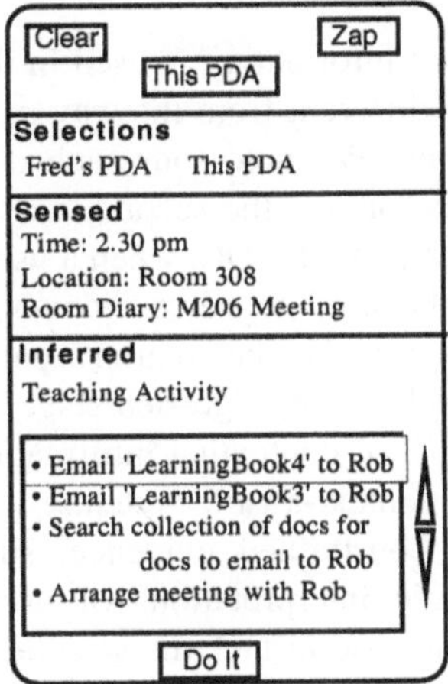

Fig. 2 A hypothetical contextually-aware interaction (See Scenario 7)

Scenario 7

Anne goes to a course team meeting and sees Rob, with whom she collaborates on both teaching and research. Anne zaps Rob's PDA with her PDA. The options offered are as follows (see Figure 2).

Email 'LearningBook4' to Rob
Email 'LearningBook3' to Rob
Search folder 'Learning Book revisions' for documents to email to Rob
Arrange meeting with Rob

Scenario 7 involves zapping between two PDAs, but this time, the colleague has twin roles, as a research collaborator *and* as a teaching collaborator. The key idea is that inferences about activities associated with the current time and with the automatically sensed location should be used to order the likely relevance of the teaching and research roles in the given context, and hence to prioritise the collection of associated documents to be offered for dissemination. In this hypothetical scenario (figure 2), Anne's location is sensed to be in a meeting room, and the booking diary associated with the meeting room shows a meeting booked with a course team of which Anne is a member. Hence the context is used heuristically to prefer the teaching role as more likely to be relevant than the research role to the interaction. Until Anne actively zaps something, these automatically sensed context objects (the location and the meeting - figure 2) are inactive in the DC system.

In all of the scenarios in this section, none of the interactions involve negotiation with the other PDA beyond extracting an object or class identity. Information from Anne's own diary, address book and desktop PC is used by Anne's DC server (see Section 6) to make the necessary inferences. In our prototype system, the DC server obtains knowledge about relevant relationships from earlier Direct Combination operations used to associate: work folders with activities (e.g. particular research projects or courses); individuals with groups; and groups with activities. These operations simulate Anne using desktop DC drag and drop operations [3] to maintain her desktop and address book. Similar mechanisms would apply, whatever sources of information were used.

The way in which contextual information is used in the implemented cases and in the hypothetical case is a little different from the way in such information is typically applied in conventional systems. We note four such differences. The first two are reflected in the implemented scenarios, the second two are not yet implemented, but follow from DC principles. Firstly, under DC, when a user selects two or more objects in the environment, this can be used to narrow down the set of intended actions, affording a substantial reduction in the search space. By applying contextual information at this late, relatively knowledge-rich stage, it can be applied in a highly targeted manner. This use in DC of contextual information at a knowledge-rich stage contrasts with many existing context-aware systems, which typically guess desired actions based on much less constrained evidence, such as automatically sensed information alone, or automatic interpretations of user's actions [2]. The second contrast is that, as is always the case in DC, the selection and execution of actions is left to the user, irrespective of whether contextual information is used or not: actions are not taken autonomously. A third difference is that under AC, context objects inferred from automatically sensed information, where context is used, are always made visible to the user (figure 2), who may opt to overrule the use of any particular item of context, or all of them. That is to say, the user always has the option of switching off the use of context in general, or particular sources of context, or explicitly discarding particular contextual inferences (e.g. inferences based on: location; calendar; address book; etc.) either until further notice, or for a given interaction. Finally, because DC creates such a knowledge rich environment, the user interface should always be able to afford the user fast ways to repair incorrect guesses on the part of the system by choosing alternatives to the presented objects. The way in which information about context is *detected* is no different in AC, but the way in which the context is *applied* appears to be different.

6 Architecture

For AC to be deployed in a given environment, various architectural elements need to be present. AC can be implemented in simple forms relatively easily. The environment needs to be made scannable in some way. We use optical barcode readers, including a thirty-foot range hand-held scanner, which are simple and effective, but may also be viewed as proxies for other more sophisticated ID technologies. The user needs a feedback device such as a PDA screen. A DC server is needed (a system which, given references to two objects, computes the options to offer the user). This may be held locally (e.g. on PDAs or other devices) or on a central server. Distributed architectures are also possible. DC servers represent objects using well-factored class hierarchies [3]. In many cases, only the class of a scanned object need be known, not its state. However, in some cases, state may needed to help infer relevant operations (e.g. when context is used). For more information on how to build DC servers and create domain models, see [3].

6.1 Security

In a full AC system, scanned objects might not reveal their identity to all enquirers; objects might not reveal their state; and some networked DC servers might be available for specialised purposes only to those with the right *subscriptions* or *credentials*. Hence, much like other computer interactions, some AC interactions are likely to be freely available to all, others available only to those who have paid, or who have the correct accreditation. Similarly, some kinds of interaction might be freely available to all *within particular locations* such as railway stations, airports, libraries, schools, universities, places of work, or supermarkets. Yet other interactions might be available only to certain *kinds of people*, for example policemen, emergency workers, maintenance workers, with access policed by specialised wands or passwords.

7 When Is DC Useful?

Because of the principle of Subsumption, there is no need for pairwise or n-fold interaction to be useful all of the time. Even if it only helps some users with some interactions some of the time, DC can be useful, since it gives a simple uniform framework for these and other interactions without getting in the way. However, it would be useful to characterise the kinds of situations to which higher order DC interactions (n • 2) are especially suited. Empirical work is needed to answer this question fully, but since DC hinges in part on reduction of search, some a priori considerations can characterise such situations in part, as we now consider. Domains can be expected to be suited to higher order DC interactions provided they have: sufficient variety of objects, sufficient variety of operations, and provided that the operations are distributed between object type pairs. For AC (over and above Desktop DC) to be applicable, a scannable environment is needed where the objects of interest are visible in the environment. Given a suitable domain, users could be expected to benefit from DC in any of the following situations: the domain is unfamiliar; there are too many commands to search quickly; the user is focussed on the environment; minimal attention situations (where time and attention are scarce for searching the user interface, and are better focused on the environment); feedback bandwidth is limited; or the task focus is specifically on combinations of objects. More generally, DC is particularly useful for all kinds of data translation and interoperability. As noted earlier, much work with mobile and ubiquitous devices involves ad-hoc tasks with novel combinations of unfamiliar resources. AC appears to be well suited to such situations.

8 Prototypes and Their Limitations

The prototypes reported here were designed for purposes of proof of concept and for refining the DC mechanism. The architecture used in the most recent prototype can support some three-fold combinations, but it is not yet clear how well this would scale to n-fold combinations in general (i.e. when four or more objects are zapped in one interaction). The domain models for our prototype DC servers have been kept minimal. We did not connect the servers to the relevant services (e.g. print queue, etc) - the user was presented with dynamically instantiated textual stubs. One of our prototypes is designed to work on Casio PDAs via front ends coded in Smalltalk using a choice of onboard DC servers or communicating wirelessly to a DC server coded in Smalltalk running on laptop. However, for reasons of practicality, the demos reported here were carried out on laptops simulating PDAs with an onboard DC server. Tagging of objects in the environment was achieved using barcodes and USB hand-held bar-code scanners, including a 30-foot range scanner. The prototype server follows the Subsumption principle, and could offer relevant actions whether zero, one, or two objects were zapped.

9 Related Work

Using pointing devices to transfer information between devices (typically computers) is not new. For example, Pick-and-Drop [7, 8] is a pioneering extension of Drag & Drop used to copy data between multiple devices via passive pens with ids. Pick and Drop can be viewed as a natural extension of Drag & Drop to mobile and ubiquitous

computing. In turn, the principle of Ambient Combination may be seen as an extension of Pick and Drop. AC offers potentially greater flexibility and expressiveness of interaction. The InfoStick [9] is an interaction device for inter-appliance computing. It may be used to pick up, and store, information items from a variety of devices, and then move them to other devices. The InfoStick may be viewed as offering a limited special case of Ambient Combination where the only available operations are
iget and *put*.

DataTiles [6] is an attractive tangible computing system using tagged transparent tiles placed on a flat display. Interactions between any two tiles are effected by physical adjacency, or by a pen gesture. The kind of interaction is determined by the tile types, although a pen gesture may be used to make this continuous or discrete in nature. DataTiles, like many tangible computing system, has the potential to be given greater flexibility and power, without loss of elegance, by applying Direct Combination in a variety of ways. For example: DC options could be visibly offered when two tiles were placed together or connected by pen. Alternatively, Subsumption would be better satisfied if DC interactions were invoked via a single additional pen gesture to invoke Direct Combination between two tiles. Or as a third alternative, a special magic lens tile could be laid on top of inter-tile borders, to reveal and allow the selection of available interactions.

From one perspective, DC may be viewed as a novel way of exploiting a *relational* approach [10] to Tangible User Interface (TUI) design, *systematically* allowing the selection of multiple objects to determine dynamically bindings between objects and computational operations. However, although DC is a generally useful technique for TUIs, DC works beyond TUIs to allow mixed interactions between virtual and real objects [6].

DC relates strongly to Direct Manipulation, and as discussed in [3], parts of DC may be variously viewed as generalisations or specialisations of DM. However, since both DM and DC can be conceptualised at several levels, which do not mutually correspond, neither idea is a superset or subset of the other.

10 Conclusions

Ambient Combination (AC) is the adaption, extension and application of Direct Combination (DC) to Mobile, Ubiquitous, Tangible and Mixed Reality Computing. In many situations, Direct Combination can reduce the amount of search, time and attention required by users to carry out actions that they do not know how to perform quickly. DC appears useful where display real estate is limited or input devices are inconvenient. It appears useful in minimal attention situations, where time spent navigating a user interface is at a premium, and in situations where users need to make unfamiliar devices work with each other. DC lets the user combat the *combinatorial explosions* inherent in these circumstances by constructing defensive *combinatorial implosions* of relevant objects. Direct Combination allows the user to exploit objects in the environment to narrow down the range of interactions. It may be viewed as a new way of distributing the user interface in the environment. DC is not a replacement for other kinds of user interaction; it is a complement: standard interaction patterns may be subsumed as special cases. When pairwise combination is applicable, it can reduce the need for memorisation and interface navigation. DC allows users to employ recognition as opposed to recall. Direct Combination seems to offer a new approach to dealing with context-aware interactions. The strength of DC arises from the twin exploitation of user's perceptual, physical and spatial situatedness in the environment, and the organisation inherent in object hierarchies. Almost all of the *elements* of Direct Combination exist in fragmented or partial form in other systems: what is new is the simple uniform framework and approach that DC affords for dealing with challenging interactions, and simple abstract architectures to make this possible. We argue that DC has a useful role to play in any future framework for mobile and ubiquitous HCI.

Acknowledgements

Thanks to: Alistair Edwards, Claudia Eckert, Martin Stacey, Randy Smith, Bill Gaver, Benedict Heal. This paper is a shorter, revised version of [6].

References

1. Pascoe, J. Ryan, N. and Morse, D. "Using While Moving: HCI Issues in Fieldwork Environments," ACM Transactions on Computer Human Interaction, vol. 7, pp. 417-437, 2000.
2. Erickson, Thomas (2002) Some Problems with the Notion of Context-Aware Computing. Communications of the ACM Feb 2002/Vol 45 No.2 pp102-104.
3. Holland, S. and Oppenheim, D. (1999) Direct Combination. In Proceedings of the ACM Conference on Human Factors and Computing Systems CHI 99, Editors: Marion Williams, Mark Altom, Kate Ehrlich, William Newman, pp262-269. ACM Press/Addison Wesley, New York, ISBN: 0201485591.
4. Weiser, M. "The Computer For the 21st-Century," Scientific American, vol. 265, pp. 66-75, 1991.
5. Binsted, Kim (2000) Sufficiently Advanced Technology: Using Magic to control the world. In Extended Abstracts of the ACM Conference on Human Factors and Computing Systems CHI 2000, pp 205-206 ISBN: 1-58113-216-6.
6. Holland, S., Morse, D.R., Gedenryd, H. (2002) Ambient Combination: The application of Direct Combination to Mobile and Ubiquitous Human Computer Interaction. Technical Report TR2002/1, Department of Computing, The Open University, Milton Keynes, MK76AA, UK.
7. Rekimoto, J., Pick and Drop: A Direct Manipulation technique for multiple computer environments, In proceeedings of UIST '87 p.31-39.
8. Rekimoto, J., Ulmer B. and Oba, H. DataTiles: A modular platform for mixed physical and graphical interactions. In Proceedings of CHI 2001. 2001 p 269-276.
9. Kohtake, N., Rekimoto, J. and Anzai, Y. InfoStick: an interaction device for inter-appliance computing, Handheld and Ubiquitous Computing (HUC '99) 1999.
10. Ullmer, B., Ishii, H., (2000) Emerging Frameworks for Tangible User Interfaces. IBM Systems Journal 39 (3&4), 915-931.

ASUR++: A Design Notation for Mobile Mixed Systems

Emmanuel Dubois, Philip Gray, and Laurence Nigay[1]

Department of Computing Science, University of Glasgow, UK
{emmanuel, pdg, laurence}@dcs.gla.ac.uk

Abstract. In this paper we present a notation, ASUR++, for describing mobile systems that combine physical and digital entities. The notation ASUR++ builds upon our previous one, called ASUR. The new features of ASUR++ are dedicated to handling the mobility of users and enable a designer to express physical relationships among entities involved in the system. The notation and its usefulness are illustrated in the context of the design of an augmented museum gallery.

1 Introduction

As defined in [7], a Mixed System is an interactive system combining physical and digital entities. Two classes of mixed systems are identified in [7]:

- Systems that enhance interaction between the user and her/his real environment by providing additional capabilities and/or information. We call such systems Augmented Reality systems.
- Systems that make use of real objects to enhance the interaction between a user and a computer. We call such systems Augmented Virtuality systems.

On the one hand, the NaviCam system [17], our MAGIC platform for archaeology [18] and our Computer Assisted Surgery system CASPER [8] are three examples of Augmented Reality systems: the three systems display situation-sensitive information by superimposing messages and pictures on a video see-through screen. On the other hand, the Tangible User Interface paradigm [13] belongs to Augmented Virtuality: physical objects such as bricks are used to interact with a computer. The design of such mixed systems, Augmented Reality as well as Augmented Virtuality, give rise to further challenges due to the new roles that physical objects can play in an interactive system. The design challenge lies in the fluid and harmonious fusion of the physical and digital worlds.

We are not aware of any existing design tools or system development methods that take into consideration the particular characteristics of mixed systems. Indeed, traditional task analysis approaches typically provide no way of including and characterising physical entities. Scenario and prototype-based design approaches overcome this limitation but don't offer the designer the ability to compare, explore or generalise design solutions. Finally, modelling languages, like UML, are well-fitted to

---

[1] On sabbatical from the University of Grenoble, CLIPS Laboratory, BP 53, 38041 Grenoble Cedex 9, France.

F. Paternò (Ed.): Mobile HCI 2002, LNCS 2411, pp. 123–139, 2002.
© Springer-Verlag Berlin Heidelberg 2002

model application functionality, but apart from users, as use case agents, no physical entities are represented or characterised.

In [10] we show that our ASUR notation can help in reasoning about how to combine the physical and digital worlds, by identifying physical and digital objects involved in the system to be designed and the boundaries between the two worlds.

In this paper we address the issue of designing mobile mixed systems by providing an extension to our ASUR notation, namely ASUR++. Mobility of a user in a mixed system requires us to consider the spatial relationships between the users and the entities involved in the mixed system. The new features of ASUR++ enable the designer to express such spatial relationships between a user and an entity. For example, using ASUR++, a condition that a user must be less than 2 meters from a specific physical object can be expressed.

The structure of the paper is as follows: We first describe the notation ASUR++. We then explain the outcomes of the notation for the design of mobile mixed systems by considering the different design phases. Having presented the notation and its usefulness, we illustrate it by considering the design of an augmented museum gallery. We compare several design solutions, all of which are described using ASUR++.

2 ASUR++ Notation

ASUR++ is a notation that supports the description of the physical and digital entities that make up a mixed system, including the user(s), other artefacts, and the physical and informational relationships among them. This description captures the characteristics of a mixed system relevant to a specific task or user activity. Relationships among the components represent informational and physical properties of both the static context and the dynamic behaviour of the system and related physical objects, during the realisation of the activity in question. The validity of the ASUR++ description of a system is thus restricted to the duration of the user's activity. Consequently, to capture a set of tasks or activities, several ASUR++ diagrams may be necessary to describe the whole behaviour of the system. For example, in the field of Computer Assisted Medical Intervention (CAMI), three tasks are commonly required: acquiring data, planning the intervention (e.g. the intervention trajectory to reproduce during surgery) and guiding the clinician during the actual intervention [20]. To describe a complete CAMI system, three ASUR++ diagrams would be required, one for each of these main user activities.

To describe the merging of physical and digital entities and relationships, ASUR++ takes into account design-significant aspects highlighted in other approaches to characterising AR systems. These existing characteristics include:

- the type of data provided to the user [3,12,16], which may be textual, 2D or 3D graphics, gesture, sound, speech or haptic, and
- the potential physical targets of enhancement, in order to combine physical and digital data [14]; the target may be users, physical objects or the environment.

To these characteristics, ASUR++ adds other factors related to the use of physical entities. ASUR++ thus combines and enriches aspects addressed in the different AR approaches.

The next section presents the two ASUR++ principles:

- Identification of the physical and digital entities involved in the systems, namely the ASUR++ components, and the identification of the exchanges between entities, namely the ASUR++ relations;
- Characterisation of ASUR++ components and relations.

2.1 Components and Relations

For a given task, ASUR describes an interactive system as a set of four kinds of entities, called components:

- Component S: computer system;
- Component U: user of the system;
- Component R: real object involved in the task as tool (R_{tool}) or constituting the object of the task (R_{object}) ;
- Component A: input adapters (A_{in}) and output adapters (A_{out}) bridge the gap between the computer-provided entities (component S) and the physical world entities, composed of the user (component U) and of the real objects relevant to the task (components R_{object} and R_{tool}). Input adapters represent all kind of sensors and may be pressure-, haptic-, sonic-, optical-, etc. sensors. This include keyboard, camera, microphones, localizers and there markers (e.g. diodes and camera). Output adapters may be any devices through which the user can perceive some information (e.g., screen, speakers, force-feedback devices, etc.)

We have identified three kinds of relationship between two ASUR components:

- **Exchange of data**: represented by an arrowed line (A→B) from the component emitter (A) to the component receptor (B), this symbolises the transfer of information between two ASUR components. For example A_{out}→U may represent the user's perception of data displayed on a screen (i.e., an output adapter).
- **Physical activity triggering an action**: a double-line arrow (A⇒B) denotes the fact that when the component A meets a given spatial constraint with respect to component B (for example, A is no further than 2 meters from B), data will be exchanged along another specific relationship (C→D). The spatial constraint and the relationship on which a transfer will be triggered are properties of this kind of relationship.
- **Physical collocation**: represented by a non-directed double line (A=B), this refers to a persistent physical proximity of two components. It might be used between any kind of components among those describing the user (U), the adapters (A_{in} and A_{out}) and the real entities (R_{tool} and R_{object}). In ASUR++ diagrams, this collocation is reinforced by a contour drawn around the components that are so collocated. This highlights groups of entities that move, or have to be multiply instantiated, if one of them is moved or multiply required. This contour is a single line contour if the set is mobile, and a double line if the set of components remains static during the interaction. This indication of grouping also makes it easier for the designer to deal with multiple instances of the collocation relationship, i.e., when more than one user or more than one instance of a physical object will be used. Multiplying the number of users or physical objects will lead to the multiplication of the components included in the contour.

Interaction with the system is thus represented by a set of relations connected to the component U, representing the user. In the following section we present several important characteristics of the ASUR++ components and their relationships.

2.2 Characterisation of Components and Relations

The characteristics described here are chosen as a first set likely to be of value in thinking about mobile mixed reality systems. They include characteristics already identified in other AR design approaches, but also include additional aspects specific to the use of real objects in the interaction. Each relation connected to the user defines a facet of the interaction, consisting of (i) an ASUR component from which information is provided for the user or to which the user provides information, and (ii) an ASUR relation between the user and this component. ASUR characteristics exist for both components and relations as presented in the Table 1 below. A more detailed description of ASUR, including these component characteristics, is given in [10].

Table 1. Characteristics of ASUR components and the relations that make up a user's different interaction facets.

Characteristics of the components	Characteristics of the relations
Perceptual/Action location: Physical area where the user has to focus in order to perceive information provided by the component or perform an action on it.	**Representation frame of reference** (only for "→"): Point of view from which information is perceived or expressed.
Perceptual/Action sense: Human sense required by the user to perceive information provided by the component (visual, audio, etc.) or to act on the component (speech or action).	**Representation language** (only for "→"): Bernsen's representation properties [5], amount of dimensions of the representation that carry relevant information.
Share: Number of users that can simultaneously access the component to perceive or provide information.	**Concept** (only for "→"): The application-significant concept about which data is carried by the relation.
	Concept relevance (only for "→"): The importance of this concept for the execution of the task.
	Triggered relation (only for "⇒"): The ASUR++ relation whose exchange of information will be triggered by the present relation.
	Spatial condition of triggering (only for "⇒"): The condition under which an exchange of data is triggered.

2.3 ASUR++ and the Design of Mobile Mixed Reality Systems

ASUR++ provides a means of describing a number of aspects of interactive mixed reality systems: aspects that are potentially significant at different stages in the development process. We have only begun exploring the use of ASUR++ for the design process and thus have not yet integrated its use into any particular design methods, nor do we yet have a mature method for its use in handling the systematic description and analysis of mixed reality designs.

Software engineering structures design and implementation into six phases: requirements definition, specification, implementation, testing, installation and maintenance [2]. ASUR++ is a design notation and can therefore be used during the requirements definition phase and the specification phase.

- Requirements definition is a formal or semi-formal statement of the problem to be solved. It specifies the properties and services that the system must satisfy for a specific environment under a set of particular constraints. Ideally requirements are defined in cooperation with the end-users.
- Specification consists of high level design (i.e., external specifications) and internal design (i.e., internal specifications). High level design is concerned with the external behaviour of the computer system. This behaviour is described in terms of functionalities as perceived by the user of the future system. For each function, valid inputs and outputs are specified as well as error conditions. Internal design determines a software organisation that satisfies the specification resulting from high level design. Internal design covers the definition of data structures, algorithms, modules, programming interfaces, etc.

For the requirements definition, an ASUR++ description may help to describe the services that the system must provide and the links between the physical and digital worlds. At this stage, the usefulness of an ASUR++ diagram will be similar to that of a UML use case diagram [19].

During the external specifications, ASUR++ is intended to provide a resource for analysts; it can be used to systematise thinking about design problems for mobile mixed systems. We will demonstrate this point in the following section. Several design solutions can be described using the same modelling approach ASUR++, enabling easy comparisons. Nevertheless we do not claim that use of ASUR++ alone is sufficient for identifying an optimal design solution. As holds true for any modelling notation, ASUR++ is a tool for the mind and a vehicle for communicating design alternatives.

3 Describing and Analysing Design Alternatives Using ASUR++

In this section we examine several different views of a design, each capturing features that can be significant during different steps of the design of a mobile mixed system design. The scenario we use is based on a system being developed as part of the City Project, one of the projects of the Equator IRC [11].

3.1 An Augmented Museum Scenario

As a vehicle for presenting ASUR++, we use an example taken from the City Project, a project developed within the Equator consortium. Based on the work of Charles Rennie Mackintosh, a Glaswegian architect of the early 1900's, the City Project has been exploring the augmentation of the permanent Charles Rennie Mackintosh Interpretation Centre, a gallery situated in the Lighthouse, an architecture and design centre in Glasgow, containing exhibits related to Mackintosh's life and work. The aim of this part of the project is to study the impact of combining multiple media to support visitors' activities, especially collaborative activities involving users in the real museum interacting with users exploring a digital version of the same museum ("co-visiting"). For the visitor to the real museum, the system being created is aimed at providing visitors with digital information tailored to visitor's current context. This information tailoring mainly relies on tracking visitor's motions in the museum and location of the exhibits. Visitor activities are thus embedded with computational capabilities. To do so, the Lighthouse has been equipped with a radio-frequency localisation system that gives the location of the visitors.

There are several services that will be provided by the system in the Lighthouse. In this paper, we consider only a single service offered to visitors: following a visit path in the museum. Clearly inspired from the City project, it does not really constitute one of the final goals of this project. We have derived this adapted scenario from the initial project to illustrate this paper. In this scenario, the considered service provides AR support to guide visitors through a pre-defined path of exhibits. A path is composed of a set of exhibits, in a given order, that the visitor has to observe. A set of paths is saved in a database. Each exhibit on the predefined path has some associated textual comments. In addition we assume that the visitor who wishes to follow a predefined path is already connected to the system and that he has already chosen a path. The main issues of this scenario are twofold: the visitor is mobile and has to be localised in the museum and the system has to know where the user is with respect to the exhibits, in order to provide the right information. Under these conditions, a visitor receives information related to:

- The path to follow: this consists of a set of textual directions and distances separating the current position of the visitor from the next exhibit of the followed path;
- The exhibits: once a visitor reaches the next exhibit along the path he/she is following, the system provides data about the exhibit not perceivable in the museum (e.g. background information about the exhibit and related items not located in the museum) according to preferences defined by the visitor (historical data, authors data, etc.).

The role of the computer system is to provide this information to the user based on his/her location relateive to the set of exhibits.

It is important to note that this example does not represent the design of an existing system, nor is it a history of an actual design development. Rather, we have chosen this scenario because it represents a realistic design problem (i.e., the design brief is a real one). However, the goals of our example scenario don't correspond to the goals of the City Project and the design alternatives that we present below are our own and don't represent any that have been developed during the City Project. Furthermore, the "augmented museum gallery" scenario has been tackled by a number of other projects as well [1, 6]. We believe that using a reasonably well-understood application makes it easier to communicate the features of ASUR++ even if the problem and our proposed solutions are neither novel nor likely to be optimal.

3.2 Abstract Description of the Scenario Using ASUR++

In terms of ASUR, the visitor is the component U, an exhibit is a component R_{object} observed by the visitor ($R_{object} \rightarrow U$) and component S includes the database that contains user paths and information related to the exhibits.

The system provides information related to the path and to the exhibit: S(path)$\rightarrow$U, S(exhibit)$\rightarrow$U. This information will be displayed according to the user's position with regard to the position of the exhibit ($R_{object} \Rightarrow U$ triggers $(R_{object} \Rightarrow U) \rightarrow U$ and $R_{object} \rightarrow U$); that is, the user will receive information relevant to the exhibit that they are near. The spatial relationship between the user and the exhibit must thus be a source of information to the system.

Figure 1 shows the resulting high-level ASUR description.

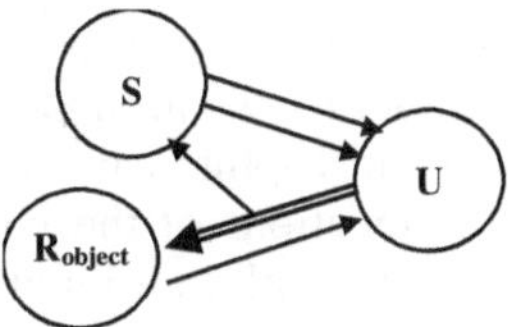

Fig. 1. Abstract description of scenario in ASUR++.

We can elaborate the abstract ASUR++ description of figure 1 by refining the relationships among the components. In the next section we begin by focussing on the relationship between system and user, examining the ASUR++ representation of two design alternatives: (1) Using one adapter to convey both kinds of information to the user, both path and exhibit, and (2) using an adapter for each kind of information.

3.3 Reasoning about Output Design Solutions

Output design refers more generally to the design of the part of a Mobile Mixed System that will provide information to the user. In the first abstract description, given in figure 1, this is identified by the set of arrows that transfer information to the user. There are two kinds of output information: that provided by the computer system, related to both the path and the exhibit, and that provided by the user's perception of the physical exhibit. The latter is fixed since it is related to the user's natural perception. Reasoning about the output design will thus focus on the output provided by the computer system, taking into account the physical realisation of the information. Three aspects will be illustrated in this section to show how ASUR++ facilitates design reasoning. These aspects are adapters elicitation, perceptual and cognitive levels.

3.3.1 Using Two Output Adapters

Adapter Elicitation Level

To follow the chosen path, the visitor must be able to perceive the guidance information provided by the system. An output adapter ($A_{out}1$) is thus required. One relationship from this component is connected to the visitor (component U), denoting the transfer of information related to the path to follow: $A_{out}1$(path)→U. Furthermore, an ASUR++ relation from the component S to the component $A_{out}1$is required because information provided by the $A_{out}1$ component is generated by the database (component S): S→$A_{out}1$. Exactly the same reasoning can apply to the transfer of information related to the exhibits, leading to the identification of a second output adapter $A_{out}2$ and the relations S→$A_{out}2$ and $A_{out}2$(path)→U. These two output adapters might be placed in the gallery infrastructure or they might be portable and carried about by the user. In what follows, we examine the latter case. The former case can also be captured using ASUR++, but space limitations prevent us from considering it in this paper. Moreover, if output adapters are placed in the gallery infrastructure, there would (probably) be a single screen for each exhibit, a less flexible solution if there are many concurrent visitors with very different interests and information needs.

The ASUR++ description shown in figure 2, represents this state of affairs, two output adapters carried by the user, by two physical collocation relations: $A_{out}1$=U and $A_{out}2$=U. It is reinforced by the contour drawn around the three components.

Now, as stated in our scenario and captured in the ASUR++ abstract description, the system has to be aware of the locations of the visitor and the exhibit in order to provide the right information. Consequently, an input adapter (A_{in}) is required to get these positions and transfer it to the computer system: U→A_{in}, A_{in}→S for the visitor's location and R_{objet}→A_{in}, A_{in}→S for the location of the exhibit. But, the component U is part of a set of components that are spatially collocated. Consequently, the relation U→A_{in} can be connected to the contour of the set rather than to the user, leaving the designer free to decide which component of this set to localise. However, since the input adapter is not further explored at this point, the ASUR++ diagrammatic representation represents this adapter as a square rather than as a circle.

Reasoning at a Perceptual Level

This analysis relies on the ASUR++ components' perceptual characteristics, including the perceptual environment (i.e. where the perception takes place), the sense(s) used, and the ability of the adapter to share the information it provides among one or several users (see Table 1).

The perceptual senses we may envision for this design are the visual and auditory senses for both adapters; in addition, the haptic sense might be used to convey the path to follow. Given that two adapters are used, different combinations of these senses might be used to realise the system. With respect to location of perception, design issues will be highly dependent on choice of modality. For example, if we consider the situation in which both information, related to the path and the exhibit, are visually conveyed via a palmtop device, s/he will have to look alternatively at the device carrying information about the path ($A_{out}2$), at the one carrying information about the exhibit ($A_{out}1$) and at the physical exhibit (R_{object}). This is an example of perceptual incompatibility, an AR ergonomic property introduced in [10], which may annoy the visitor. Projecting information into the same visual field as the actual exhibit may lead to a partial occlusion of the real exhibits, and is probably not a good solution in a museum context, but would be extremely interesting in an Computer Assisted Surgery system for example, where a surgeon may wish to have a permanent view of the patient and to perceive collocated guidance information for the surgical tools s/he is manipulating [8].

Other aspects may also have to be considered, regarding the context of use of the system being designed. The gallery is likely to have a number of visitors and visitors often operate in groups. The choice of audio for a personal output adapter might be more intrusive than a visual adapter, disrupting social interaction among other visitors. However, audio might also promote "co-visiting".

Finally, taking into account the number of users that should be able to access to the information leads us to consider three cases: restricting to one user only, allowing a group of users to access to the data or to broadcast the information to every visitor present in the same place. Again, the context of use of the system will greatly impact on the choice of one of these possibilities. For example, if we choose to limit access to the data to one person only, the consequence is that a group guided by a leader is not possible because the members of the group can't read the data related to the exhibit.

Note that the use of ASUR++ does not offer a way of resolving these design choices (that remains a question of usability evaluation and/or the use of appropriate guidelines), but it does provide a means of expressing the aspect of the system to which the choices apply, viz., the physical realisation of the output adapters between system and user.

Reasoning at a Cognitive Level

This analysis is based on ASUR++'s characterisation of the language and the frame of reference of the representation conveyed by a relation (see table 1).

The languages that may be used to express information about the exhibit include text, graphics (2D or 3D) and speech. In addition, path information may be conveyed by sounds (non-speech audio) or tactile stimulation. The resulting possible combinations are of course highly dependent on the choices made in the previous phase, concerning the human senses the adapter will exploit.

The usability of particular representational combinations also has to be considered. This can be assessed, of course, via interaction design patterns, analytic evaluation in terms of ergonomics principles and/or psychological theories modelling the cognitive processes of the users, or empirical user studies. For example, if we consider that the path is provided using a textual language rather than with graphics, the user has to interpret the presented textual information in terms of her/his physical 3D environment. This interpretation introduces a cognitive discontinuity (an AR ergonomic property introduced in [10]) which, in this case, may complicate the task for the user.

The frame of reference of the representation of the information related to the exhibit must be presented from a user's point of view so that s/he can access it. The path may be expressed in different frames of reference: a user-centred point of view (e.g., "turn right at the urn") or in a global reference scheme (e.g., a map). The impact of choosing either one or the other is not immediately apparent and again will need observational studies, design patterns or analytic studies, in collaboration with usability professionals or psychologists.

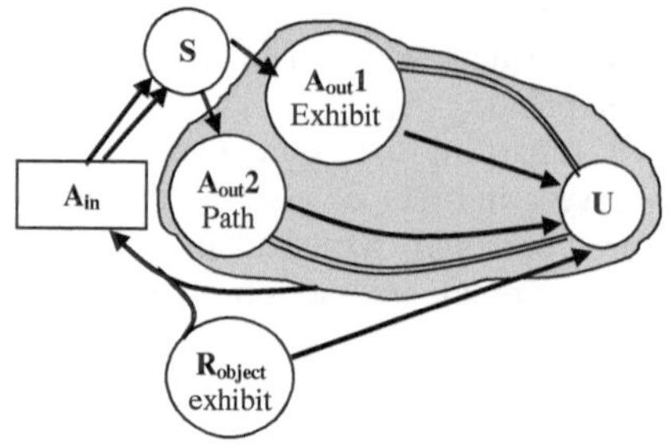

Fig. 2. Partial ASUR++ description of the scenario using two output adapters.

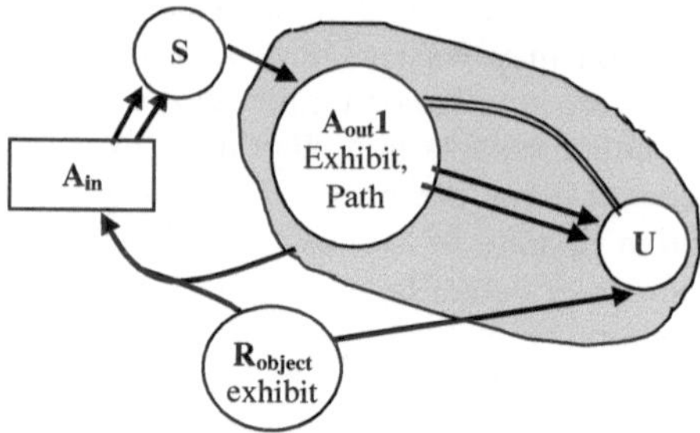

Fig. 3. Partial ASUR++ description of the scenario using one output adapter.

3.3.2 Using Only One Output Adapter

Adapter Elicitation Level

The only differences between this ASUR++ description and the one presented in Figure 2, is (i) the use of a single output adapter and (ii) the existence of two ASUR++ relations between the user and this adapter. These two relations indicate that the adapter provides to the user information related both to the path and to the exhibits.

Using only one output adapter instead of two will have an impact on the design possibilities identified in the following phases of the ASUR++ based reasoning process.

Reasoning at a Perceptual Level

The limitation to one adapter restricts the possible design solutions and forces trade-offs. First of all, haptic feedback is no longer a viable alternative since it unlikely to be suitable for information related to the exhibits. Significant compromises will also have to be made if either audio or visual techniques are used on their own.

Reasoning at a Cognitive Level

Consider the case of a visual output adapter. The likely possible languages are either text or graphics. If text is used, there remains the problem of a cognitive discontinuity when conveying path information textually, but presenting all the information (path and exhibit) via the same representation might be considered to offer a form of coherence in the output interaction. An observational study might be conducted to assess this hypothesis.

More importantly, using the same adapter for both information streams may interact badly with the physical properties of the adapter. For example, it may be difficult to present all the relevant information concurrently via a palm-sized display. Once again, the context of use of one adapter is proven to be important to take into account when envisioning a design solution.

Thus far we have explored the design space of system's output to the user. We now focus on the design of input, that is, the ways the computer system will get information from the physical world and from the user. Note that the input aspects of the user's interaction with the system is only a subset of the whole input design.

3.4 Reasoning about Input Design Solutions

As shown in figure 1, the system needs to be aware of the spatial relationship of the visitor to the exhibits. The most direct design solution is to utilise two input adapters, one dedicated to localising the exhibit, while the second is dedicated to localising the visitor. We describe this solution in the next section. For the purposes of our example, we assume the "single adapter" output design.

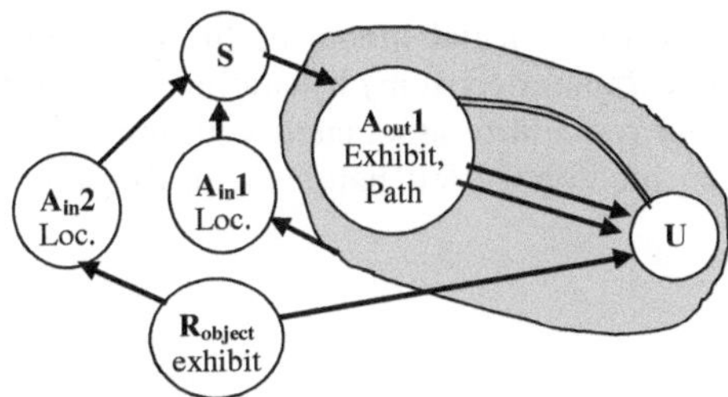

Fig. 4. ASUR++ description of the scenario using one output adapter and two input adapters.

3.4.1 Using Two Input Adapters

In order to be aware of the visitor's location in the museum an input adapter ($A_{in}1$) is required to retrieve the position of the visitor in the museum ($U \rightarrow A_{in}1$) (more exactly the set of collocated ASUR++ components that includes the user), and to transfer the position to the computer system ($A_{in}1 \rightarrow S$). In addition, a second input adapter ($A_{in}2$) is required to locate the exhibit ($R_{object} \rightarrow A_{in}2$) and transfer the location to the computer system ($A_{in}2 \rightarrow S$). The ASUR++ description of the overall system using one output adapter and two input adapters is presented in the figure 4.

Considering the concept of localisation, the system has to deal with two inputs: the one related to the user's location and the one related to the exhibit's location. Matching these two sources of information may be a problem for the computer system and is similar to a discontinuity problem on the user's side. Solutions to address this problem can be driven by the solutions envisioned when a discontinuity problem is identified on the user's side. It would thus be better to:

- Use only one reference scheme in which to encode the location information provided by the adapters (similar to a cognitive discontinuity problem)
- Track only one entity (similar to a perceptual discontinuity problem)

Addressing the first kind of problem is relatively easy. One global reference scheme may be used. The second problem is harder to address. On the user's side, this led us to group the two output adapters into only one. We explore this solution in the next section.

3.4.2 Using One Input Adapter

In this present case, grouping the two adapters may be achieved by using one of the following mechanisms:

- Avoiding the need of the relationship between the exhibit and the input adapter, or of the relationship between the user and the input adapter.
- Grouping the exhibit with the user or the exhibit with the input adapter;

Avoiding the Need of a Relationship

Let us first consider the localisation of the exhibit. A solution could be to use a static model of the positions of the exhibits. In fact this is achievable by adding a field in the exhibit database holding the location of the exhibit in the museum. Consequently, having the position of the visitor in the museum is sufficient to find in the database the exhibit which has the nearest coordinates and thus to display the right information. To represent the existence of a virtual model of the physical exhibit, we refine the ASUR++ component S (computer System) by adding a decoration to the S node: V-R_{object} (virtual model of the real entity associated with the component R_{object}). The new ASUR++ diagram is presented on the left-hand side of figure 5.

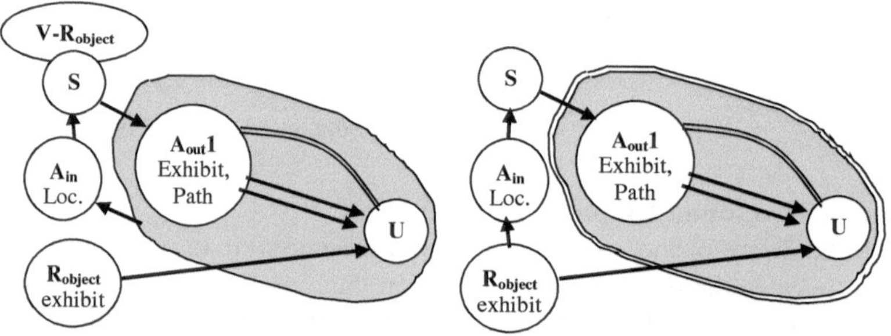

Fig. 5. Complete ASUR++ description of the scenario using one output adapter and one input adapter when avoiding the localisation need of the exhibit (left) or of the visitor (right).

Avoiding the need of the visitors' localisation could be achieved in two ways: either the display of information is time dependant or the user is static and the exhibit moves in front of him. In fact, in the first case, time dependent display of information is similar to providing the computer system with a virtual model of the visitor's motion based on the time. But, the visitor might be rapidly lost if he spends more time than planned in front of an exhibit. This solution is thus quite risky. The second solution seems to be more reliable. Its technical realisation is another question. However, in this futuristic situation the user and the devices s/he is carrying would be static and the exhibits would automatically pass in front of the visitor. The ASUR++ diagram representation of this design variant is presented on the right-hand side of figure 5.

Grouping Mechanism

The role of the grouping mechanism is to physically link a component with an input adapter, so that when this adapter sends information to the system about another component, the system also knows where the information comes from.

One way of implementing this mechanism consists of installing near an exhibit the input adapter responsible for the visitor's localisation. This is represented by a physical collocation relationship between the exhibit and the input adapter ($R_{object}=A_{in}$).

The set of components that make up the exhibit and the input adapter remain static. The system can determine the visitor's position by associating the visitor's identity with the exhibit to which the input adapter is collocated. Thus the system can display the right information to the user. To be localised, the visitor, or more exactly the set of component that includes the visitor, has to come near the exhibit. In terms of ASUR++, when the visitor comes near the exhibit, it triggers the transfer of information from the visitor to the input adapter. The following relations emerge: $U \Rightarrow A_{in}$ and $U \rightarrow A_{in}$. The ASUR++ diagram of this system is shown in the left part of figure 6. Examples of devices that might play the role of the component A_{in} as described here are motion detectors or an rfid tag and sensor. In the later case, the relation denoting the transfer of information between the user and the input adapter requires the addition of the rfid emitter to the set of components carried by the visitor. The relation between the "visitor set" and the "adapter set" is $U \rightarrow A_{in}$. The designer has to think about which component of the "visitor set" to embed: the visitor or the output adapter.

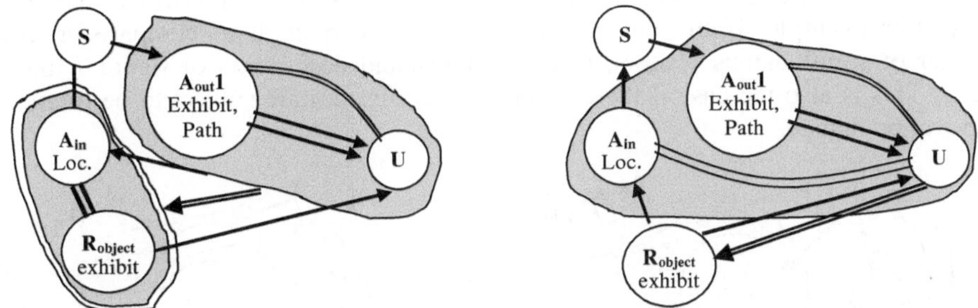

Fig. 6. Complete ASUR++ description of the scenario using one output adapter plus one input adapter grouped with the exhibit (left) or with the visitor (right).

Another way of applying this mechanism is to group the input adapter with the visitor. The information provided to the system by the adapter refers here to the localisation of an exhibit, given that the relationship between the adapter and the visitor is known and fixed. In this case, the input adapter has to be physically collocated with the user (A_{in}=U) and is added into the mobile "visitor's set". When the visitor approaches the exhibit, it triggers the exchange of information between the exhibit and the input adapter responsible for the localisation of the exhibit: $U \Rightarrow R_{object}$ and its associated triggered relation $R_{object} \rightarrow A_{in}$. The right side of figure 6 illustrates this alternative. A candidate A_{in} component for this version would be an rfid on the exhibit or, more elaborately, a camera with an image processing module added in the computer system to automatically recognise the exhibit in front of the camera.

3.5 Scalability of Design Solutions

Consider the design solutions shown in figure 6. Both of them satisfy the system's functional requirements. So far we have only considered the system with respect to a single user. However, the context of use of this mobile mixed system is a museum, which may involve multiple visitors and, of course, a number of exhibits. Thus, reasoning at a larger scale means in this case considering the existence of several users and exhibits at the same time. Describing the large-scale system with ASUR++ will result in multiple U components (visitors) and multiple R_{object} components (exhibits). Given that these components are organised as collocated sets, an ASUR++ description will be based on the use of several of these sets, generating as many as necessary to characterise the system at the new scale.

When the input adapter is collocated with the exhibit (left part of figure 6), multiple exhibits will result in multiple input adapters, one for each exhibit. Multiple visitors will require multiple output adapters. The left side of figure 7 shows this ASUR++ description. On the other hand, when the input adapter is collocated with the user, multiple exhibits have no influence on the devices to connect to the system, but multiple users result in the need for multiple adapters for input and output as illustrated on the right side of figure 7.

On the basis of these large scale descriptions, it is possible to assess alternative solutions by considering aspects such as implementation complexity or cost. Indeed, the description reveals the number of required adapters for input and output and also indicates whether exhibits must be modified or users equipped with devices to wear or carry. For example, if the number of exhibits is very high in comparison to the number of simultaneous visitors, then the right-hand description of figure 7 may be better. This is also the case if the exhibits of the museum are subject to be frequently removal or change.

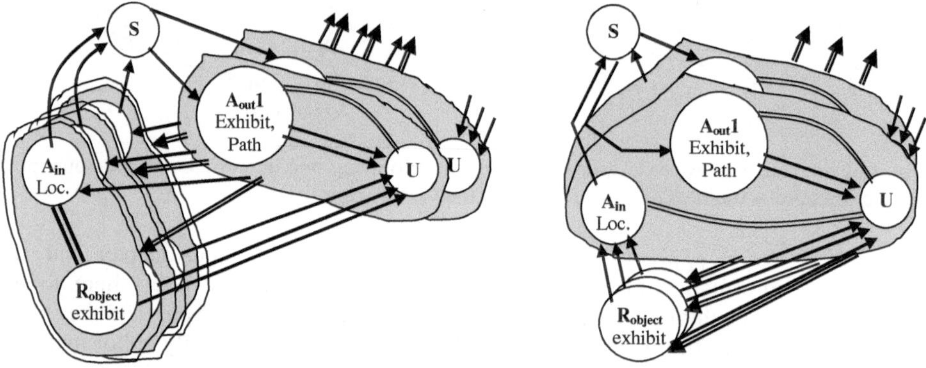

Fig. 7. ASUR++ description of the large scale version scenario using one output adapter and one input adapter when grouping the input adapter with the exhibit (left) or with the visitor (right).[2]

4 Conclusions and Future Work

We have presented ASUR++, a notation for the design of mobile mixed systems. ASUR++ is an extension of ASUR, our earlier notation dedicated to the design of

[2] For sake of clarity, the ASUR++ relation between the second user and the input adapter or exhibits are only partially represented.

mixed systems. We have presented and analysed several design solutions for an augmented museum gallery, expressed using the ASUR++ notation.

In this paper we do not claim that one can, using ASUR++ alone, identify the optimal design solution. As holds true for any modelling notation, ASUR++ is a tool for the mind. As we pointed out in a previous study [4], "Like a screwdriver, a modelling approach concentrates force (of reasoning) in the appropriate area; it does not mean that there is no role for the artisan and no element of skill and judgement involved." As a consequence, we do not claim that the various design solutions developed for our example scenario are the best ones, or that we have explored the entire design space. Indeed we cannot prove that the described solutions do cover all possible perspectives on design. In addition, it is important to point out that different individuals may achieve different results than the ones described in this paper with the same modelling technique.

ASUR++ is intended to provide a resource for analysts. It can be used to systematise thinking about design problems for mobile mixed systems. We demonstrated this point in the paper. Several design solutions have been described using the same modelling approach, enabling easy comparisons. The notation, with its underlying semantics, encourages the analyst to think about design issues in a particular way: In particular ASUR++ prompts the analyst:

- to study the spatial and other physical relationships amongst the entities involved in the system: physical objects, adapters and users,
- to study the scalability of the design solutions.

One further research avenue that we have begun to explore is use of ASUR++ modelling in conjunction with other modelling approaches. As we have shown in [4] the use of multiple modelling techniques extends the range of perspectives on the design problem. Diverse notations can work in concert and in a complementary fashion to identify and propose corrections to design flaws. For example in [9] we have established links between ASUR diagram and a software architecture model and in [10] we have explained how ergonomic properties can be assessed based on an ASUR diagram.

Another research avenue involves identifying recurrent ASUR++ diagrams that can be generalised and applied across different application domains. Such diagrams might describe reusable interaction design patterns for mobile mixed systems. Furthermore, such interaction design patterns expressed using ASUR++ may then be translated in terms of software architectural patterns, such as the ones we presented in [15], providing assistance with realising the implementation of the patterns.

Finally, we are currently engaged in investigating how ASUR++ fares in actual mixed system design activity. The results of these empirical studies will inform further development of the notation and the method of its use.

References

1. Abowd, G., D., Atkeson, C., G., Hong, J., Long, S., Kooper, R., Pinkerton, M.: Cyberguide: a mobile context-aware tour guide. Wireless Networks Vol. 3, **5**, Special issue: mobile computing and networking. Kluwer Academic Publishers, Hingham, MA, USA (1997) 421-433.

2. ANSI/IEEE Standard 729-1983. Software Engineering Standards. IEEE, New York, (1989).

3. Azuma, R., T.: A survey of Augmented Reality. In Presence: Teleoperators and Virtual Environments Vol. 6, 4, (1997) 355-385.

4. Bellotti, V., Blandford, A., Duke, D., MacLean, A., May, J. and Nigay, L.: Interpersonal Access Control in Computer Mediated Communications: A Systematic Analysis of the Design Space. Human-Computer Interaction, Vol. 11, 4. Lawrence Erlbaum (1996) 357-432.

5. Bernsen, O.: Foundations of multimodal representations. A taxonomy of representational modalities, in Journal Interacting with Computers, Vol. 6, 4, (1994), 347-371.

6. Cheverst, K., Davies, N., Mitchell, K., Friday, A., Efstratiou, C.: Developing a context-aware electronic tourist guide: some issues and experiences. Conf. Proc. of CHI'2000. Netherlands. ACM Press, NY, USA, (2000) 17-24.

7. Dubois, E., Nigay, L., Troccaz, J., Chavanon, O., Carrat, L.: Classification Space for Augmented Surgery, an Augmented Reality Case Study. In Conf. Proc. of Interact'99 (1999) 353-359.

8. Dubois, E., Nigay, L., Troccaz, J., Carrat, L., Chavanon, O.: A methodological tool for computer-assisted surgery interface design: its application to computer-assisted pericardial puncture. In Westwood, J. D. (ed.): Conference Proceedings of MMVR'2001. IOS Press (2001) 136 - 139.

9. Dubois, E.: Chirurgie Augmentée : un Cas de Réalité Augmentée ; Conception et Réalisation Centrées sur l'Utilisateur. PhD Thesis University of Grenoble I, France (2001) 275 pages.

10. Dubois, E., Nigay, L., Troccaz, J.: Assessing Continuity and Compatibility in Augmented Reality Systems. To appear in Int. Journal on Universal Access in the Information Society, Special Issue on Continuous Interaction in Future Computing Systems. Springer-Verlag, Berlin (2002).

11. Equator Interdisciplinary Research Consortium. http://www.equator.ac.uk/

12. Feiner, S., MacIntyre, B., Seligmann, D.: Knowledge-Based Augmented Reality. Communication of the ACM, Vol. 7 (1993) 53-61.

13. Ishii, H., Ullmer, B.: Tangible Bits: Towards Seamless Interfaces between People, Bits and Atoms. In Conference Proceedings of CHI'97. ACM Press (1997) 234-241..

14. Mackay, W.E., Fayard, A.-L., Frobert, L., Médini, L., "Reinventing the Familiar: an Augmented Reality Design Space for Air Traffic Control", In Proceedings of CHI'98, LA, (1998) 558-565

15. Nigay, L., Coutaz, 1997, J.: Software architecture modelling: Bridging Two Worlds using Ergonomics and Software Properties. In Palanque, P., Paterno, F. (eds.): Formal Methods in Human-Computer Interaction, Springer-Verlag, Berlin Heidelberg New York (1997) 49-73.

16. Noma, H., Miyasato, T., Kishino, F.: A palmtop display for dextrous manipulation with haptic sensation. In Conference Proceedings of of CHI'96. ACM Press (1996) 126-133.

17. Rekimoto, J., Katashi N.: The World through the Computer: Computer Augmented Interaction with Real World Environments. In Proceedings of UIST'95. ACM Press (1995) 29-36.

18. Renevier, P., Nigay L.: Mobile Collaborative Augmented Reality, the Augmente Stroll. In Little, R. Nigay, L. (eds): Proceedings of EHCI'2001, Revisited papers, LNCS 2254. Springer-Verlag, Berlin Heidelberg New York (2001) 315-334.

19. Stevens, P., Pooley, R. Using UML: Software Engineering with Objects and Components. Addison-Wesley.

20. Troccaz, J., Lavallée, S., Cinquin, P.: `Computer augmented surgery`, Human Movement Science 15, (1996) 445-475.

Designing LoL@, a Mobile Tourist Guide for UMTS

Günther Pospischil, Martina Umlauft, and Elke Michlmayr

Forschungszentrum Telekommunikation Wien, Tech Gate Vienna,
Donau-City-Straße 1, A-1220 Wien, Austria,
{pospischil, umlauft, michlmayr}@ftw.at

Abstract. Modern lifestyle corresponds with high personal mobility. People want to work or use leisure-time applications while on the road. Modern mobile communications systems allow to meet these requirements for the first time. Advanced new features like user positioning allow sophisticated applications that are not possible in the fixed Internet or traditional cellular networks. Still, application development for the Mobile Internet is a complex task. Users have special demands because of the mobile environment. Stringent technical constraints are imposed by mobile networks and mobile devices.
In this paper we present a prototype of a mobile application for UMTS. It is called LoL@ (Local Location Assistant) and implements a tourist guide for users in the city of Vienna. We discuss user interaction and interface design, design process, and technical solutions used to implement the application. Because of the initial lack of powerful PDAs, currently a laptop is used as terminal.

Keywords: Location Based Service, Electronic Guide, Mobile Internet, Semi-Formal Requirements Specification.

1 Introduction

Currently, Internet applications and applications for mobile communications systems are pretty separated. This is caused by the vastly differing technical constraints: different terminals, data transmission speeds, and operating environments. However, technology improves and advanced mobile systems, like the Universal Mobile Telecommunications System (UMTS) or public wireless network access points, will be available in the near future. Together with the evolution of handheld computers and cellular phones, a convergence of both scenarios can be expected [1]. We think that there will be two important groups of mobile applications, supporting mobile business and mobile leisure time usage.

Mobile business applications, like online banking and trading (m-commerce), file access, mail, or calendar, are quite similar to the well known Internet/Intranet applications. They are commonly used with a Laptop or a powerful Personal Digital Assistant (PDA). On the other hand, mobile leisure applications provide entirely new possibilities, like mobile multiplayer games and video messaging

F. Paternò (Ed.): Mobile HCI 2002, LNCS 2411, pp. 140–154, 2002.
© Springer-Verlag Berlin Heidelberg 2002

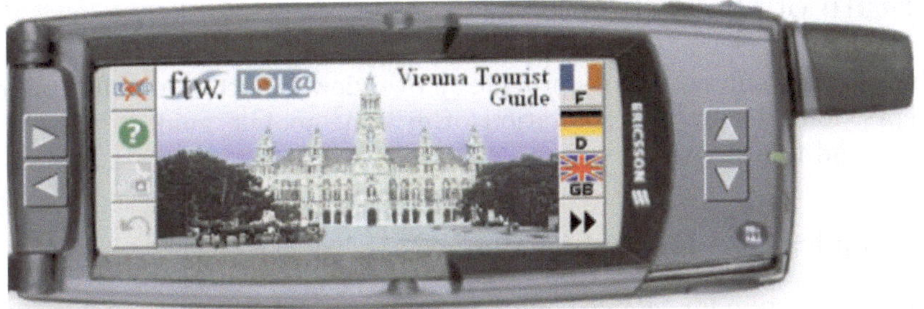

Fig. 1. LoL@ – The Local Location Assistant

(mobile entertainment), or shopping and other tools for leisure time use. These applications may be enhanced with a location component. Access will usually happen with a smart phone or a PDA.

About 18 months ago, we started to develop LoL@, the Local Location Assistant (Fig. 1). It is a prototype of a location based application for the General Packet Radio Service (GPRS) or UMTS. Finally it will run on a GPRS/UMTS-enabled PDA, currently a laptop is used as terminal. LoL@ is designed for tourists, providing predefined or user defined tours in the city center of Vienna. The application can either be used while walking through the city or in the hotel room. In the first case, the system offers interactive navigation between the sights and an electronic tour diary.

The specification of LoL@ was based on two major ideas: to provide value to the user by exploiting the capabilities of new mobile communications systems (GPRS, UMTS) and to study the technical feasibility of such an application in the field. A market survey analyzing user's needs was out of the scope of our project. Therefore, we based our assumptions of tourist behaviour on literature [2,3] and currently available tools, like written tour guides [4] and electronic navigation tools [5,6]. The technical constraints influenced the detailed system design – the chosen programming languages, interfaces, and components. More details can be found in [7] and [35].

We exploit several ergonomic criteria for efficient and fast user interactions to achieve a positive user experience. Especially we consider use of known metaphors, multi-modality, and avoidance of application modes (see Sect. 5.1).

Additionally we have chosen several technical solutions to improve application performance even over slow links. A part of the application logic resides in the terminal, network initiated data transfers (push communication) are possible, and special care was taken with data flow design.

In Sect. 2, we discuss the state of the art of software development in mobile communications environments. Constraints for mobile Internet applications are presented in Sect. 3, followed by an overview of LoL@ in Sect. 4. The design process and technical considerations are shown in Sect. 5. The paper ends with an overview of our future work (Sect. 6) and summary and conclusions in Sect. 7.

2 State of the Art

In the following subsections we provide an overview of current mobile application concepts, followed by a presentation of existing mobile guides, showing their features and limitations.

2.1 Applications for Mobile Phones

The Global System for Mobile communications (GSM) provides circuit switched data connections with 9.6 kbit/s. System extensions, like High Speed Circuit Switched Data (HSCSD) or the packet switched GPRS technology, facilitate data transmission with up to 57.6 kbit/s. UMTS will provide up to 384 kbit/s, although a single user will typically obtain only 64–144 kbit/s [15]. Applications are based on specific toolkits and protocol stacks, like the SIM toolkit [11] and the Wireless Application Protocol (WAP) [12]. A WAP application can be accessed with a simple browser and over slow links.

Unfortunately, WAP is not really suitable for complex applications because of several limitations. The GUI of WAP applications is very limited due to the small, monochrome screens and the small numeric keyboard (Fig. 2). WAP does not provide advanced user interaction primitives like graphical buttons or selection lists. Multimedia features are currently very restricted; it is not possible to include audio or video streaming into WAP applications. Finally, WAP browsers support a simple script language but no compiled code (Applets).

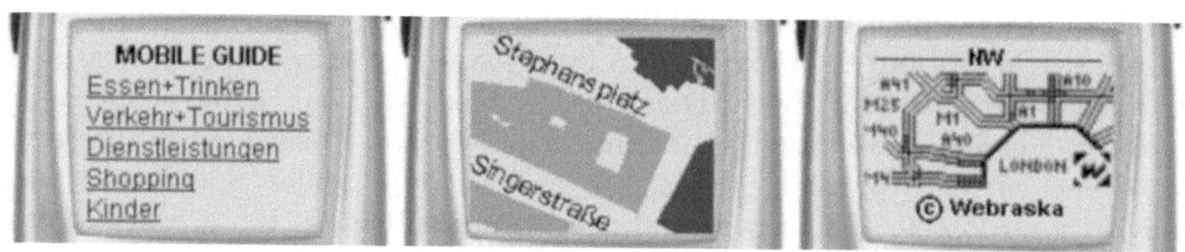

Fig. 2. Mobilkom's Mobile Guide (left, middle) and Webraska Traffic Information System (right)

2.2 Existing Mobile Guides

There is a variety of mobile guides for various devices and operating systems, including research projects like GUIDE [2] or Cyberguide [3] and commercial products.

Cyberguide is primarily designed for indoor usage. It provides locally stored data, a short-range infrared localization system, and bitmap images showing the layout of buildings. In GUIDE, a hypertext-centric presentation of information, based on locally available data, was chosen. Initial data downloads are done via a wireless LAN. The terminals have to be rented at a registration desk. In contrast, LoL@ is based on widespread mobile technology: mobile phones and Laptops/PDAs.

A WAP-based guide is provided by the Austrian GSM operator Mobilkom (Fig. 2, left and middle) [14], providing menu-driven selections and a static map around the destination. Other prominent examples are Palmtop Software's TomTom [5] and Webraska (Fig. 2, right) [6]. After selecting a starting point and a destination, a map shows the route between both locations. Additional features, like textual route descriptions or search functionality and in some cases, automatic positioning features based on GPS are provided.

The shortcomings of the existing systems are the relatively simple user interfaces and the simple maps. The existing applications do not provide multimodal interaction or multimedia information. Better integration with the network infrastructure could provide additional positioning features (e.g. based on the UMTS location architecture) and system awareness (e.g. to cope with varying data rates due to handovers or transmission errors).

3 Constraints for Mobile Internet Applications

Developing applications for the Mobile Internet is a challenging task because of technical and user interaction constraints.

Technical constraints arise primarily from the low data rates of mobile networks. Additional problems for real-time applications like streaming audio or video are the varying transmission bandwidth and delays (jitter). These effects are caused by network interworking functions and the mobile radio channel.

The limited resources of mobile devices cause additional technical constraints. PDAs running PalmOS 4 [16] or Windows for PocketPC 2002 [17] provide between 32 and 64 MB of memory and colour screens with up to $320 * 240$ pixels. User interaction is based on pen-input using text recognition, a virtual keyboard, and simple GUI primitives (Fig. 3). For LoL@ we assume a device with a color display with $320 * 120$ pixels and pen-input.

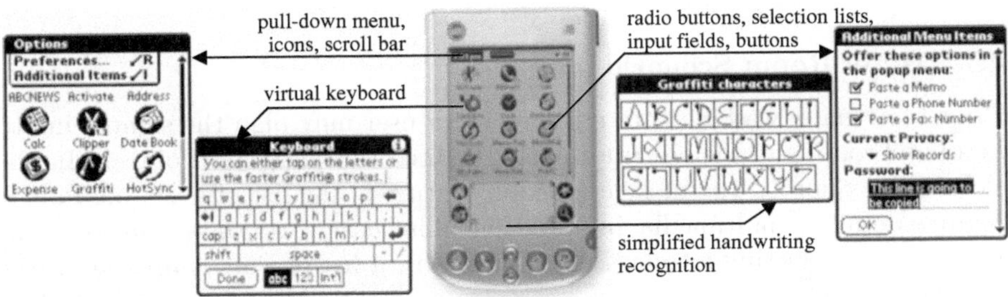

Fig. 3. Simple PDA GUI with handwriting recognition and virtual keyboard (left) and GUI primitives (right)

Typically, PDA applications are not designed for permanent network connectivity. They often support the download of Web-Sites for offline browsing [13]. In the near future, a convergence between PDAs and cell phones can be expected [1].

The *user interaction constraints* arise to some extent because of the technical limitations which have to be compensated partly by the users. For instance, because of the small screen size it is often not possible to show all relevant information simultaneously. The resulting hierarchical information structure causes higher demands on the user's short term memory. Additionally, the user is in a potentially uncomfortable environment (rain, cold, frightening neighborhood). Also, the user might be stressed because he/she is lost and has an urgent need for information, e.g. to find the nearest shop before it closes.

Because of these facts and the relatively high prices for mobile data services, users demand efficient applications. When developing the user interface, it is necessary to consider the limited keyboards and displays. Additional constraints are created by the environment, like direct sunlight which reduces screen brightness and contrast considerably.

4 LoL@ – A Prototype of a GPRS/UMTS Mobile Guide

To study the possibilities of advanced mobile applications, a prototype of an interactive tourist guide is being developed at Forschungszentrum Telekommunikation Wien (FTW). The application is called LoL@, the Local Location Assistant. It provides a map of the inner city of Vienna in two zoom levels (overview and detail) plus textual/multimedial information screens. Three usage scenarios are considered:

- A user walks through the city (the most important scenario, see Sect. 4.2).
- A user retrieves sightseeing information from a hotel room (see Sect. 4.1).
- A user accesses tour information and personal information after finishing the tour (see Sect. 4.3).

LoL@ is based on conventional Internet software technology and user interface paradigms, extended by concepts to improve usability for the mobile domain (see Sects. 5.1 and 5.3).

4.1 Hotel Room Scenario

Prior to actually walking through the city, a user may plan the sightseeing tour in a hotel room and request information related to the sights (Points of Interest, PoI). Consequently, the most important feature is the list of available PoIs and related detailed multimedia information. The usage workflow can be seen in Fig. 4. After selecting a tour (Figs. 4.a and b), a list of PoIs is presented to the user (Fig. 4.c).

After selecting a PoI in this list, the PoI information menu is shown (Fig. 4.d). This screen provides four options, namely "Information", "Details", "Virtual Visit", and "My Data".

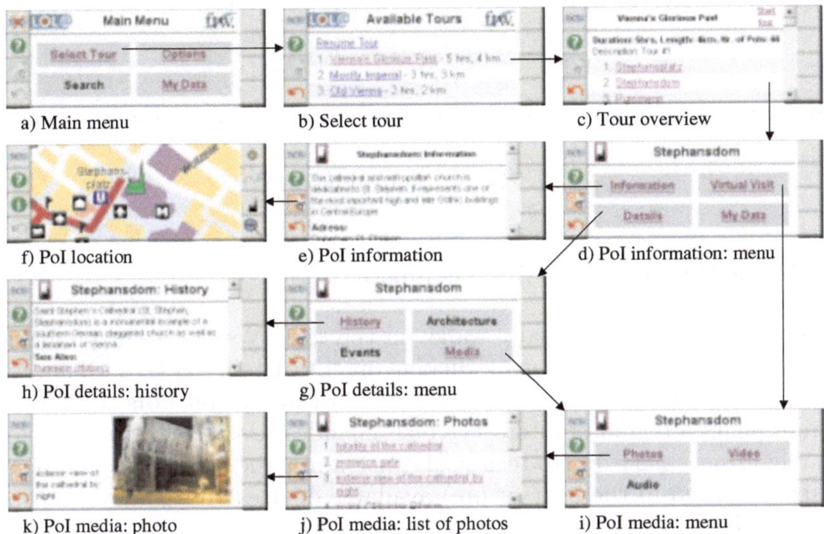

Fig. 4. Multimedia information in LoL@

"Information" gives an overview of the current PoI; it includes a brief description, address and contact information, as well as opening hours and entrance fees (Fig. 4.e). The "Map" button can be used to see the location of the current PoI in the detail map (Fig. 4.f).

Selecting the *"Details"* button brings up the details menu, shown in Fig. 4.g. It gives access to detailed historical information (including links to related PoIs, see Fig. 4.h), architecture description, a list of events (e.g. special exhibitions), and multimedia data. An example of multimedia data, related to the PoI "Stephansdom" is shown in Figs. 4.j and k.

The *"Virtual Visit"* button (Fig. 4.d) is used to access a pre-selected list of multimedia information for this PoI, grouped into three categories (Fig. 4.i). Alternatively, a user may access a similar screen with the "Media" button in the details menu (Fig. 4.g). In the latter case, no information filtering is done, i.e. all available multimedia information for this PoI is shown. In a future version of LoL@, the "Virtual Visit" menu will include user specific information prefiltering according to his/her interest profile. The basic profile may be set in the application preferences. Additional details can be created automatically by monitoring user behavior and through recommendations by other users. Such a concept is already available at some Internet e-commerce portals, for example the Amazon web shop [19]. In this case, user patterns are created based on shopping behavior.

The *"My Data"* item is used to access the personal tour diary, described in Sect. 4.3.

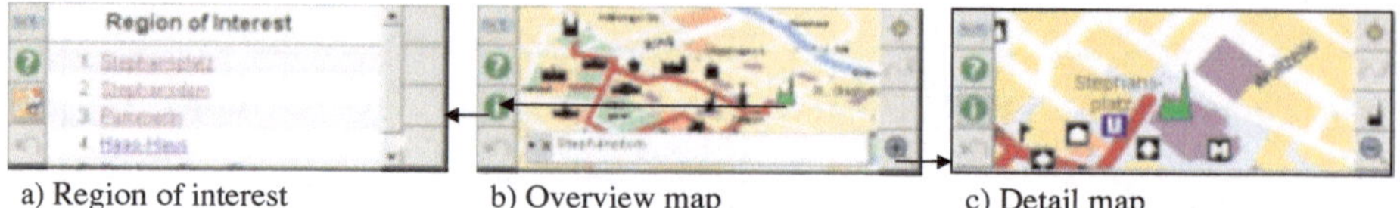

Fig. 5. Map hierarchy

4.2 Walk through the City Scenario

In this scenario, a user actually walks on the tour. We use a "map-centric" approach: after starting the tour by clicking on "Start tour" (Fig. 4.c), an overview map shows the complete tour area and the most important PoIs (Fig. 5.b). Each PoI represents a region of interest, which includes several sights (Fig. 5.a). This hierarchy is necessary due to technical constraints: the screen size and resolution of currently (and in the near future) available devices are too low to display all details in the overview map.

A user can manipulate objects by selecting them in the map and invoking the desired action by pressing a button. For example, when a user selects a PoI, then clicks on the information symbol , the information screen (Fig. 4.d) for this PoI is brought up.

The other possible options are to zoom into the detail map (Fig. 5.c) or turn on positioning. The detail map is also used for routing, i.e. to guide the user from his/her current position to the next PoI on the tour. All visited PoIs are automatically included in a server-based tour diary. A user may additionally upload private data, e.g. photographs retrieved from a digital camera.

When positioning is turned on, the position of the user is displayed in the map. Different location methods and environmental conditions may cause considerable variations in the measurement accuracy. Especially for walking people, an accuracy of 100 m has a significant impact. Therefore we decided to avoid the well-known cross-hairs symbol. Fig. 6 shows the graphical metaphor we developed to represent the uncertainty of the location estimate: the user position is shown as a disc with a radius that corresponds to the current accuracy; color and opacity of the disc fade from the middle to the edge. Unfortunately our map software [20] currently does not allow to show such a symbol. Therefore, currently the accuracy information is only available by clicking the user position symbol (cross-hairs, shown in Fig. 7.b). This displays location accuracy and measurement time in a textbox (similar to a tool tip).

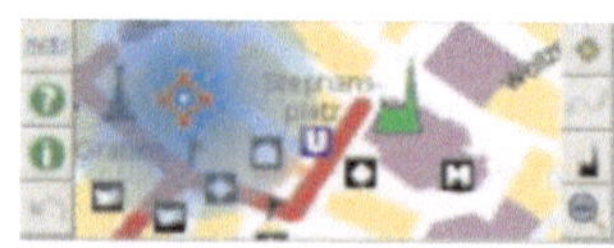

Fig. 6. Representation of location uncertainty

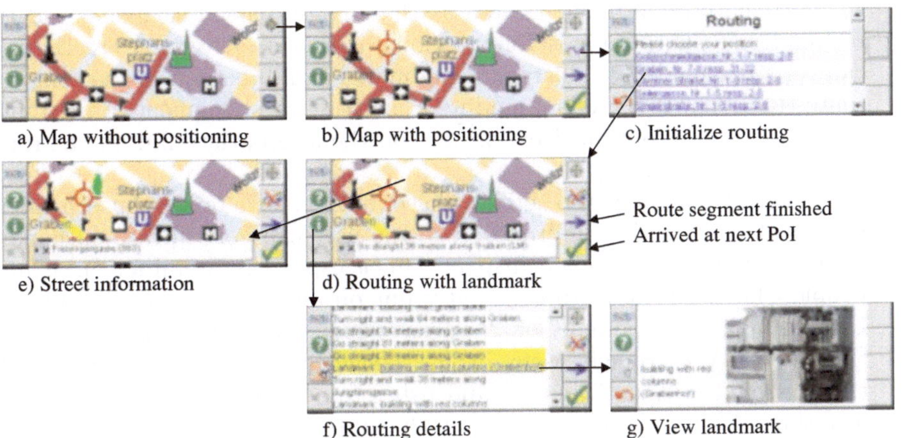

Fig. 7. Routing

Additionally, routing can be used to find the PoIs by navigation with interactive maps. LoL@ uses a hybrid routing concept, combining automatic user positioning with GPS and some user interaction. Additional location methods, based on radio system characteristics (cell ID) and radio signal propagation are currently being studied [7]. Similar to [3], a micro-positioning concept based on short-range Bluetooth beacons is foreseen. Because of the limited accuracy of the current location methods, the user has to select or confirm his/her location in some cases. The workflow for routing is shown in Fig. 7.

- After activating positioning, the location server requests a location estimate from the terminal and performs some calculations (e.g. coordinate transformations). The result is sent back to the terminal to be displayed in the map (Figs. 7.a and b).
- Routing always starts with the determination of the accurate user position. If a user has confirmed arrival at a PoI previously, and the location measurement data does not violate this assumption, it is assumed that he/she is still located at this PoI. If the measurement data does not match the location of the PoI an interactive positioning procedure is performed. This procedure is also used if there is no current PoI because the user has just started the tour. In such situations, the location system calculates a location estimate and an accuracy value. A list of street names within the accuracy radius is presented to the user who selects his/her current location (Fig. 7.c).
- After having identified the starting point of the route, the route to the next PoI is calculated. Alternatively, a user may also select an arbitrary location in the map and activate routing to this point. The only difference is that routing between two PoIs is enhanced by Landmarks, i.e. prominent points between the PoIs (Fig. 7.d).
- Now the user walks along the calculated route and confirms arrival at every route segment and eventually at the next PoI (Fig. 7.d). The map shows

the current route segment as well as the already passed segments and the remainder of the route in different colours (Fig. 7.e). Alternatively, the user can get a textual list of all streets, landmarks and directions of the complete route (Fig. 7.f). If there is a landmark at the current segment of the route, its details can be obtained by clicking on the appropriate link in the route description (Fig. 7.g).

– Optionally, a user may also activate voice routing, i.e. a text-to-speech engine that reads the routing information so the user can concentrate on looking at street signs instead of the display. This is especially useful for far-sighted people who would have to constantly put on and off their reading glasses. Even though the experience of [2] warns of using too much spoken output we think that it is extremely useful in this special case. Users can keep e.g. their thumb over the "Next" button and listen to one line of routing directions at a time. After pressing "Next" (which can be done without looking at the device), the next line of directions will be read. Since only one short sentence is read to the users each time we do not think that the amount of spoken information will be overwhelming. This assumption will be verified by an upcoming usability study (see Sect. 6).

4.3 Accessing Information after Finishing the Tour

After having completed a tour, a user may wish to access the tour diary, e.g. from a home PC after returning from the trip or letting friends access their diary over the web. By default the tour diary, which is stored at the application provider's server, is accessed directly within LoL@ (Figs. 8.a and b). Optionally, a complete tour diary may be downloaded to a PC after user authentication. In this case, a better graphical representation exploiting the superior displays and processing power of a PC can be used. Also, all linked objects are embedded directly into the diary (Fig. 8.c).

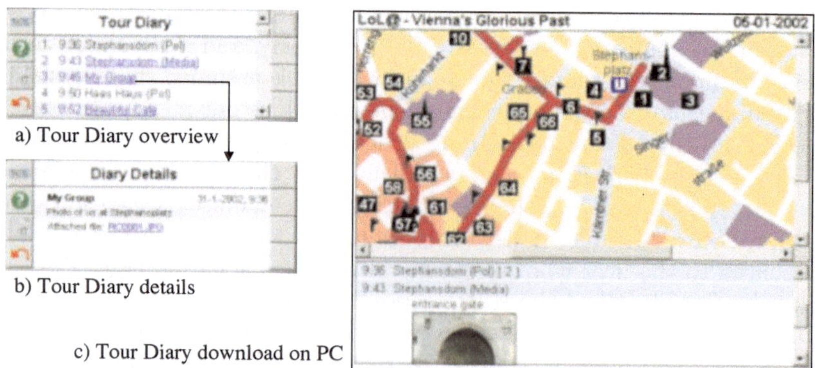

a) Tour Diary overview

b) Tour Diary details

c) Tour Diary download on PC

Fig. 8. Tour diary

5 Design Process and Technical Realization

5.1 General User Interface Considerations

The LoL@ user interface was developed following the recommendations of [18]. A design process (see Sect. 5.2) influenced by [21] that focused especially on the user's needs was used. The following design issues were considered in particular:

– Use of known metaphors: LoL@ uses two main metaphors - the map metaphor and the browser metaphor with hypertext links to present textual and multimedial information. Another important metaphor developed for LoL@ is the position symbol described in Sect. 4.2 which also gives an indication of the location accuracy.
– Navigation based on two map scales: To overcome the limitations of the small display, LoL@ uses an overview map for general orientation. A hierarchy of sights is introduced: PoIs are grouped into regions of interest which are displayed in the overview map. Major sights, like St. Stephen's Cathedral, are displayed using dedicated symbols while minor PoIs, like a café, are displayed with a category symbol only. Maps are dynamic and show user location, already visited, and remaining parts of the tour. In addition, the map is augmented with tool tips showing PoI and street names.
– Multi-modal interactions: Users can interact with LoL@ by clicking on icons, hypertext links, and buttons, by selecting from menus and through spoken commands. We have chosen a server based, speaker-independent voice recognition system [22]. A set of pre-defined voice commands can be used as "short cuts" to perform specific actions, e.g. to access the tour diary.
– Avoidance of "modes" as much as possible. The map screen and textual screens, for instance, can be seen as two differing "views" rather than application modes. If, for example, a new PoI is selected in the textual screen, this selection is also relevant for the map screen.
– Careful colour design: This constraint is typically relevant for elderly far-sighted people or disabled (colorblind) users. In our scenario, it gains additional importance because of bad display contrast in direct sunlight.

5.2 Design Process

We had no possibilities to conduct a market survey, so no input from "real" end users was available, except from a few foreign students. The reason for this limitation was the tight time frame and the initial technical focus of our project. As a consequence, we did not have all resources for system design as recommended in [21]. However, about half of the people involved in the human-system interface design were not computer scientists, but domain experts from the Institute of Cartography and Reproduction Techniques at Vienna University of Technology and colleagues from other disciplines. This team was selected to define the basic system layout without thinking too much about implementation details before finishing the interaction design [23]. The computer scientists gave input regarding the technical feasibility and the system definition process, enabling an integrated GUI and functional design as recommended by [24].

Initial requirements were taken from literature [2,3,25] and enhanced in several brainstorming sessions with the domain experts. Based on these, the main scenarios as described in Sect. 4 were iteratively designed together with the domain experts.

A very early prototype of the human-system interface was produced using Macromedia Flash [26] software and presented to the whole project team including project management. Refinements of this prototype were produced using proposed "screenshots" drawn on paper, together with a functional overview.

The detailed requirements specification was developed in a spiral-model fashion with several rapid iterations over the course of about 4 months by a core team of 3 domain experts and 2 computer scientists. To track changes, a central document [27] was created. It documented human-system interactions in a semi-formal style, using graphical elements and natural language. For each screen the user would see, a preliminary "screenshot" was drawn accompanied by a table listing all possible actions (Fig. 9). This specification included pre-conditions and post-conditions for all interactions. We defined the current system state, user interaction, and the resulting application state-change together with the following screen. For the main usage scenarios we additionally defined flow diagrams to provide a better overview of the interactions.

Screen 210 - DETAIL MAP / SELECTION: ANY PoI A PoI is selected (active). The symbol of the currently active PoI is displayed in the symbol field. For POIs without a symbol (secondary POIs) a pictogram denoting the category of the POI is displayed. A list of pictograms is given in the Introduction (Sec. 3.1).	
[Positioning]	Enable positioning, try to locate user. 5 possibilities for next screen: • 210p - DETAIL MAP / LOCATION: WITHIN RoI *or* • 212p - DETAIL MAP / LOCATION: OUTSIDE TOUR *or* • 211p - DETAIL MAP / SELECTION: ANY PoI, indicate that current position is outside of map • Error Message „Positioning not successful" (if locationing failed) *or* • „You are *XX* km outside the tour area. Please move to *Address of selected PoI*"
[i]	Next Screen: 300 - PoI INFO MENU for active PoI.
[Zoom Out]	Next Screen: 201 - OVERVIEW MAP / SELECTION: ANY RoI, selected RoI contains active PoI.
[Click on PoI]	Next Screen: • if PoI is not selected: 210 - DETAIL MAP / SELECTION: ANY PoI, select clicked PoI, deselect current PoI. • if PoI is selected: 212 – DETAIL MAP / SELECTION: NONE, PoI is deselected.
[Click any place]	Deselect current PoI. *OPTIONAL:* next screen is 211 - DETAIL MAP / SELECTION: ANY PLACE, select clicked place.

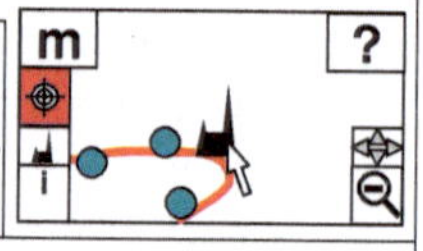

Fig. 9. Human-system interaction specification with pseudo "screenshot"

This semi-formal approach was very helpful because it was easily understood by the non-technical people in our team, who were not familiar with object-oriented analysis or UML diagrams [28]. Still, the system behaviour was described accurately enough to allow an implementation by programmers who were not involved in the system design.

5.3 Technical Realization

The technical realization of LoL@ was based on a functional decomposition. We identified several functional blocks, related to specific tasks:

- *Multimedia:* We use off-the-shelf products for streaming audio and video, namely the Apple Darwin Streaming Server and the Apple Quicktime Player [29] in the terminal.
- *Location, Navigation and Maps:* We have a map viewer applet in the terminal, together with location measurement hardware and software. Coordinate transformations are done by a location server, map preparation and route calculations are performed by a routing server (Geomedia Web Enterprise [30]). Additional details are published in [31] and [32].
- *Speech processing:* Voice transmission is done with a Voice-over-IP solution based on the Session Initiation Protocol (SIP) [33] and the GSM voice codec. Speech recognition is implemented with the Nuance Speech Toolkit [22].
- *Content server:* An SQL database is used for textual content, multimedia data is stored in the file system of the content server.
- *Application logic:* The application logic is split between JAVA applets in the terminal and JAVA servlets [10] in an HTTP servlet engine, content preparation is based on XML technology [8].

In principle we assume permanent network connectivity of the terminal. All content data is stored in a content database and prepared on demand by JAVA servlets. However, current mobile communications systems offer only moderate data throughput. Therefore – and in contrast to traditional Internet applications where usually all the logic resides in the server – we decided to split the application logic into a server part and a terminal part, thus reducing the amount of transferred information: simple interactions, e.g. the pressing of a button, can be handled directly within the terminal. Network interactions are only required if complex interactions are invoked (e.g. activate routing) or if new information has to be fetched from the server. This concept improves system response time.

The terminal part of LoL@ is implemented on top of a Web-Browser to minimize the memory footprint of the application. LoL@ can use the browser functionality to display hypertext information; multimedia streaming and map display is possible with browser plug-ins and 3rd-party JAVA applets.

Because of the selected terminal architecture it is quite straightforward to use JAVA servlets for the server side of the application logic. Connectivity is provided using the standard HTTP protocol [34]. We use flexible templates for content preparation, based on XML [8] and XSL [9] technology. These templates provide a means to model human-computer interaction flow in machine-readable and human-understandable form. They allow the integration of input data from various data sources with different access methods and data formats. Different output formats, like HTML or WML, can be supported easily by using appropriate XSL stylesheets. A detailed discussion of the content preparation system can be found in [35].

6 Future Work

We have almost finished system implementation and are currently planning an extensive trial campaign that will be performed during summer 2002. These trials will be used to evaluate technical aspects of the system, e.g. location accuracy and data throughput, as well as usability issues. The usability study will be conducted by students attending a user interface design and usability course. The results will be used to improve LoL@. Currently we plan several system enhancements:

- Add instant messaging: This extension will allow communication among a group of tourists, e.g. to notify all group members about interesting sights or a leaving bus. A location based component may be included to obtain the location of other group members.
- Extended context and profile awareness: Tour details will be sorted w.r.t. date and time, e.g. to consider opening hours of museums. Additionally we will include information pre-filtering in the "Virtual Visit" menu to present information according to the user's interest profile.
- Click-to-call: By clicking on a link within LoL@, a voice call can be established. This mechanism will allow to integrate ticket booking or room reservations into our application.
- Unfortunately the software environment of current PDAs does not provide all functionality that is required for the map viewer applet and the speech software yet. Therefore we use a Laptop for the demonstration of LoL@, although the processing capabilities of current PDAs would be sufficient.

7 Summary and Conclusions

We have presented LoL@, the Local Location Assistant, a UMTS tourist guide for the inner city of Vienna. After providing an analysis of existing software environments in the mobile communications community, we gave an overview of the functionality of LoL@. The chosen design methodology, user interface concepts, and the technical considerations for implementation have been discussed in detail.

Although we did not perform the field trials yet, we can already see that map presentation on smart phones with small displays is feasible. Even routing systems can be implemented on such devices. Considering the varying positioning accuracy, an interactive routing concept will give the best user satisfaction.

The chosen semi-formal design and specification methodology has turned out to be a very efficient means to structure the communication between engineers and users. The resulting specifications are quite understandable for non-technicians, still they are accurate enough to implement a medium-scale project.

Acknowledgements

This work is supported within the Austrian competence center program ***Kplus*** and by the member companies of ftw (Alcatel Austria, Connect Austria, Ericsson Austria, Kapsch AG, Mobilkom Austria AG, Nokia Austria, and Siemens Österreich). Special thanks are addressed to our colleagues at the Institute of Cartography and Reproduction Techniques at the Vienna University of Technology.

References

1. Forman, D., "Mobile Convergence – Ultraportables come together", *LAPTOP Magazine*, p. 52–60, August 2000.
2. Cheverst, K., et al., "Experiences of Developing and Deploying a Context-Aware Tourist Guide: The GUIDE Project". In *Proc. Sixth Annual International Conference on Mobile Computing and Networking*, p. 20–31, ACM, 2000.
3. Long, S., et al., "Rapid Prototyping of Mobile Context-Aware Applications: The Cyberguide Case Study". In *Proc. 2nd ACM International Conference on Mobile Computing and Networking (MobiCom '96)*, p. 97–107, ACM, 1996.
4. Polyglott Tourist Guides, `http://www.polyglott.de`, 2002.
5. Palmtop Software, "TomTom CityMaps", `http://www.palmtop.nl/palm/citymaps.html`, 2002.
6. Durocher, J. M., "WAP and Mobile Location? Spawning a Mobile Information Revolution". In *Proc. IBC Content and App. for the Mobile Internet 2000*, 2000.
7. Anegg, H., et al., "LoL@: Designing a location based UMTS application", *E&I Elektrotechnik und Informationstechnik*, 2:48–51, Springer Verlag Wien, 2002.
8. Bray, T., et al., "Extensible Markup Language (XML)", W3C Recommendation, `http://www.w3.org/TR/REC-xml`, 2000.
9. Froumentin, M., "Extensible Stylesheet Language (XSL)", W3C XSL Working Group, `http://www.w3.org/Style/XSL`, 2002.
10. Davidson, J., et al., "JAVA Servlet Specification V.2.3", Sun Microsystems, `http://www.sun.com`, 2000.
11. 3GPP, "USIM Application Toolkit (USAT)", 3GPP Technical Specification, `ftp://ftp.3gpp.org/Specs/2001-03/R1999/31_series/31111-340.zip`, 2001.
12. Wapforum: "Wireless Application Environment Specification", WAP Forum, `http://www.wapforum.org`, 1999.
13. Harris, D., "AvantGO: Optimizing Web Pages for Handheld Devices". In *Proc. IBC's second annual Mobile Software Forum*, 1999.
14. Mobilkom Austria, "Mobile Guide", `http://wap.a1.net`, 2002.
15. UMTS Forum, "Enabling UMTS/Third Generation Services and Applications", UMTS Forum Report 11, `http://www.umts-forum.org`, 2000.
16. Palm, Inc., "Hardware Comparison Matrix", `http://www.palmos.com/dev/tech/hardware/compare.html`, 2001.
17. De Herrera, C., "Pocket PC 2002 Comparison", `http://www.cewindows.net/wce/30/ppc2002/ppc2002comp.htm`, 2001.
18. Preece, J., "Human Computer Interaction", Addison-Wesley, 1994.
19. Amazon Web Shop, `http://www.amazon.com`, 2002.
20. GISquadrat Active CGM JAVA Map Viewer, `http://www.gisquadrat.com`, 2001.
21. Guida, G. and Lamperti, G., "AMMETH: A Methodology for Requirements Analysis of Advanced Human-System Interfaces". *IEEE Trans. on Systems, Man, and Cybernetics - Part A: Systems and Humans*, 30(3): 289-321, IEEE, May 2000.
22. Nuance, "Software for a Voice-Driven world", `http://www.nuance.com`, 2002.
23. Gentner, D. R., Grudin, J., "Why good Engineers (Sometimes) create bad Interfaces". In *Proc. Empowering People: Human Factors in Computing Systems. Special Issue of the SIGCHI Bulletin*, p. 277–282, ACM, April 1990.
24. Van der Veer, G., Van Vliet, H., "The Human-Computer Interface is the System". In *Proceedings of the 14th Conference on Software Engineering and Training*, p. 276–286, IEEE, 2001.

25. Cheverst, K., et al., "Providing Tailored (Context-Aware) Information to City Visitors". In *Lecture Notes in Computer Science 1892*, p. 73–85, Springer Verlag Berlin, Heidelberg, 2000.
26. Macromedia, "Flash Product Overview", Macromedia Inc., `http://www.macromedia.com/software/flash/productinfo/product_overview/`, 2001.
27. Pospischil, G., et al., "Report 4.4.b: Interactive Human Interface", Technical Document, Forschungszentrum Telekommunikation Wien, `http://www.ftw.at`, 2001.
28. Object Modelling Group, "Unified Modelling Language Specification, V. 1.3", Object Modelling Group, `http://uml.shl.com/docs/UML1.3/99-06-08.pdf`, 1999.
29. Apple Quicktime Streaming, `http://www.apple.com/quicktime/products`, 2002.
30. Intergraph Corp., `http://www.intergraph.com/gis/gmwe/default.asp`, 2001.
31. Pospischil, G., et al., "LoL@: A UMTS Location Based Service". In *Proc. International Symposion on 3G Infrastructure and Services*, Greece, 2001.
32. Friedrich, B., et al., "Visualisierungskonzepte für die Entwicklung kartenbasierter Routing-Applikationen im UMTS-Bereich". In *Proc. Symposium für Angewandte Geographische Informationsverarbeitung (AGIT) 2001*, `http://www.agit.at/papers/2001/brunner_friedrich_EAR.pdf` (in German), Austria, 2001.
33. Handley, M., et al., "SIP: Session Initiation Protocol", IETF RFC 2543, `http://www.ietf.org`, 1999.
34. Fielding, R., et al., "Hypertext Transfer Protocol – http/1.1", IETF RFC 2068, `http://www.ietf.org`, 1997.
35. Michlmayr, E., "Flexible Content Management for the LoL@ UMTS Application", Masters Thesis, Vienna University of Technology, 2002.

Location-Aware Shopping Assistance: Evaluation of a Decision-Theoretic Approach

Thorsten Bohnenberger[1], Anthony Jameson[2], Antonio Krüger[1], and Andreas Butz[3]

[1] Department of Computer Science, Saarland University
[2] German Research Center for Artificial Intelligence (DFKI)
[3] EyeLed GmbH

Abstract. We have implemented and tested a PDA-based system that gives a shopper directions through a shopping mall on the basis of (a) the types of products that the shopper has expressed an interest in, (b) the shopper's current location, and (c) the purchases that the shopper has made so far. The system uses decision-theoretic planning to compute a policy that optimizes the expected utility of a shopper's walk through the shopping mall, taking into account uncertainty about (a) whether the shopper will actually find a suitable product in a given location and (b) the time required for each purchase. To assess the acceptability of this approach to potential users, on two floors of a building we constructed a mock-up of a shopping mall with 15 stores. Each of 20 subjects in our study shopped twice in the mall, once using our system and once using a paper map. The subjects completed their tasks significantly more effectively using the PDA-based shopping guide, and they showed a clear preference for it. They also yielded numerous specific ideas about the conditions under which the guide will be useful and about ways of increasing its usability.

1 Introduction

Mobile systems offer great potential for providing novel services to users. But this potential can be fulfilled only if the systems are designed in a thoroughly user-centered way. User-related considerations can range from fine-grained problems in dealing with novel input and output modalities to high-level considerations concerning the way in which a system fits into users' customary patterns of activity.

Systems for mobile commerce are a case in point. Especially when equipped with location-awareness, they in theory have a good deal to offer their users. But mobile commerce systems have so far largely been rejected by consumers, even by those who were initially eager to try them out.

Part of the problem lies in the technical limitations of current wireless technology, which lead to long waits and frequent interruptions of connections, in addition to clumsy interaction with small devices (cf. [10]). One remedy is to aim for designs that work well with the current limited technology, checking with users to see whether they really do work well enough.

Another type of problem concerns designs that do not adequately take into account the conditions under which mobile systems are used (e.g., while the user is in motion and/or is simultaneously engaging in some other activity). Only realistic tests conducted early enough in the design process can reliably prevent such problems (see, e.g., [9,6]).

F. Paternò (Ed.): Mobile HCI 2002, LNCS 2411, pp. 155–169, 2002.
© Springer-Verlag Berlin Heidelberg 2002

In this paper we report on an effort to apply a user-centered strategy in the design and development of an intelligent shopping guide based on artificial intelligence techniques of decision-theoretic planning.

The rest of this Introduction describes the basic ideas of the shopping guide, and the next section explains how the current prototype was realized. The remaining sections describe and discuss a study that we conducted with 20 users in a semirealistic setting with a view to assessing and improving the usefulness and acceptability of the shopping guide.

1.1 Combining Product Location with Navigation Support

Location-aware mobile commerce systems that aim to bring together customers and products[1] have so far fallen into two main categories (see [5] for more extensive discussion and examples).

1. Product location services: A user $\mathcal{U}$ queries her system $\mathcal{S}$ about the availability of a particular type of product near her current location (see, e.g., the ADAPTIVE CLASSIFIED SERVER of AdaptiveInfo; http://www.adaptiveinfo.com).
2. Location-dependent alerting services: When $\mathcal{U}$ arrives at a given location, she is automatically notified about nearby products that are known to be of special interest to her.

In both of these schemes, the location of $\mathcal{U}$ is seen as being determined by other factors, and the job of $\mathcal{S}$ is to help $\mathcal{U}$ do as well as possible in or near that location.[2] Consequently, $\mathcal{U}$ may discover that no especially desirable products are available near her current location even though there exist more attractive ones at some location that she might just as well have visited.

The alternative approach to be examined here takes into account the fact that people often have considerable freedom in determining what particular locations to visit. To take an example that involves a small geographic region, consider a user who wants to walk through a large shopping mall on the way to some other destination and would like to pick up a few items along the way – providing that she can do so without investing too much extra time. Instead of taking the shortest route through the shopping mall and finding out about products that happen to be available along that route, she may well be willing to take a longer route that leads her by places where she is more likely to find the desired products. Suppose now that $\mathcal{U}$ is to be guided through the shopping mall by a system running on a handheld device: It makes sense for the system to apply techniques similar to those used in the more sophisticated route planning systems for automobiles, which choose a route according to a mixture of criteria such as driving time, gas consumption, and scenic attractiveness.

On the other hand, the planning problem in the shopping mall has an important difference from the typical route planning problem: the need to deal with the uncertainties

[1] To facilitate exposition, we will use the term *products* even when what is really meant is products, services, and combinations of both. The symbol $\mathcal{S}$ will denote the system under discussion, while $\mathcal{U}$ will refer to a user of that system.

[2] A system may provide navigation support to guide $\mathcal{U}$ from her current location to the exact nearby place where she can obtain the recommended product; but the current location is still treated as given.

that are inherent in almost any attempt to match customers with products. For example, if $\mathcal{U}$ is looking for a specific book, it is in general hard for a system to know with certainty in advance whether a given bookstore has that book in stock. If $\mathcal{U}$'s interest in books is more indefinite (e.g., "Something amusing to read on the plane"), there will be uncertainty for another reason: Even if $\mathcal{S}$ has exact information on the available books, $\mathcal{S}$ cannot predict with certainty if $\mathcal{U}$ will find one of them that fulfills her requirements.

As is explained in [2], $\mathcal{S}$ can deal adequately with this uncertainty only if it computes not a fixed plan but a *policy* in the sense of decision-theoretic planning. A policy can specify, for example, that if $\mathcal{U}$ does not find a suitable book in Bookstore *A* she should make a slight detour to try the nearby Bookstore *B*.

One of the issues to be addressed in the current paper is whether this type of decision-theoretic planning can actually lead to more effective shopping in the context of a working system. And even if the shopping is more effective, how do users evaluate it subjectively? Although we will address this question within a mock-up of a shopping mall, the basic method is applicable on very different scales, such as shopping within a city.

A second main issue that is specific to smaller scenarios concerns the usability of a particular solution to the problem of providing precise location-awareness indoors. We test a relatively simple, robust technology that makes use of the infrared ports found on many handheld devices.

2 Realization of the Shopping Guide

For concreteness, we will explain the workings of the shopping guide with reference to the hardware used and the (fictitious) shopping mall that we set up for the user study. Consider a user $\mathcal{U}$ who enters a mall with the intention of buying five products, if possible: a CD, some bread, a newspaper, a PDA, and a plant. As the partial map in Fig. 1 indicates, products of each of these types are available in one or more stores. But when deciding which stores to visit, $\mathcal{U}$ needs to take into account the facts that (a) it is not certain that she will find a truly suitable product even in a potentially relevant store; and (b) stores differ in the time that it typically takes to wait in line to pay for a product.

The goal is to allow $\mathcal{U}$'s PDA to determine at each point in time which store $\mathcal{U}$ should visit next. Since the necessary planning (described below) is computationally expensive, it is performed on a stationary computer at which $\mathcal{U}$ enters a specification of the desired products through some suitable form of interaction (e.g., menu selection on a touch screen). The computer then computes a *policy* for moving through the shopping mall. The policy is compact enough to be transferred to $\mathcal{U}$'s PDA, and it contains all the information that the PDA-based application requires.

As Fig. 2 indicates, the PDA guide should at any moment direct $\mathcal{U}$ toward the next store by displaying a navigation instruction on the PDA screen; $\mathcal{S}$ will determine the current location and orientation[3] by receiving an infrared signal from one of a number of beacons affixed to the walls, and $\mathcal{S}$ will know what products $\mathcal{U}$ has already found, because $\mathcal{U}$ will check each one off after purchasing it.

[3] A beacon transmits signals only within a small angle. Therefore, receiving signals from a beacon means pointing with the PDA's infrared interface in the direction opposite to that in which the beacon transmits signals.

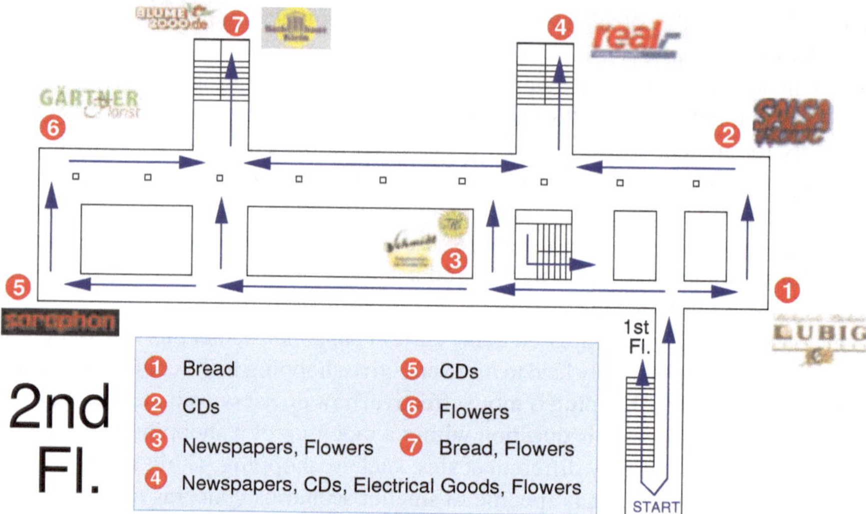

Fig. 1. Part of one of the maps representing the shopping malls for the user study. Icons represent stores, arrows indicate the directions $\mathcal{U}$ is permitted to walk to

Fig. 2. The hardware of the PDA-based shopping guide (Left: Example display on the PDA, which receives location signals from a beacon (middle). Right: Close-up of part of the display in the situation where the user has arrived at a relevant store. $\mathcal{U}$ checks off the items found in the current store and clicks "Okay" to proceed with the next navigation recommendation)

2.1 Hardware

The particular hardware configuration employed in the user study (see Fig. 2) gives an idea of the typical hardware requirements: The specific PDA was a Compaq iPAQ 3620 Handheld, which can receive signals from the small infrared beacons via its built-in IrDA interface. We mounted 30 of these beacons at decision points (i.e. turns, crossings, stairs) in the building in which the study was conducted, each at a height of about 2.5

meters. The beacons emit a 16 bit ID twice a second, and they have a transmission range of about 7 meters. Running on batteries, they do not need access to the power supply system or to a data network.

The PDA software for the shopping guide is implemented in TCL/TK. When a beacon signal is received, the browser displays the corresponding navigation recommendation according to the precomputed navigation policy, taking into account both the beacon's location and the products that $\mathcal{U}$ has checked off on the shopping list shown on the right-hand part of the screen. If a beacon ID corresponding to a store location is received, the items offered in that store are highlighted in the shopping list.

2.2 Decision-Theoretic Planning

The techniques used for the decision-theoretic planning are closely related to those described in [2] and [7] (see [1] and [3] for discussions of the more general family of decision-theoretic planning techniques). Since these techniques are not the focus of the present paper, they will be described only in enough detail to permit an overall understanding of how the shopping guide works.

The result of the planning process is a navigation policy that allows $\mathcal{S}$ in each situation to recommend the action with the highest expected utility for the user. We use a Markov decision process (MDP) to model the dynamics of the shopping trip. Our basic assumptions are the following: (a) $\mathcal{U}$ moves from a starting point to a destination (entry and exit of the shopping mall) via distinct positions in the building. (b) On her way she wants to find as many items of her shopping list as possible in a relatively short time. (c) In each of the stores she has a chance to purchase some of the items needed. (d) In each of the stores in which she finds an item, she will spend a certain amount of time waiting in line to pay for it.

We encode this information in the state features and the transitions between the states as follows: A state is defined by (a) the position of an infrared transmitter, (b) a list of the desired items and the items already purchased, and (c) a flag indicating the status of $\mathcal{U}$'s shopping procedure at that position – a feature that is significant only if there is a store at the position in question. This flag has one of four values: (1) NEUTRAL, indicating that $\mathcal{U}$ has not yet checked if any of the desired articles is available at the current store; (2) NOT FOUND, indicating that none of the articles needed is available in the store; (3) FOUND, indicating that $\mathcal{U}$ has determined that at least one of the articles needed is available; and (4) BOUGHT, indicating that $\mathcal{U}$ has bought at least one article.

Figure 3 shows an example of the system dynamics. Suppose $\mathcal{U}$ approaches a junction located at position 2. Whether or not she has already purchased article A (corresponding to the two states shown for position 2), $\mathcal{S}$ has the choice between recommending LEFT or RIGHT, which would lead $\mathcal{U}$ to position 6 or position 3, respectively. Each of these transitions has a cost that depends on the time it will take $\mathcal{U}$ to reach the next position.

Suppose that $\mathcal{U}$, who has not yet purchased article A, approaches the store at position 3, which offers article A. In this case, $\mathcal{S}$ can recommend RIGHT to have $\mathcal{U}$ pass by the store and go ahead toward position 4. But $\mathcal{S}$ can also recommend that $\mathcal{U}$ enter the store and LOOK FOR ARTICLE A. In this case, $\mathcal{S}$ needs to consider what $\mathcal{U}$'s chances are of finding article A in this store. Moreover, looking for article A is associated with a cost: the usual time needed to check if article A is available or not. If $\mathcal{U}$ finds article A in the store,

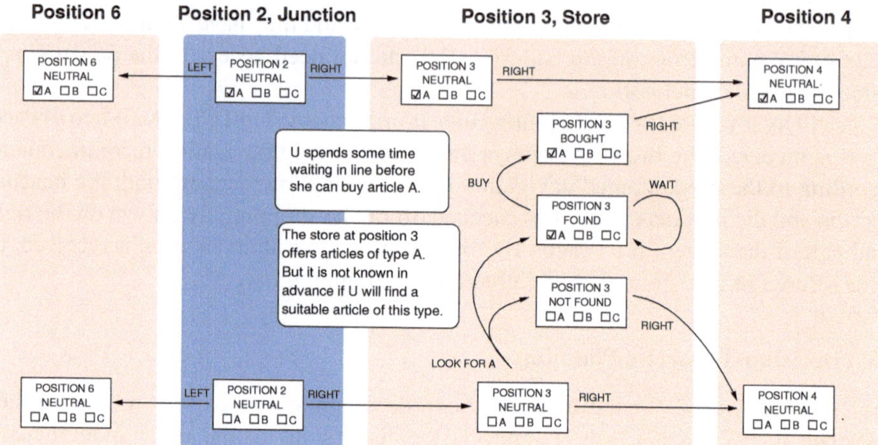

Fig. 3. Illustration of the modeling underlying the decision-theoretic planning employed in the shopping guide (Explanation in text)

it remains uncertain how long she will have to wait in line to pay for it. This waiting is modeled in a simple fashion by a circular WAIT-transition which is associated with a particular probability of the form $(n-1)/n$ for an integer n. This model would be entirely accurate if (a) the cashier required a constant amount of time to deal with each customer, (b) there were always n customers waiting, and (c) the cashier always chose the next customer at random from the n who were waiting. These assumptions will seldom be satisfied exactly, but they do make it possible to approximate a broad range of more realistic waiting processes. Once $\mathcal{U}$ has finally BOUGHT article A, $\mathcal{S}$ will give the next navigation recommendation: In this case the only option is RIGHT.

The Markov decision process described here is *fully observable*: The system always knows with certainty $\mathcal{U}$'s position, the items needed and the items already purchased. The rewards for the goal states, which are associated with the position at the mall exit, are a function of the number of articles that $\mathcal{U}$ has bought during her shopping trip.

Given this representation of the planning problem, we can use a standard algorithm, such as value iteration (see, e.g., [3]), to compute the navigation policy. This computation takes only a few seconds, since the number of state features is small in the problem at hand.

The modeling underlying the decision-theoretic planning can in principle take into account a variety of factors other than the ones included in our current implementation, such as the difficulty of following each instruction correctly (cf. [7]). The main constraints on the modeling are due to complexity considerations, as will be discussed at the end of the paper.

3 User Study: Method

The overall purpose of the user study was to check the objective effectiveness and subjective acceptability of the shopping guide just described, with regard to both its planning methods and its hardware solution.

So as to have a point of comparison for the results with the PDA guide, we also had subjects perform the same basic task with a conventional paper map (see the example in Fig. 1). The map was designed to be at least as helpful as the best maps that can be found in shopping malls: It was portable (in contrast to the frequently found wall-mounted maps, which burden users' memory), and it showed all of the relevant information that could reasonably be expected to be found on a paper map. This information did not include probability distributions for waiting times or for the likelihood of finding suitable items in particular categories.

3.1 Mockup of the Shopping Malls

One configuration of 30 beacons was installed on two floors of the main Computer Science building at Saarland University. Two alternative shopping malls (of equal navigational complexity), each comprising 15 stores, were defined, so that each subject would be able to perform a distinct shopping task with each of the two types of guidance. Figure 1 shows the upper floor of one of these malls. It was not feasible to decorate the building with physical representations of stores, but the subject could always see from the map or the PDA display when she had reached a given store.

In addition to the physical constraints on movement enforced by the architecture of the building, we added some further constraints that might be found in a more complex setting: Some segments of hallways were designated as "one-way passages", as is indicated by the arrows in Fig. 1. Subjects using the map were instructed not to walk in the wrong direction on these segments, and the instructions given by the PDA guide likewise obeyed these constraints. Analogous constraints are found in real shopping malls in the form of (a) escalators that run in only one direction at a given location; (b) entrances through which shoppers are not allowed to exit, and vice-versa; and (c) one-way arrows drawn in aisles within stores. Since we designated a relatively large number of one-way segments in our malls (partly for technical reasons that will be discussed below), our malls were presumably untypically challenging in this respect.

Experiments on consumer behavior in which subjects are only pretending to buy products are frequently criticized on the grounds that the subjects' motivation is untypical. But in the present study, the focus was not on the buying of products but rather on the process of navigating to find the products.

3.2 Shopping Tasks

Each subject used the paper map and the PDA guide in two successive trials.[4] The subjects' task in each case was to buy as many products as possible from a shopping list with 5 categories: bread, a plant, a CD, a newspaper, and a PDA.

Each of the following combinations and temporal orderings was employed with 5 of the 20 subjects:

- PDA in mall 1; map in mall 2

[4] A within-subject design was chosen over a between-subject design because of (a) our expectation that individual differences would be large; and (b) our desire to hear the comparative comments of subjects who had used both types of shopping guidance.

- PDA in mall 2; map in mall 1
- map in mall 1; PDA in mall 2
- map in mall 2; PDA in mall 1

Therefore, neither learning effects nor differences between the two malls could bias the overall pattern of results.

3.3 Subjects

We recruited 20 subjects (15 of them female). Of these subjects, 14 were students, 3 were employed persons, 1 was a homemaker and 1 was an unemployed person. Only 2 of the subjects had used a PDA before, neither of them for a navigation task. Most of the subjects knew what a PDA is and what it is normally used for. The subjects' reward was made up of (a) a fixed amount of money (15 German marks) plus (b) an additional performance-dependent amount: They received 2.00 marks for each product they managed to "buy", but 0.01 mark was deducted for each second that they spent shopping. (This cost of time was explained by analogy to the parking costs incurred during shopping.)

3.4 Procedure

Each subject performed the tasks individually under the guidance of an experimenter. First, the experimenter explained the nature of the tasks. Then he familiarized the subject with the paper map (e.g., how the paths between stores were visualized) and with the use of the PDA guide (e.g., how to hold the device to ensure good reception of the beacons' signals). During the explanation of the reward policy, the subject was instructed not to race through the building in order to maximize her financial reward but rather to walk at a normal pace, obeying the constraints shown on the map (when using the map) or following the PDA guide's directions.

In each of the two trials, the subject began at the mall entrance and was followed around by the experimenter. When the subject reached a store, the experimenter took on the role of the sales clerk at that store: He informed the subject about the availability of the desired product(s); and if at least one was available, he determined how long the subject would have to wait in line to buy it, enforcing the waiting time with a stopwatch.

After both trials had been completed, the subject filled in a questionnaire and was interviewed briefly.

4 User Study: Results

4.1 Objective Results

As is illustrated in Fig. 4 (left), the PDA guide did enable subjects to complete their shopping task 11% faster on the average than with the map, the difference being statistically significant according to a paired t-test ($t(19) = 2.65, p < .05$). This difference in itself may not seem to constitute a practically important advantage, but it seems more noteworthy when we consider two factors:

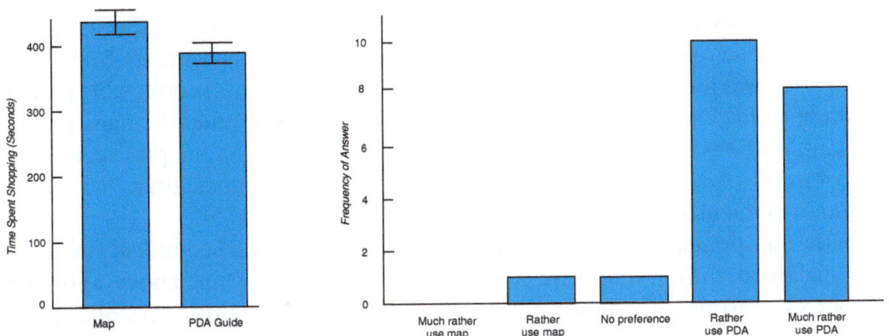

Fig. 4. Left: Mean values for the quantitative performance index for each of the two types of shopping guidance. Each bracket indicates the standard error of the corresponding mean. Right: Frequency distribution of answers to the question "Which method would you rather use if you had to perform more shopping tasks like the ones you performed in this study?"

- In the map condition, subjects were allowed to study the map as long as they wanted before starting to shop, and this study time was not counted. In the PDA condition there was nothing to study, and the subjects set off immediately.
- Whereas all subjects were familiar with paper maps, none had previously used a PDA navigation aid like the one in this study, and only 2 had used PDAs at all. So the time for the PDA condition includes any time that subjects might have required to get used to using the PDA guide and to dealing with the infrared signals from the beacons.

The map and PDA conditions were just about identical in terms of the number of products that subjects found. It is not surprising that there is no reliable difference here: The number of products found is a relatively coarse-grained variable, the possible values being just the integers between 0 and 5; and the situation was set up in such a way that in general most of the desired products could be found, so there was little room for differences.

It is unclear how well these quantitative results would generalize to other concrete scenarios and to qualitatively different users. Still, the results for this study demonstrate that the PDA guide can fulfill its promise of enabling shoppers to complete their task more effectively despite the two handicaps just listed.

4.2 Subjective Results

While the objective effectiveness of the PDA guide may be a necessary condition for its ultimate acceptance, it is not a sufficient condition. An important question is how attractive subjects would find the prospect of using such a guide in real shopping (bearing in mind the fact that they had no opportunity to experience it in truly realistic conditions). As an assessment of overall preference, the questionnaire included a question that asked which method they would prefer to use if they had to perform further shopping tasks like the ones they had just completed. Figure 4 (right) shows that the subjective preference is much clearer than the objective advantage. It is possible that some subjects were biased in favor of new technology or wanted to please the experimenter (cf. [8]).

It might be suspected that this difference (and perhaps the objective speed difference as well) were due to specific inadequacies in the design of the paper map. But subjects' responses to the relevant questions made it clear that they saw fewer correctable problems with the map than with the PDA. Nevertheless, the usage of multiple maps put up in important locations of the building (and designed according to principles such as those in [4]) instead of a map carried along by the subjects might facilitate the subjects' orientation in the map condition.

On the questionnaire, subjects were encouraged to add verbal comments to their ratings. In addition, the questionnaire contained open questions asking about (a) problems and sources of irritation with the map and the PDA guide; and (b) ways of improving the PDA guide. Since in many cases closely related comments were made to different questions, we summarize them here not in terms of the individual questions but in terms of the overall pattern of ideas that emerged.

4.3 Reasons Given for Preferring the PDA Guide

Saving Time and Effort. The most obvious advantage of the PDA guide is that the shopper can typically complete her tasks in less time than with a map. As one subject mentioned, there is also a corresponding savings of effort in moving around the shopping mall – an advantage which some users may find appealing even if they are not concerned about time, for example if their mobility is for some reason limited or if the mall is much larger than the one used in this study. These comments confirm that subjects recognized the superior ability of the PDA guide to deal with complexities such as uncertainties about product availability and waiting time – despite their unfamiliarity with systems of this sort and with the underlying technology. The comments also confirm that the potential advantages conferred by the PDA guide's knowledge and reasoning were not seen by subjects as being outweighed by usability problems or unfamiliarity.

Less Cognitive Effort and Frustration. A less objectively measurable advantage is the reduced need for mental exertion with the PDA guide. As one subject put it, "You don't have to plan where you're going to go, or think for yourself." When using the map, most subjects recognized the advisability of planning ahead, though it would not have been feasible for them even to attempt the type of decision-theoretic planning that is performed by the PDA guide. A typical strategy was to

- plan a few steps ahead, assuming that the products they wanted to get in the nearby stores would be available; and
- replan whenever an product proved unavailable.

In addition to leading in general to suboptimal solutions, even this simple strategy requires more mental exertion than simply following the advice given by the PDA guide. Moreover, users found it frustrating to have to abandon part of a plan that they had already spent some effort devising.

Less Need for Thinking about Orientation. A small number of subjects pointed out that with the map, they continually had to ensure that the map was oriented correctly and to keep in mind where they were. By contrast, the PDA guide simply gave directions relative to their current position and orientation; it can take into account the direction the user is facing to a certain extent because each beacon transmits signals only within a limited angle.[5] Our subjects' problems when navigating with the map are similar to those discussed in a more general analysis by Thorndyke and Hayes-Roth ([11]) about problems that people may have with orientation under different conditions.

4.4 Perceived Drawbacks of the Current PDA Guide

Even subjects who were on the whole enthusiastic about the PDA guide mentioned some limitations, some of which can be eliminated through improvements in its design.

Feeling of Not Having the Big Picture. Although, as was just mentioned, some subjects appreciated not having to think about their orientation and location, for some subjects this consequence was associated with a drawback: They felt almost as if they were being led blindfolded through the shopping mall – a feeling that was evidently not entirely satisfactory to them even if they were convinced that the system was leading them in an optimal way. Proposals made by subjects for improving their overall feeling of orientation included some ideas that are familiar from GPS-based navigation systems:

- Display overview maps of (parts of) the mall.
  For example, while $\mathcal{U}$ is walking from one position to the next, an overview map showing her position within (some part of) the shopping mall could be displayed.
- Make it possible for $\mathcal{U}$ to be given longer sequences of directions.
  Instead of always being led from one beacon to the next, subjects sometimes prefer longer sequences of instructions (e.g., the graphical equivalent of "Go straight ahead, then take the next right and then the next left"). In addition to giving users a greater sense of knowing where they are going, such sequences can save time, because they reduce the number of occasions on which $\mathcal{U}$ needs to acquire information from a beacon and consult the PDA. On the other hand, longer sequences place heavier demands on $\mathcal{U}$'s working memory, and they lead to a greater likelihood that some instruction will be executed incorrectly. This type of tradeoff was investigated experimentally in a different setting in [7]: This study showed that finding an optimal instruction sequence length is a task that can be handled naturally within the framework of decision-theoretic planning used here.
- Give advance announcements of upcoming stores.
  Some subjects wanted to know not only what direction they should proceed in but also what they could expect to find there. For example, if the PDA displayed along with the arrow a message like "Saraphon (CDs), 40 m", $\mathcal{U}$ could utilize the walking time to prepare mentally for the task of looking for a suitable CD.

In addition to the advantages already mentioned, users who have more than the minimally necessary amount of information about their surroundings are in general better able to cope with problems with the technology, which will be discussed next.

[5] Where uncertainty about orientation was possible, the PDA display included an identifiable landmark, such as a flight of stairs, located in the immediate vicinity of the user's current position.

Need to Attend to the PDA and to the Beacons. The current realization of the PDA guide requires users to attend to a certain extent to the handheld device and/or to the beacons even while walking: $\mathcal{U}$ needs to know when she has reached the range of the next beacon, so that she can attend to the next instruction. $\mathcal{U}$ may therefore

1. look out for the physical beacons themselves,
2. aim the PDA explicitly at a beacon, and/or
3. keep looking at the PDA to see when a new instruction appears.

Both the first and second activities can be made unnecessary if the beacons are simply made more powerful, so that their signals reach the PDA whenever the user comes into the relevant area.[6] The third activity can interfere with $\mathcal{U}$'s walking, for example increasing the danger of colliding with other persons or stumbling on stairs. An obvious solution, mentioned by some subjects, is the use of acoustic stimuli, which could be delivered via a loudspeaker or a headset. The acoustic output can range from (a) acoustic signals that alert $\mathcal{U}$ to a change in the display to (b) the use of speech for all navigation instructions.

5 Further Issues

The aspects of the shopping guide that could be tested in the present study passed their tests fairly well, and the improvements that proved desirable can be realized without much difficulty. But the goal of deploying the system in real shopping situations raises some issues that were not (fully) addressed in the present study.

5.1 Increasing the Volume of Products Purchased

One important potential advantage of the PDA guide was *not* demonstrated in the results of the present study: the advantage that shoppers might find more products that they want to buy. But this advantage might well be achieved in situations with somewhat different properties than the situation of the present user study:

First, if the time available is more severely limited (e.g., because stores are about to close), shoppers with the slower map method will have to stop shopping before they have exploited all possibilities. In this case, shopping times will tend to be equalized, with the difference appearing in the number of products found.

Second, shoppers often have a choice between shopping or not shopping (e.g., on the way to their gate in an airport) or between shopping in mall *A* or shopping in mall *B*. If a given shopping mall, by providing a shopping guide, offers the prospect of getting the job done more efficiently and with less cognitive effort, this mall has an advantage over competing malls in attracting customers to shop there in the first place.

5.2 Applicability to Spontaneous Shopping Behavior

Especially recreational shopping is often highly spontaneous: A shopper may suddenly abandon any plans she may originally have had in order to pursue some attractive new

[6] This technical solution is currently being implemented by the company that supplied the beacons. But as this solution does not overcome the problem if $\mathcal{U}$ points interface at the floor, other technologies, such as Bluetooth, might be employed to indicate the presence of a transmitter regardless of the position of the PDA.

option. Yet in situations where it is very hard to predict what the shopper might do, it is hard to specify an accurate model as a basis for decision-theoretic planning. The approach presented here is therefore most applicable when the shopper has particular goals that she wishes to achieve within a limited amount of time.

Our approach does not, however, presuppose that the shopper is mechanically tracking down specific products on a shopping list with maximal efficiency. If $\mathcal{U}$ is allowed to specify vague product descriptions like "something to read" or "a present for my 3-year-old daughter", $\mathcal{S}$ will lead $\mathcal{U}$ to relatively promising stores, and $\mathcal{U}$ can be as creative as she likes in deciding what (if anything) to buy there.

It is possible to extend the model that underlies the decision-theoretic planning to account for possible events such as $\mathcal{U}$'s spontaneously stopping to spend time at stores that she passes (see [2] for further examples of possible extensions). The proportion of shoppers whose behavior is adequately fit by the model's assumptions will depend on the sophistication and the accuracy of the modeling effort, an issue to be discussed next.

5.3 Complexity of Decision-Theoretic Planning

Decision-theoretic planning based on Markov decision processes has an inherently high computational complexity. Since in the present study the focus was not on computational issues, we made – and enforced – some assumptions about the reality being modeled which ensured that planning would be computationally tractable. For example, the introduction of "one-way passages" made it impossible for subjects to visit any store twice, thereby making it unnecessary for the MDP models to include information about the stores that $\mathcal{U}$ has already visited.

For the modeling of real shopping situations, it will be necessary to deal with models which have not been selected primarily because of their tractability. A great deal of research has been done in recent years on ways of making decision-theoretic planning more tractable in certain classes of situations (see, e.g., [3]). We are examining how these approaches can help with the type of planning problem involved here.

5.4 Learning of Parameters

For the modeling, a great many specific quantitative parameters need to be specified, concerning matters like walking and waiting times and the probability of finding a given type of product in a given store. In our study, we were able to construct a reality that corresponded to our model. For real use, one approach would be to estimate the parameters initially on the basis of a more or less thorough analysis of the shopping mall. If this initial specification is accurate enough to ensure that the system is of some use to shoppers, further data can be obtained from the PDAs of shoppers who have used the system, provided that they are willing to take a few seconds to allow data about their searches to be (anonymously) transmitted back to the central system. In addition to supporting estimates of the parameters considered in this paper, this method would open up the prospect of introducing some form of social navigation or collaborative filtering.

6 Conclusions

We have provided an initial demonstration of the feasibility and acceptability to users of several aspects of a new approach to intelligent location-aware shopping guidance. The method is characterized by:

- an integration of navigation assistance with product location;
- the exploitation of decision-theoretic planning techniques that perform computations too complex for humans to approximate mentally; and
- the use of robust location-awareness technology which is already available at reasonable cost.

Our study with 20 users in a semirealistic situation yielded a number of insights into the reasons why users find the system acceptable – and the conditions on which this acceptability depends (e.g., unfamiliarity with the environment or time pressure). It also turned up some problems and limitations, along with ideas for overcoming them.

Studies of the use of the shopping guide with real shoppers in a real shopping mall will undoubtedly bring to light new challenges that have not been addressed in the present paper. Still, the intermediate results yielded by our study seem likely to prove useful in connection with both (a) more realistic application scenarios of the shopping guide and (b) other types of mobile system that raise similar issues.

Acknowledgements

The research of the first and third authors was supported by the German Science Foundation (DFG) in its Collaborative Research Center on Resource-Adaptive Cognitive Processes, SFB 378, Projects B2 (READY) and A4 (REAL). Beacons for the study were supplied by Eyeled GmbH (http://www.eyeled.de).

References

1. J. Blythe. Decision-theoretic planning. *AI Magazine*, 20(2):37–54, 1999.
2. T. Bohnenberger and A. Jameson. When policies are better than plans: Decision-theoretic planning of recommendation sequences. In J. Lester, editor, *IUI 2001: International Conference on Intelligent User Interfaces*, pages 21–24. ACM, New York, 2001.
3. C. Boutilier, T. Dean, and S. Hanks. Decision-theoretic planning: Structural assumptions and computational leverage. *Journal of Artificial Intelligence Research*, 11:1–94, 1999.
4. R. P. Darken and B. Peterson. Spatial orientation, wayfinding, and representation. In K. Stanney, editor, *Handbook of Virtual Environment Technology*. Erlbaum, Mahwah, NJ, 2001.
5. S. Duri, A. Cole, J. Munson, and J. Christensen. An approach to providing a seamless end-user experience for location-aware applications. In *Proceedings of the First International Workshop on Mobile Commerce*, pages 20–25, Rome, 2001.
6. A. Jameson. Usability issues and methods for mobile multimodal systems. In *Proceedings of the ISCA Tutorial and Research Workshop on Multi-Modal Dialogue in Mobile Environments*, Kloster Irsee, Germany, 2002.
7. A. Jameson, B. Großmann-Hutter, L. March, R. Rummer, T. Bohnenberger, and F. Wittig. When actions have consequences: Empirically based decision making for intelligent user interfaces. *Knowledge-Based Systems*, 14:75–92, 2001.

8. T. K. Landauer. Behavioral research methods in human-computer interaction. In M. Helander, T. K. Landauer, and P. V. Prabhu, editors, *Handbook of Human-Computer Interaction*, pages 203–227. North-Holland, Amsterdam, 1997.
9. N. Sawhney and C. Schmandt. Nomadic Radio: Speech and audio interaction for contextual messaging in nomadic environments. *ACM Transactions on Computer-Human Interaction*, 7:353–383, 2000.
10. M. Singh, A. K. Jain, and M. P. Singh. E-commerce over communicators: Challenges and solutions for user interfaces. In *Proceedings of the First ACM Conference on Electronic Commerce*, pages 177–186, Denver, 1999.
11. P. W. Thorndyke and B. Hayes-Roth. Differences in spatial knowledge acquired from maps and navigation. *Cognitive Psychology*, 14:560–589, 1982.

handiMessenger: Awareness-Enhanced Universal Communication for Mobile Users

Stacie Hibino[1,†], Audris Mockus[2,†]

[1]Eastman Kodak Company, 100 Century Center Court, Suite 600
San Jose, CA 95112 USA
hibino@acm.org
[2]Avaya Labs Research, 233 Mt. Airy Road, Rm 1C05
Basking Ridge, NJ 07920 USA
audris@mockus.org

Abstract. Successfully contacting colleagues while away from the office is especially difficult without information about their availability and location. HandiMessenger is a service designed to facilitate opportunistic communication by tightly integrating awareness and contact capabilities into a wireless, unified messaging and awareness application. Users can securely access intranet email, instant messages, and other messages from a handheld mobile device (e.g., a wireless PDA). Simultaneously, they are presented with awareness information about the sender to enable a reply in the most appropriate mode, given the situation of the participants. For example, they can read email and reply by initiating a phone call, if the sender is available by phone. Analysis of handiMessenger usage shows that users do request awareness information while inspecting message headers or content, and that users also, at times, respond to a message with a different type or in a different mode than the original message.

1 Introduction

Cell phones, pagers, wireless PDAs, and laptop computers are designed to help people stay connected in today's mobile society. The use of these devices alone, however, does not overcome some of the common communication problems that many mobile corporate workers face on a day-to-day basis. These users may play constant "phone tag," since they often do not know when their colleagues are available, where they are located, or which devices are accessible to them. Enterprise users also deal with several problems in retrieving messages, such as email, when away from the office. For example, they usually need to deal with securely getting through a firewall. Currently, such access may require specialized hardware for added security purposes (e.g., a secure token generator, a Research in Motion (RIM) Blackberry device). Enterprise users may also experience difficulty in accessing urgent messages in certain social contexts (e.g., in meetings).

[†]Work conducted at Bell Labs, Lucent Technologies.

F. Paternò (Ed.): Mobile HCI 2002, LNCS 2411, pp. 170–183, 2002.
© Springer-Verlag Berlin Heidelberg 2002

HandiMessenger is an awareness-enhanced universal communication service for mobile enterprise users. This prototype service is designed to facilitate group and personal communications when away from the office. Users can easily access intranet web pages or their intranet messages from their wireless handheld device (e.g., such as a Palm VII or a WAP-enabled cell phone). Such access can be done unobtrusively in a meeting situation. Users can also request awareness information and use tap-to-contact links for quick and easy communication with colleagues. These awareness and contact features are tightly integrated with users' messages and message headers to provide opportunistic communication. That is, a user can check messages, view integrated awareness information to see if and how the message sender is available, and simply tap an embedded link to initiate communication with that sender. The usage logs confirm that such scenarios do occur in practice.

The following types of communication are supported within handiMessenger: intranet email, group instant messages, and click-to-dial on mobile phone devices. We also support a new type of message, namely ConnectIcons [1]. A ConnectIcon is an electronic correspondence that includes a brief message line, presence information about the sender and recipients, communication link(s) specified by the sender (e.g., email, phone, and/or instant messaging), and optional URL attachments. In addition, we recently added third-party call control to setup calls between two IP-based H.323 phones, calls between H.323 phones and standard PSTN phones, and calls between two IP-based Session Initiation Protocol (SIP) phones.

HandiMessenger is implemented as a web-based service with server-side processing and is thus easily delivered to a variety of users' existing handheld devices. We tested handiMessenger with a variety of mobile devices, including several Palm-based devices, PocketPC device with wavelan (802.11) card, and WAP phone with GPRS service.

HandiMessenger also provides a mobile complement to TeamPortal [1], a Java-based portal designed to aid teams in communication and coordination. TeamPortal provides a visual GUI to calendar and awareness information, click-to-contact links, connectIcons and group instant messaging (IM). Thus, when a calendar entry or group IM is updated through handiMessenger for example, this update is automatically propagated to users logged in to TeamPortal.

The main contribution of this work is a service that provides opportunistic communication by transparently integrating awareness and communication capabilities into a messaging application. More specifically, we have: embedded awareness information and communication functions into the display of message headers and content, calendar, and corporate directory listings; added persistent instant messages to the family of unified messages; incorporated an infrastructure to securely access intranet information from wireless, web- and internet-capable handheld devices; incorporated third-party communication initiation where appropriate; and used a commercial communication infrastructure to initiate voice calls between various types of endpoint devices (including standard telephone and voice over IP devices—i.e., PSTN, H.323, and SIP). Analysis of handiMessenger usage logs shows that users most frequently use the service just to obtain the latest message headers, or to obtain headers and then retrieve a message, thus indicating the utility of the awareness and other information provided on the message header page. The logs also show that users sometimes replied to a message in a different mode from the received message and that they also made awareness requests from the context of both message headers and message content.

The remainder of this paper is divided into five additional sections. In the next section, we present sample scenarios on how handiMessenger can be used. We then describe the system architecture and key functionality. This is followed by a discussion of handiMessenger usage and user feedback. Finally, we discuss some related work and present our conclusions.

2 Sample Scenarios

A common situation when handiMessenger is useful is when one or more colleagues travel out of town on business (e.g., to attend a conference). In the corporate world, users typically need to get through a firewall to access their messages. For many users, this could mean using a local network and setting up a virtual private network (VPN) connection, or traveling with larger devices such as laptops and going through a series of several steps to connect to the corporate intranet (e.g., using an 800 number, entering access codes, etc.). Although devices such as the RIM Blackberry and services such as AvantGo provide secure email and limited intranet web access from a wireless handheld device, these applications and hardware platforms do not provide access to universal messaging nor do they integrate awareness information.

Assume Diane is attending the CHI'2001 conference. Using a Palm VII, she can simply raise the antenna of the device, start handiMessenger and specify the type and number of messages to retrieve. She is presented with a list of her messages along with available awareness information of the message senders. She decides to click on an email from Dave because she sees that he is currently logged into TeamPortal. Dave wants to know how the conference is going. Diane decides to respond by tapping on Dave's office phone icon to initiate a voice call between her hotel phone and Dave's office.

3 System Description

3.1 Overview

HandiMessenger is a web-based service, the core of which is implemented as a set of Java servlets. The "home page" for the handiMessenger service is a web page that has been installed onto a user's wireless PDA (e.g., a Palm Query Application (PQA) installed on a Palm VII). Devices that do not support installation of PQAs, bookmark a reference and each time the initial page is downloaded or retrieved from the device cache. The top of the handiMessenger home page is presented in Figure 1.

Starting with the handiMessenger home page, users can access the service to check various types of intranet messages or initiate various modes of communication. For example, users can check corporate email, team-based instant messaging (IM), and other types of specialized messages such as connectIcons. They can also use handiMessenger to examine awareness information about various team members and initiate communication through handiMessenger's simple "tap-to-contact" links.

Fig. 1. Top part of the handiMessenger home page

3.2 Architecture

The handiMessenger architecture is presented in Figure 2. A handiMessenger request from a wireless PDA is transmitted over the air to a nearby cell tower and then relayed through a proxy of the wireless service provider (e.g., service providers such as OmniSky or Palm.net) to the handiMessenger gateway (i.e., external server). Inside of a corporate intranet, handiMessenger pools for requests from the gateway via an https tunnel through the corporate firewall (using a corporate https proxy).

When a request is received, handiMessenger authenticates the user, processes the request, and uploads the result to the gateway using the https protocol. At the gateway, the HTML content is filtered to rewrite all intranet references. The results are then sent back to the user's device. The following basic handiMessenger requests are currently supported:

- getMessageHeaders,
- getMessage, composeMessage and sendMessage,
- getCalendar and setCalendar,
- getAwareness, and
- getCorporateDirectoryListing.

When processing a user's request, handiMessenger uses standard protocols to gather information from various messaging and awareness servers and/or other corporate databases or URLs as is necessary to fulfill the request. User authentication to these other servers are set up by the user ahead of time, during a one-time provisioning session. This provisioning is similar to specifying mail server and login information for Netscape Messenger.

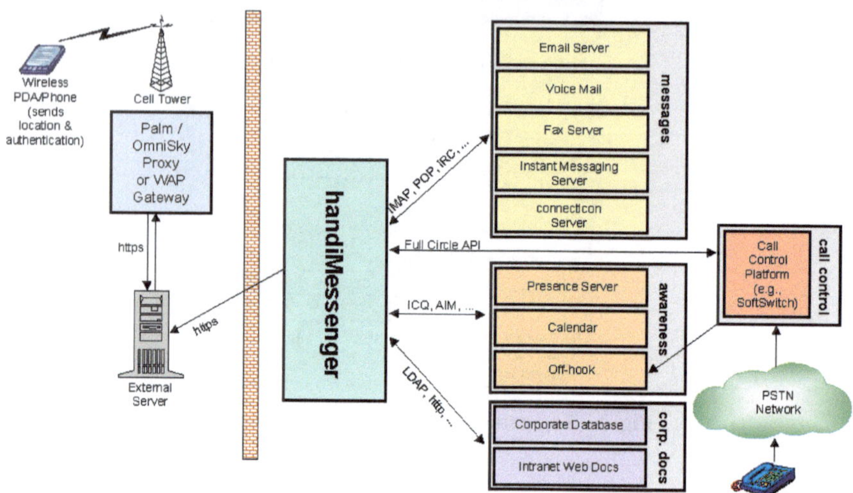

Fig. 2. handiMessenger Architecture

3.3 Checking Messages

When using handiMessenger to check for messages, users can select the type of messages to retrieve, along with the maximum number of messages to check. A list of message headers is then presented to the user, sorted by the order in which they were most recently received. This is similar to a list of email headers presented when reading email, only all selected types of message headers are included in the list (i.e., in addition to email). On the left of each message header in the list, an icon is used to indicate the type of message. These are the same icons displayed on the handiMessenger home page and shown in Figure 1. A sample list of message headers displayed by handiMessenger is shown in Figure 3.

The message headers summarize typical message information such as:

- *Date:* the date or time for the message. If the message was received today, only the time is shown, in order to save space. For email, the date displayed is the date that the email was sent. For group instant messaging (IM), the date of the last chat line is displayed.
- *Subject:* the subject of the message. Similar to the Palm OS email reader, the subject line is truncated to a few letters to save screen space while still providing a hint of the message subject. For group IM, the subject line includes the user ID and beginning text corresponding to the last chat line added.
- *From:* an indication of who sent or updated a message. If awareness information about a sender or group is available, an abbreviated version of the presence information involving the last device used is displayed underneath their email address or group ID. This enables users to see at a glance who might be available for them to contact immediately.

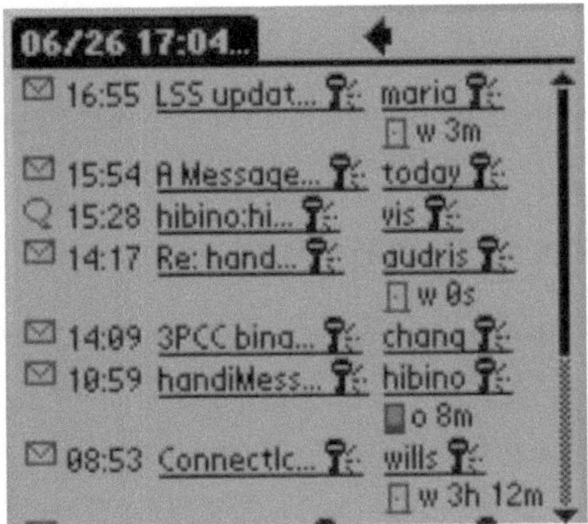

Fig. 3. Message headers shown by handiMessenger

Users can click on a message subject to retrieve the body of the message. They can also click on a sender's name to get more detailed awareness information about that person. The "key" icon at the end of each link indicates that the link is a secure https link. This icon is automatically added by the Palm OS web clipping application.

When handiMessenger retrieves the contents of a message, the system annotates the message with awareness information and tap-to-contact links. Thus, when users read a message, they can see if the sender is currently available and easily contact them via any one of the sender's available devices. Users are thus not constrained to reply to a message in the format in which it was sent, but they can easily mix various communication modes (e.g., phone call in response to an email message) according to their preference and the party's availability.

3.4 Checking Awareness and Initiating Communication

HandiMessenger provides access to awareness outside of checking messages. From the handiMessenger home page, users can request awareness about registered individuals or groups, or look up individuals from the corporate database. Based on the information presented, users can then contact the desired party or parties via handiMessenger's tap-to-contact links. For example, they can click on a phone icon to initiate a phone call with someone.

An example of awareness information about the handiMessenger group is shown in Figure 4. The top of the page provides iconic links for initiating the following types of group contact: email, group IM, connectIcon, and conference call. Thus, clicking on the envelope email icon will bring up an email composition page with the emails of all group members inserted into the "To:" field; clicking on the talk bubble group IM icon will display the persistent group IM session, and so forth.

Following the group contact icons, group members are listed alphabetically by user ID. When group awareness is requested, all members are displayed. When individual awareness is requested, only that person's information is presented. In either case, the individual's awareness information is presented in the same format. Each member's email address is displayed next to their ID. This is followed by a list of their available devices, sorted by when they were most recently used. Each device line includes:

- an icon indicating the type of the given device or application (a door icon represents TeamPortal),
- a letter indicating the location of that device (w=work, h=home, o=other)
- abbreviated time information to indicate how long ago the device was updated (e.g., 10s=10 seconds),
- information about the current status of the device.

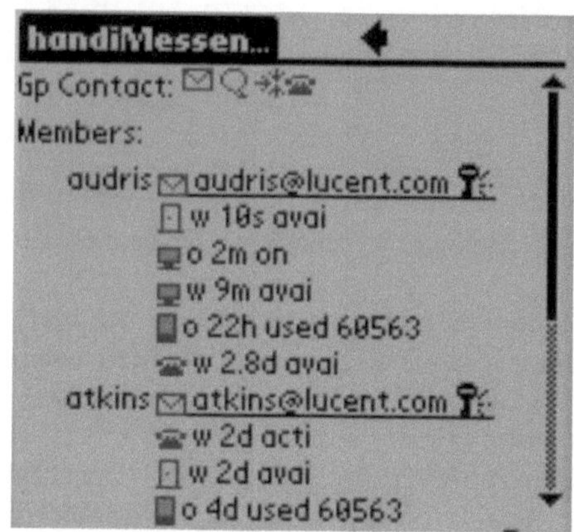

Fig. 4. Sample group awareness information displayed by handiMessenger

In the case of PDA devices, zip code information of the closest tower to where the device was last used is also presented, when available. When combined with the access time indicated, the zip code can indicate whether a group member is currently in or out of town, and could thus influence a decision on how to contact that member.

3.5 Implementation and Current Status

As mentioned above, the core of handiMessenger is implemented as a set of Java servlets. For Palm OS-based devices, the handiMessenger icons and home page are installed as a PQA application directly onto the user's wireless PDA. Individual user information and provisioning are saved in an LDAP database. We currently support access to email, group instant messages, and connectIcons. We also support third party call control over traditional (PSTN) and IP-based phones such as H.323 and SIP endpoints. The call control features related to traditional phones require additional communications setup, such as a Lucent SoftSwitch, however.

handiMessenger has been running since April 2001. During the summer of 2001, we tested its use within our research group and gathered some additional feedback from some business unit partners.

4 Usage and User Feedback

4.1 handiMessenger Usage

A typical usage pattern within handiMessenger is to:

- list message headers,
- retrieve the body of one or two messages,
- reply or compose a short message.

Based on PDA constraints and bandwidth, mobile users are not so interested in reading the details of all of their messages, but more likely to look for messages with higher priority or those requiring some sort of timely response. Since most PDAs are not very conducive to composing long messages, users tend to either keep their communications brief or turn to a device with a higher bandwidth (i.e., the phone).

4.2 Analysis of Usage Logs

We collected usage logs at the server side, where we recorded user id, user location (ZIP code), request URL, and the content of the request. We did not collect information at the client (PDA) side so we can only reconstruct over-the-air request sequences, and not navigation between application pages on the PDA itself. That is, pages are cached on the PDA, so a user could get message headers, get a message, use the back button to review the cached headers, and then request to get a different message; such an interaction would result in three (not four) over-the-air requests.

In this section, we discuss preliminary results of handiMessenger usage based on logfile analysis of trial usage of the service. Our goal was to understand basic usage patterns and to use this information to refine the handiMessenger interface. The voice communication capabilities of handiMessenger were available only at the very end of the period, so we do not have sufficient data to analyze their usage. To insure uniformity of the data, we consider only the period before the voice call capabilities (via 3rd party call setup) were introduced.

Types of handiMessenger Requests Made. From April until August of 2001 there were 1428 over-the-air requests by ten users made from 34 different zip codes. Figure 5 summarizes the distribution of the most frequent requests.

The most frequent handiMessenger request was getMsgHeaders; the next frequent requests were getMessage, followed by getAwareness, composeMessage, sendMessage, getDirectoryListing, and getCalendar. This indicates that users were mainly concerned about obtaining their latest message headers and reading some of their messages. Also, since full awareness is automatically presented with getMessage requests, users did not need to make as many separate getAwareness requests. Out of the 337 total getMessage requests, 251 were to read email, 56 to see instant messages, and 30 to retrieve connectIcons.

Sequences of handiMessenger Interactions. We analyzed sequences of handiMessenger interactions in order to better understand how the service has been used. We first extracted sequences of interactions by breaking the full sequence of a user's accesses into subsequences at the points where the delay between two accesses was more than five minutes. In our data, there is no significant difference between the number of sequences when using a delay of one to nine minutes. There were 333 such sequences of user requests, with about 75% of the sequences shorter than four accesses. The frequency distribution of sequence lengths is shown in Figure 6.

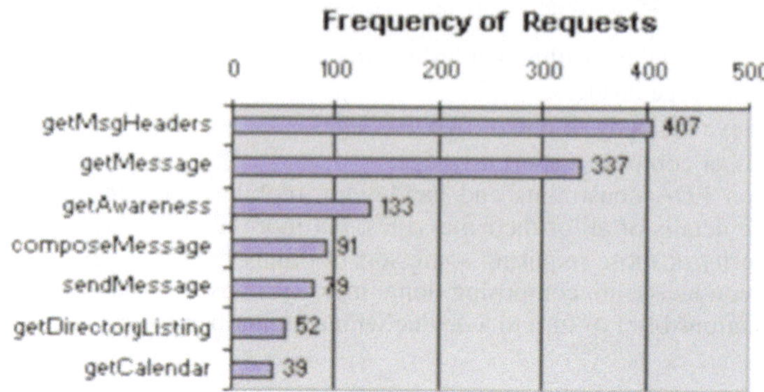

Fig. 5. Distribution of most frequent handiMessenger requests from April to August 2001

Sequences of handiMessenger Interactions. We analyzed sequences of handiMessenger interactions in order to better understand how the service has been used. We first extracted sequences of interactions by breaking the full sequence of a user's accesses into subsequences at the points where the delay between two accesses was more than five minutes. In our data, there is no significant difference between the number of sequences when using a delay of one to nine minutes. There were 333 such sequences of user requests, with about 75% of the sequences shorter than four accesses. The frequency distribution of sequence lengths is shown in Figure 6.

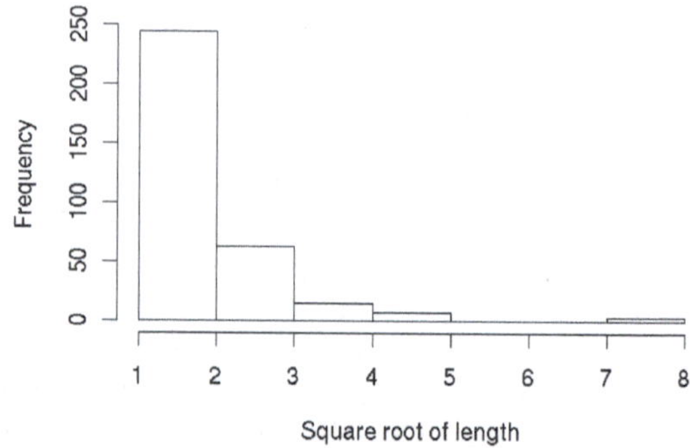

Fig. 6. Frequency distribution of session lengths, by square root of length. The majority of sessions only made one to three over-the-air requests.

One-step sequences accounted for 26% (these were mostly getMsgHeaders requests) and two-step for another 21% of all sequences. This means that roughly half of the time, users make only one or two requests (where getMsgHeaders, or getMsgHeaders followed by getMessage were the most frequent among these). Based on this, we hypothesized that users are not relying on handiMessenger as a primary messaging program (e.g., for reading *all* email), but they are likely just using it to check if any new messages require more immediate attention. Direct feedback from users confirmed this hypothesis.

Table 1 summarizes transition probabilities for handiMessenger sequences. The Table shows that after reading headers, users are likely to:

- read messages (59% of the time),
- get headers again (30% of the time), or
- check the sender's full presence (11% of the time).

Table 1. Transition probabilities between frequent handiMessenger requests

	To:			
From:	**getHdrs**	**getMsg**	**getAwr**	**compose/ sendMsg**
getHdrs	0.30	0.59	0.11	0.00
getMsg	0.07	0.61	0.09	0.23
getAwr	0.14	0	0.42	0.44

The logs show that when users get headers again, they are either refreshing the header listing (e.g., to check for any new messages), changing the number of headers retrieved (e.g., to get more headers), or requesting a different collection of message types to be shown (e.g., accessing instant messages and connectIcons instead of just email). When users check the presence of a message's sender, they are retrieving the full presence information available for that sender, versus the minimal presence information that is automatically displayed with the message header. The zero in the right column of Table 1 reflects the fact that it was not possible to respond without looking at the message content first. While we do not know whether or not the decision to respond to a message was taken based on the presence information, the 11% of the time users chose to investigate presence information in more detail suggests that it is worth providing it at the message header level.

Table 1 also shows that after reading a message via a getMessage request, users are 61% likely to place a subsequent request to read a message. Further analysis shows that 70% of time it is a request for a different message and 30% of the time it is a request for the same message, but with complete content. Users are 23% likely to compose or send a message after reading a message (e.g., to reply to email or add to an IM session). They may also check presence or refresh headers after reading a message, but this occurs less frequently. This indicates that on a low-input-bandwidth device such as PDA, composing text messages is not frequent (note that the data does not reflect tap-to-call functionality, though).

Based on Table 1, we also see that detailed awareness information (from "getAwareness") is used as a step to initiate a communication of the most appropriate mode (44% of the time) or to check awareness of several people (42% of the time). This supports our conjecture that making presence available in multiple contexts may increase the likelihood of communication.

Replying to Messages. We also analyzed the usage logs to identify how users reply to the messages they receive. Email is the most likely response to email, although there were instances when a connectIcon or an instant message was used to reply. The same holds true for instant messages (i.e., an instant message is the most likely response to an instant message). For connectIcons, the most likely response was an instant message (it makes no sense to send a connectIcon back). Although there were relatively few cases of such response sequences overall, we do see that users do not feel that their response messages need to be of the same type as the originating messages they received.

We believe that the addition of click-to-dial and third-party call control will be frequently used by handiMessenger users in the future. We expect this to be especially true in situations where users need to provide longer replies to messages. Current PDAs are not conducive to lengthy text input, and thus users are less likely to use them for composing long messages. This is also in line with a study of IM in the workplace, which showed that IM often transitions to a wider bandwidth communication channel [5]. However, the switch to wider-bandwidth devices such as the phone is also limited to particular contexts. We would not, for example, expect users in a meeting situation to interrupt an ongoing meeting by initiating a phone call. The study of such usage patterns thus needs to be conducted in additional detail in the future.

4.3 User Feedback

Six users completed a user survey on the utility of handiMessenger, its current features, and possible future features. The users indicated the utility of easy intranet email access, the value of an integrated service, and the desire for future features. Details about their feedback are provided in the remainder of this section.

Easy Access to Intranet Email. An important feature of handiMessenger is secure access to corporate email from outside of a firewall. Although other commercial products support this, many of these have hardware or software constraints at the client or server level. The RIM Blackberry, for example, constrains access to email from a particular device. Users who already own popular Palm-based devices may be reluctant to switch to another device or unwilling to carry an additional one. Some services also require a particular email server, such as Microsoft Exchange. Many corporate end-users do not have control over their email servers, which are typically maintained by a centralized systems support team.

HandiMessenger only requires the availability of a standard POP3 or IMAP interface from an existing email server. The service also can be accessed from a number of handheld devices, including Palm OS-based devices and PocketPC devices. HandiMessenger has been used from a variety devices, including: Palm V or HandSpring Visor with OmniSky modem, Palm VII and VIIx, Kyocera SmartPhone, Compaq iPaq with waveLAN card, and Motorola Timeport 260 phone with WAP over GPRS service.

No Need for Context or Application Switching. Although users appreciated easy access to email, they also liked the ability to select the types of messages to retrieve, along with the ability to reply via various modes of communication. In handiMessenger, users do not need to switch to different applications to access different types of messages. Also, the integrated awareness information provides insight into how one might successfully contact the message sender. Finally, the tap-to-contact links provide easy access to alternative communication modes.

Pending Issues and Requests for Future Features. While users were generally pleased with the existing handiMessenger features, they also had some concerns and requests for future features. Some of these requests included the following:

- ability to control awareness information for privacy,
- desire to receive notification about incoming messages,
- text shortcuts to improve efficiency in text input (e.g., for writing an email or IM),
- ability to add filters to focus on specific messages,
- translators for attachments (e.g., such as for a Microsoft Word file attached to an email message).

5 Related Work

The explosion of instant messaging, the convergence of voice and data networks, and the fast-paced evolution of wireless devices are opening up new avenues for coordinating and staying in touch with colleagues. Although awareness research has been going on for years (e.g., especially related to video awareness such as in [2]), it has only recently begun to go mobile (e.g., see [4, 7]).

Hubbub supports mobile instant messaging with novel sound features [3]. Short musical tunes are used to indicate when users are present, as well as to communicate short messages. Users can login on a desktop or send and receive instant messages from a Palm OS device with a designated IP. Hubbub also allows users to login to the system from as many places as they wish (e.g., home or office PC, or Palm device) and the system automatically "follows" users according to their current or most recent interaction with the system—thus updating presence information appropriately. Hubbub does not integrate other types of group communication, however.

iMobile is a proxy-based platform for mobile services [6]. The platform provides a messaging-based service where messages are relayed between various mobile devices or between information servers and these mobile devices. The approach can be thought of as using an instant messaging window (e.g., such as that found in AOL) to either send messages to other users or to treat any message as a kind of shell command. Thus, when users open an instant messaging session with an iMobile agent, they can send messages that request information such as the quote of a stock, airline flight information, or the weather. Email can also be accessed in a command-line like fashion. The key feature of the iMobile platform is that new endpoint devices or new services can easily be added without changing the overall service logic. Although iMobile does not currently support a service such as handiMessenger, we believe that handiMessenger could be ported to run on top of the iMobile platform.

Awarenex [7] is perhaps the mobile messaging application that is most closely related to handiMessenger. In Awarenex, users start with awareness information in the form of a contact list (i.e., a kind of "buddy list"). Using this list, users can click on an individual in the list to get more detailed contact information. The user is then presented with the "Contact Locator," where they can get information about an individual's likely location, based on usage of their various registered devices (e.g., office phone, mobile PDA, home computer, etc.). We display similar information within handiMessenger on our individual and group awareness screens. From the Contact Locator of Awarenex, users can click on an email address to send email to the person, or click on a device to initiate either a phone call or instant messaging session with the person. We also support these types of click-to-contact links within handiMessenger. Awarenex, however, does not support the *retrieval* of various types of messages such as email. Awarenex also does not support the notion of groups or persistent group chat. All instant messaging is done one-to-one within Awarenex, versus one-to-many. We thus feel that handiMessenger provides better support for team-based communications.

Internet services such as Yahoo are moving towards integrating limited presence information with email and group correspondence. For example, when users view one of their Yahoo emails, a smile icon indicates whether another Yahoo user is logged into Yahoo Messenger (Messenger includes chat). However, this presence information does not provide more detailed awareness nor is it displayed in the context of email headers.

6 Conclusion

handiMessenger is a novel service that provides mobile access to awareness-enhanced communication from a wireless PDA. Users can access intranet email, group instant messages, and connectIcons without having to switch contexts from one application to another. Awareness information about a message sender supports users in making opportunistic connections with their colleagues. Users are not constrained to the mode of communication in which they are accessing a message, but can easily follow tap-to-contact links to initiate alternative modes of communication. Thus, for example, users can (and do) access a connectIcon and reply via phone or email.

Analysis of usage logs shows that users most frequently used handiMessenger to obtain the latest message headers, or to obtain headers followed by obtaining the message content, thereby using message headers as a filter to determine if they have to take any urgent action in replying to a message or in contacting a currently available colleague. The logs also showed that users have taken advantage of other unique features of handiMessenger, including requesting awareness information after viewing message headers, or replying to messages in a different mode from the received message.

Users of the handiMessenger service appreciated easy, mobile access to corporate email, the ability to stay better connected with their team members, and the consistent access to different types of messages. They did, however, also express concern for privacy issues and requested additional features such as message notification, filtering, and translations for attachments. We plan to address some of these issues in the future.

References

1. Colbert, R. O., Compton, D. S., Hackbarth, R. L., Herbsleb, J. D., Hoadley, L. A., and Wills, G. J. (2001). "Advanced Services: Changing How We Communicate." *Bell Labs Technical Journal, 6(1)*, 211-228.
2. Dourish, P. and Bly, S. (1992)."Portholes: Supporting Awareness in a Distributed Work Group." In *CHI'92 Conf. Proc.*, NY: ACM Press, 541-547.

3. Isaacs, E., Walendowski, A., and Ranganathan, D., (2001). "Hubbub: A wireless instant messenger that uses earcons for awareness and for sound instant messages." In *CHI 2001 Conference Companion*, NY: ACM Press.
4. Milewski, A.E. and T.M. Smith. (2000). "Providing presence cues to telephone users." In *Proceedings of the ACM 2000 Conference on Computer Supported Cooperative Work*. NY:ACM Press, 89-96.
5. Nardi, B.A., Whittaker, S. and E. Bradner. (2000). "Interaction and outeraction: instant messaging in action." In *Proceedings of the ACM 2000 Conference on Computer Supported Cooperative Work*. NY: ACM Press, 79-88.
6. Rao, C-H.H., Chen, Y-F.R., Chang, D-F., and M-F Chen. (2001). "iMobile: A Proxy-Based Platform for Mobile Services." Presented at *First ACM Workshop on Wireless Mobile Internet (WMI 2001)*, Rome, Italy.
7. Tang, J.C., Yankelovich, N., Begole, J., Van Kleek, M., Li, F. and J. Bhalodia. (2001). "ConNexus to awarenex: extending awareness to mobile users." In *CHI 2001 Conference Proceedings*. NY: ACM Press, 221-228.

An Empirical Study of the Minimum Required Size and the Minimum Number of Targets for Pen Input on the Small Display

Sachi Mizobuchi[1,3], Koichi Mori[1], Xiangshi Ren[2], and Yasumura Michiaki[3]

[1] Nokia Research Center, Nokia Japan,
2-13-5 Nagata-cho, Chiyoda, Tokyo 100-0014 Japan,
{sachi.mizobuchi, koichi.mori}@nokia.com
[2] Kochi University of Technology,
185 Miyanokuchi, Tosayamada-town, Kami-gun, Kochi 782-8502 Japan,
ren@info.kochi-tech.ac.jp
[3] Keio University,
5322 Endo, Fujisawa, Kanagawa, 252-8520 Japan,
{sachim, yasumura}@sfc.keio.ac.jp

Abstract. Two experiments were conducted to compare target pointing performance with a pen (stylus) and with a cursor key. on small displays. In experiment 1, we examined participants' performance of target pointing with both input methods at different target sizes. It was found that pen operation is more erroneous than key based operation whn target size is smaller than 5 mm, but at a target size of 5 mm, the error rate decreased to the same level as for key input. In experiment 2, we examined the effect of the number of targets. The results showed, with a target size of 5 mm, the pen could point to targets quicker than with key input, when the distance to the target exeeds a path length of 3 steps.

1 Introduction

1.1 Emerging Mobile Devices and Input Device

A great variety of handheld devices are coming onto the market. Some have advanced features compacted into very small sized hardware. For example, Fig. 1 shows a multimedia device released in the end of 2000 in Japan ("Eggy" by NTT DoCoMo, produced by Sharp) . It provides users with a message editor, web browser, address book, drawing tool and other applications on 2-inch display. Users interact with the device with seven hard keys or with a pen (stylus).

In small size display devices, objects are often made very small, so it can be contained in one screen. As a result, target are so small that it makes pen interaction more difficult. Another way is to enlarge objects, however, this might reduce the amount of objects shown in one screen, which requires users to scroll more to have a glance to the same amount of imformation. This kind of trade-off is the common problem that we face to in the user interface design process.It is

F. Paternò (Ed.): Mobile HCI 2002, LNCS 2411, pp. 184–194, 2002.
© Springer-Verlag Berlin Heidelberg 2002

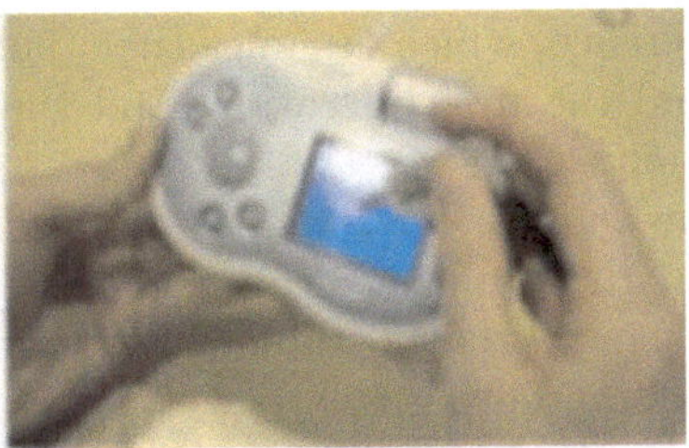

Fig. 1. A small sized display device which enables users to interact with a key and with a pen

required to find a balanced condition under which maximum number of object can be displayed, and the accuracy of user interaction is assured.

In order to maximize the capability of the device, we need to understand the characteristics of different input methods and utilize them in a sufficient way.

The cursor key is one of most widely-adopted input methods for current handheld devices. It enables users to control the focus point on the display. The operation is simple, and it is a big advantage that users can operate it with only one hand.

The pen is another quite highly-adopted input device for handheld devices, especially for PDAs (Personal Data Assistants). Pen operation is considered to be more intuitive than a cursor key, since it allows direct manipulation. Some handheld devices provide both cursor key and pen input. However, "both input methods are available" does not always mean "both input methods guarantee a satisfying performance". It is still unclear which device is most effective for what kind of task. Our study aims to understand characteristics of each device, so that we can design user interfaces more effectively.

1.2 Previous Studies of Pen

Previous studies have tried to reveal characteristics of interaction using pen input. Mackenzie et al.[1] and Uoi et al.[2] conducted experiments on drugging operations. Both of their results showed that pen input is quicker than mouse input.

Ono [3] and Kato and Nakagawa [4] highlighted that participants' pointing performance with a pen was affected by the target direction. They found that moving right and lower-right for right-handed people was inferior to moving in the opposite directions. Kato and Nakagawa also found the number of manipulations with a pen is greater in tasks where precision is required. Their results showed higher improvement with a pen in performance when the subject knows the next manipulation than with a mouse.

Ono [5] found it harder to change the operation for pen input than with mouse input once operation is started. He also found that, with a mouse, the pointing time in a combination of tasks is almost equal to the sum of the times for each task, but with a pen, it is shorter. He introduced a model to explain the difference, which assumes that there is an overlap between the user's mental work and the first operation, and that this overlap makes a pen easier to use than a mouse.

Sears and Shneiderman [6] tested three pointing devices: touch screen, touch-screen with stabilization, and mouse, each with a different target sizes. Their results showed that a stabilized touch-screen was effective for reducing the error rates when pointing to a target.

Ren and Moriya [7] investigated the effects of pointing strategies on target pointing tasks in various target sizes, distances and directions. They found significant interaction between the pointing strategy and target size.

In this study, we investigate characteristics of pen input in comparison to cursor key input, whereas most xof previous studies have compared pen to mouse input. We used target pointing task, which is one of most basic interaction style.

1.3 Experiments and Goals

We conducted two experiments to compare target pointing performance with a pen and with a cursor key. Our goals are to make following questions clear;

1) What is minimum required size of a target? (Experiment 1)

2) What is minimum number of targets at which performance with a pen is superior to that with a cursor key? (Experiment 2)

Besides, we were interested in whether the effect of target direction on pen pointing performance can also be observed on handheld device. Thus we also examined;

3 Does the position of the target affect the pointing performance? (Experiment 1)

2 Experiment 1: Effect of Target Size

In this experiment, we investigated minimum required size of a target for a pen.

In general, the target becomes more difficult to point to as target size becomes smaller. However, the operation of cursor key is the same when target size on the display changes. Thus, it can be predicted that pointing performance with a pen declines as target become smaller whereas performance with a cursor key is not affected by the target size.

2.1 Method

Design. The experiment used a two-conditions, within-group design. The independent variables were target size (2, 3, 4, 5 mm) and input method (pen and cursor key). The dependent variables were pointing time and error rate.

Participants. Ten people (male 9 female 1) participated in the experiment. Ages ranged from 19 to 60 years. All of them were right-handed, and had normal eyesight. None of them had prior experience of using a PDA..

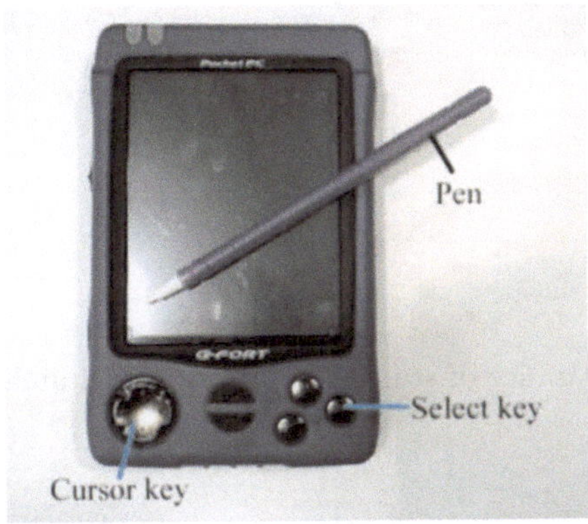

Fig. 2. Device used in the experiment

Table 1. Specification of used device

OS	Microsoft® Windows®CE Version3.0
CPU	VR4122 150MHz
Memory	32MB
Display	65,536 cobr 240×320 dots TFT LCD
Display size	610×810 mm
Size and weight	W85×H135×D25.5mm 300g

Apparatus. The testing program ran on G-FORT, a PDA produced by NTT DoCoMo and Casio. It has color LCD of 610*810mm (240*320 pixels). In addition to a touch screen and pen, it had 4-way (up, down, left and right) cursor key on the left bottom, and a select key was assigned on the right bottom of the device.

As stimuli, we provided 6*6 cells with 1.5mm of white-space between them on the screen. The size of each cell varied from 2mm to 5mm. Under the key condition, a starting cell appeared with bold frame in one of 4 corners (it was balanced between conditions), and a target cell was highlighted in gray (Fig. 3). Pointing time was measured after a cursor key was firstly pressed until a select key was pressed. Under the pen condition the first subject only saw a start cell with bold frame. When users touched the cell with the pen, the target cell highlighted in gray (Fig. 4). Pointing time was measured after the pen left the surface of start cell until it touched on the surface of target cell.

Procedure. Participants jointed the test session in turn. An experimenter gave instructions how to operate the pen and the cursor key, and allowed participants to practice pointing to targets of 4 mm for 20 trials. After a practice session, each participant was tested under all conditions in different order. One condition consisted of 50 trials.

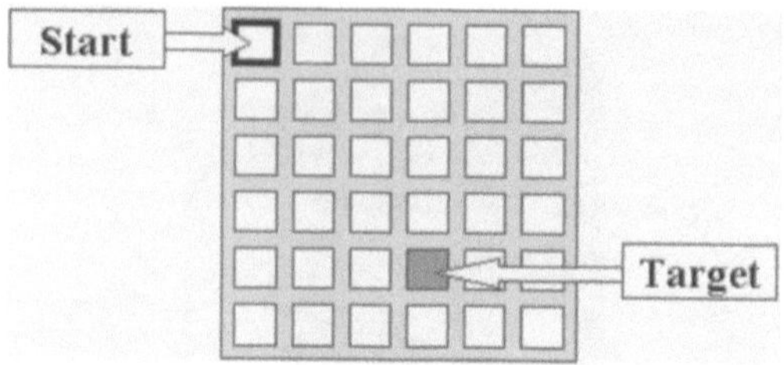

Fig. 3. Appearance of start cell and target cell under key condition

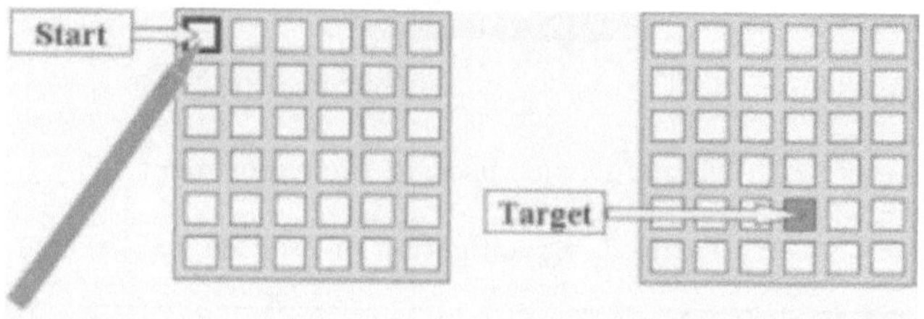

Fig. 4. Appearance of start cell and target cell under pen condition

They were seated at a table holding the device in their hand.

(Pen condition) Participants were instructed to hold the device with their left hand on table, and held a pen with their right hand.

(Cursor condition) Participants were instructed to hold the device with both hands, and operated cursor key and select key with their thumbs. The cursor moved from one cell as a participant pressed a key once. It stopped when a cursor reached the edge of the matrix.

Under both conditions, the participants were instructed to point the as target quickly and accurately as possible. There was no feedback for a hit or a miss.

Interactions were automatically logged to the device.

2.2 Results

Pointing Time. The results of pointing time by target size and input method can be seen in Fig.5.

The analysis of variance of pointing time showed significant main effect of input method ($F_{1,7992}$=2789.45, $p < .001$), significant main effect of target size ($F_{3,7992}$=14.48, $p < .001$), and significant interaction between the input method and target size. Pen input was quicker than key input at every target size, and the effect of target size was stronger with the pen than with the key. The results of post hoc test (Tukey HSD) showed the pointing time when target size is 4 and 5 mm is significantly quicker than that when target size is 2 and 3 mm ($p < .001$).

The average length between the center of start cell and the center of target cell in mm is shown in the table.

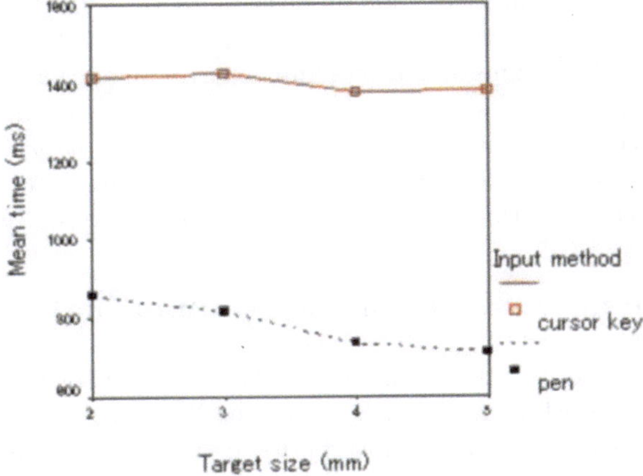

Fig. 5. Pointing time by target size and input method

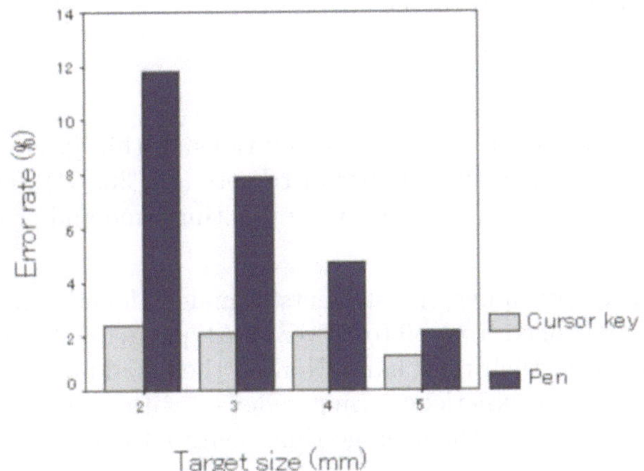

Fig. 6. Error rate by target size and input method

Error Rate. The results of the error rate are shown in Figure 6.

The analysis of variance of error rates showed the main effect of the input method ($F_{2,9}$=36.16 $p < .001$). Pen input made more errors than key input. However, as the target size became larger, the error rate decreased. Though the error rate of pen was significantly higher than key at the size of 2-4mm, there was no significant difference at the size of 5mm ($F_{1,9}$=1.23, $p > .05$).

Effect of Starting Position. We compared pointing time with a pen by four starting positions (upper right: UR, upper left: UL, lower right: LR, lower left:

LL). The analysis of variance of starting position by target size showed significant main effect of starting position ($F_{3,3717}$=3.0, $p < .05$). Post-hoc test results showed that pointing time of UL is significantly longer than UR (HSD=23.3, $p < .05$). UL can be considered to require movement from upper left to lower right. The inferiority of moving toward the lower right corner is also shown in previous studies [3, 4]. One of the possible reasons that target is more difficult to be found in the direction because it is likely to be covered by hand.

3 Experiment 2: Effect of the Number of Targets

In experiment 2, we examined the effect of the number of targets on pointing performance with both pen and with key input. We compared performance with a pen and that with a cursor to move between the same positions.

It is expected that cursor key will be strongly affected by the number of target between starting point and target because it will be the sum of required number of click. On the other hand, pen will be less affected by the number of targets because it does not require additional steps when the number of targets between starting point and target increases.

3.1 Method

Design. The experiment used a two-conditions, within-group design. The independent variables were the number of targets (16, 36, 64) and input method (pen, key). The dependent variables were pointing time and error rate

Participants. Twelve university students (7 male, 5 female) participated in the experiment. Ages ranged from 19 to 26 years. All participants were right handed and held the pen with their right hand. None of them had problems with eyesight. Six of them had prior experience using a device with pen interaction, ranging from 6 months to 2 years, the average time being 1.13 years.

Apparatus. The device used in this experiment was the same as that in experiment 1. As stimuli, we presented 5mm squares (cells) in a matrix with 1.5mm of white-space between them. The number of displayed cells was varied between 16 (4*4), 36 (6*6), and 64 (8*8). To simulate smaller size display, we prepared three plastic covers and put one of them in accordance to the number of cells, so that it surrounded whole cells.

Procedure. We took the same procedure as in experiment 1. Each participant was tested under all conditions in different order.

3.2 Results

Pointing Time. The results of the pointing time can be seen in Fig. 8.

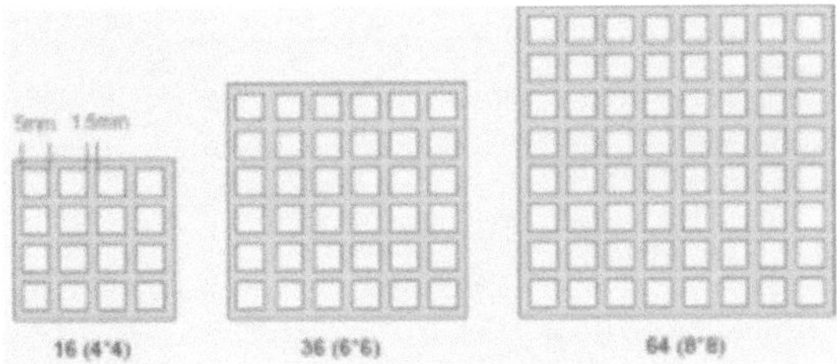

Fig. 7. Three conditions of the numnber of targets

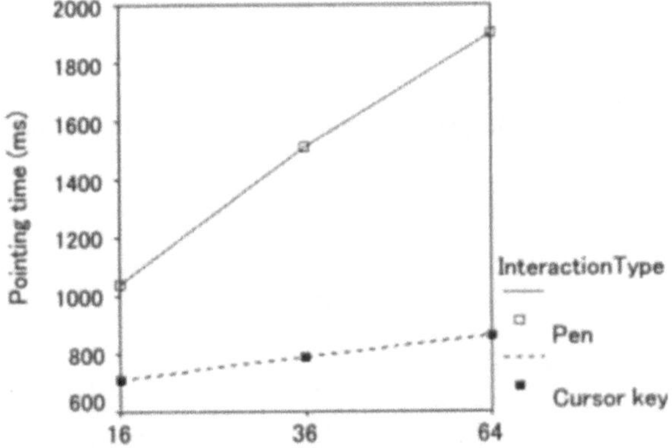

Fig. 8. Pointing time by the number of targets and input method

The analysis of variance of pointing time shows a significant main effect of the number of targets ($F_{2,3593}$=220.98, $p < .001$), a significant main effect of the input method ($F_{1,3593}$=1235.43, $p < .001$) and a significant interaction between the number of targets and input method ($F_{2,3593}$=109.21, $p < .001$). Pointing time was quicker with pen input compared to key input in all conditions, and the difference became larger as the number of targets increased.

Fig.9 shows the pointing time with a key and with a key by path length that is the minimum steps for a cursor to move to the target. Though it is not relevant for the pen, we plotted the pen's result with the same unit. It allowed us to compare the pointing time to the same target between pen and cursor. The pointing time of the pen and cursor are equal when the path length is 2, and pen input is quicker when the path length is more than 3. The difference is bigger as the path length becomes longer.

Error Rate. Error rate in this experiment was quite low as shown in Fig. 10, and there was no significant difference between two input methods ($F_{1,11}$=0.28, $p > .05$). This result is consistent to the result in experiment 1.

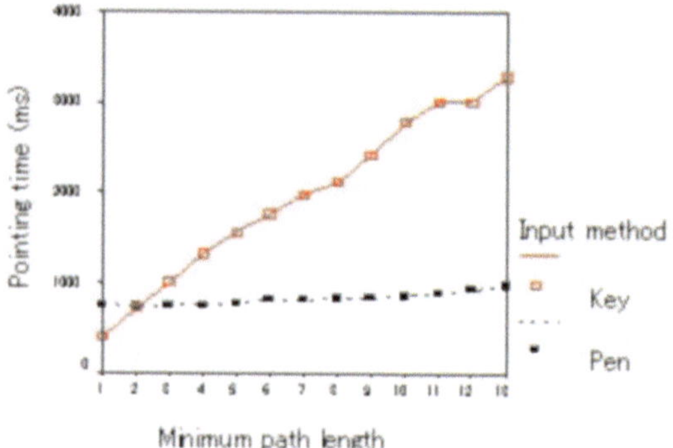

Fig. 9. Pointing time by path length and input method

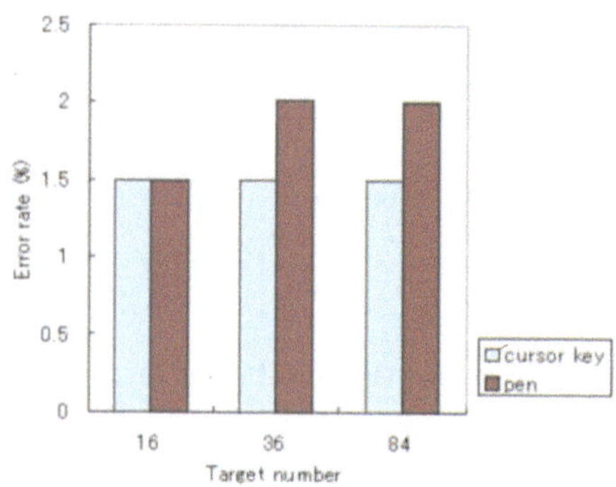

Fig. 10. Error rate by the number of targets and input mehtod

4 Discussion and Conclusions

In this study, we compared target pointing performance with two widely-adopted input methods for handheld devices: pen and cursor key. We conducted two experiments to find the minimum required target size and minimum number of target for a pen to work better than cursor key in target pointing.

4.1 The Minimum Required Size of Target

In experiment 1, we compared that the target pointing time and error rate of pen and of cursor key input for different target sizes. We found that pen input was quicker than cursor key input, but generally made more errors. The results show fairly clearly that error rates of pen input became lower as the target size became larger. It declined to the same level as that of cursor key input when the target size became 5mm. From this result, we could say that target size should be at least 5 mm for pen input to be as accurate as a cursor key for pointing tasks.

4.2 The Minimum Number of Targets at which Performance with Pen is Superior to that with Cursor Key

In experiment 2, we compared target pointing time and error rates for pen input and cursor key input for a different number of targets. We found that pen input is quicker than the cursor key input when the path length is greater than 3 steps at the minimum required target size (5mm) found in experiment 1.

4.3 The Effect of Target Direction

The analysis of the effect of the starting point showed that pointing time from upper left to the area 90 degrees lower right was longer than the other directions, as proposed in previous studies. This result suggests that user interface designers should limit the use of target buttons or icons in the lower right-hand part of a touch-screen, for right-handed users using pen input

It the reason for the inferiority of "from upper left to lower right" movement is still unclear. One possible factor is the ease of finding the target. For right-handed users, the lower-right area is likely to be covered by their own hand. It also might be affected by human motor system of the wrist and fingers. Further investigation will be needed to reveal what causes the difference in performance for target direction.

4.4 Future Work

From this study, we got practical findings about minimum required size of target and effective target number for pen. The results of this study will provide a basic reference for further investigation into the operation of pen input and cursor key input.

In order to refine user interface, several factors that affect pointing performance need to be considered as well. For example, the performance will be affected by dominant hand, and by user's posture; it will be different when a user operates handheld device sitting in chair compared to standing in a street. Some factors can improve performance as Brewster [8] pointed out. He compared target pointing performance on PDAs using with sound feedback and without sound feedback as the condition, and found that sound feedback improved performance and subjective evaluation by participants.

Though we did not analyze our data in this study in terms of performance models such as Fitts' law, it will be useful to investigate the applicability of Fitts' law to this kind of task.

Acknowledgments

We thank Miika Silfverberg for his advice and designing the target pointing test, Jun Imai and Yuriko Shimizu for their help with conducting the experiments, Jan Chipchase and Fumiko Ichikawa for the review.

References

1. MacKenzie, S., Sellen, A. and Buxton, W., A Comparison of Input Devices in Elemental Pointing and Dragging Tasks, in Proceedings of.CHI '91 Conference on Human Factors in Computing Systems, ACM Press (1991), 161-166
2. Uoi H., Shinoda M.,Yamamoto Y., Tsujino Y., Tokura N., An Experimental Comparison of Pen-Tablet and Mouse in Selecting Two Targets, IPSJ SIGNotes Human Interface Abstract No.043 – 005 (1992), 82-99
3. Ono, M. An experimental study of a pen's user interface, in proceedings of the 41st meeting of SIG-HI of Information Processing Society of Japan (IPSJ) (1992) 93-100
4. Kato, N. and Nakagawa, M. A study of operability of pens for designing pen user interface, IPSJ journal, Vol. 39, No. 5 (1998) 1536-1546
5. Ono M., Performance Modelling for Pen Input, in proceedings of the 47st meeting of SIG-HI of Information Processing Society of Japan (IPSJ) (1993) 1-6
6. Sears A., Plaisant C. and Shneiderman B., A new era for high precision touchscreens, Advances in Human-Computer Interaction, Vol.30, Ablex, Norwood, NJ (1992) 1-33
7. Ren, X. and Moriya, S., Improving selection performance on pen-based systems: A study of pen-input interaction for selection tasks, ACM Transactions on Computer Human Interaction (ToCHI), Vol.7, No.3 (2000) 384-416
8. Brewster, S. A., Sound in the interface to a mobile computer, in Proceedings of HCI International'99, Lawrence Erlbaum Associates, NJ, (1999) 43-47

KSPC (Keystrokes per Character) as a Characteristic of Text Entry Techniques

I. Scott MacKenzie

Dept. of Computer Science
York University
Toronto, Ontario, Canada M3J 1P3
+1 416 736 2100
smackenzie@acm.org

Abstract. *KSPC* is the number of keystrokes, on average, to generate each character of text in a given language using a given text entry technique. We systematically describe the calculation of *KSPC* and provide examples across a variety of text entry techniques. Values for English range from about 10 for methods using only cursor keys and a SELECT key to about 0.5 for word prediction techniques. It is demonstrated that *KSPC* is useful for a priori analyses, thereby supporting the characterisation and comparison of text entry methods before labour-intensive implementations and evaluations.

1 Introduction

An important research area in mobile computing is the development of efficient means of text entry. Interest is fueled by trends such as text messaging on mobile phones, two-way paging, and mobile web and email access. Coincident with this is the continued call in HCI for methods and models to make systems design tractable at the design and analysis stage [5]. This paper addresses these two themes. We propose a measure to characterise text entry techniques. It is calculated a priori, using a language model and a keystroke-level description of the technique. The measure is used to characterise and compare methods at the design stage, thus facilitating analyses prior to labour-intensive implementations and evaluations.

2 Keystrokes per Character (KSPC)

KSPC is an acronym for *keystrokes per character*. It is the number of keystrokes required, on average, to generate a character of text for a given text entry technique in a given language. That *KPSC* ≠ 1 for certain text entry techniques has been noted before [e.g., 1, 7, 8]. The contributions herein are (i) to systematically describe the calculation of *KSPC*, (ii) to provide examples of *KSPC* over a wide range of text entry

F. Paternò (Ed.): Mobile HCI 2002, LNCS 2411, pp. 195–210, 2002.
© Springer-Verlag Berlin Heidelberg 2002

techniques, some with $KSPC > 1$, others with $KSPC < 1$, and (iii) to demonstrate the utility of *KSPC* as a tool for analysis.

Although we write "keystrokes", *KSPC* also applies to stylus-based input, provided entry can be characterised by primitives such as strokes or taps.

3 Qwerty Keyboard (*KSPC* = 1)

The ubiquitous Qwerty keyboard serves as a useful baseline condition for examining *KSPC*. First, considering just lowercase letters, we note that the Qwerty keyboard is *unambiguous* because each letter has a dedicated key. In other words, each keystroke generates a character of text. Given this, we conclude the following for the basic Qwerty keyboard:

$$KSPC = 1.0000 \tag{1}$$

Of course, the value edges up slightly if we consider, for example, shift key usage. However, this is a good start. The process is not as simple for other keyboards and techniques, however. There are two central requirements in computing *KSPC*. The first is a clear keystroke-level description of the entry technique. This we provide as each technique is presented. The second is a language model. This is needed to normalize *KSPC*, so it is an "average", reflecting both the interaction technique and the user's language.

4 Language Model

A language model is built using a representative body of text — a *corpus*. We used the British National Corpus (ftp.itri.bton.ac.uk/bnc) which contains about 90 million words.[1] Since the corpus is rather large, we used forms more easily managed, yet suited to our needs. We used three such forms of the corpus.

4.1 Letters and Letter Frequencies

In the most reduced form, we work only with letters. The result is a small file (< 1KB) with 27 lines, each containing a letter (SPACE, a-z) and its frequency. The frequencies total 505,863,847. The first five entries are

---

[1] To test for a "corpus effect", all our *KSPC* figures were also computed using the Brown corpus [3]. The largest difference was 1.07%, with no change in the rank order of figures in Table 1. Our word-, digram-, and letter-frequency files for both corpora are available upon request.

```
_ 90563946
a 32942557
b 6444746
c 12583372
d 16356048
```

The SPACE character ('_') is the most prevalent, representing about 18% of all input in text entry tasks (see also [10]). Punctuation and other characters were excluded for a few reasons. In particular, they are problematic because the method of entry is not standardized and is typically dependent on the implementation, rather than on the technique per se.

If the keystrokes for an entry technique are appended to each entry in the letter-frequency file, then *KSPC* is computed as follows:

$$KSPC = \frac{\sum(K_c \times F_c)}{\sum(C_c \times F_c)} \tag{2}$$

where K_c is the number of keystrokes required to enter a character, $C_c = 1$ (the 'size' of each character), and F_c is the frequency of the character in the corpus.

Although small and efficient, the letter-frequency form of the corpus is useful only if the entry of each character depends only on that character. If the keystrokes also depend on neighboring characters (as demonstrated later), then a more expansive form is required.

4.2 Digrams and Digram Frequencies

The digram form of the corpus contains $27 \times 27 = 729$ entries. Each contains a digram — a two-letter sequence — and its frequency. The frequency is the number of times the second letter occurred immediately following the first. The frequencies again tally to 505,863,847. The five-most-frequent digrams are

```
e_  18403847
_t  14939007
th  12254702
he  11042724
s_  10860471
```

Equation 2 still applies if the digram table is used. The only difference is that summation occurs over 729 entries, rather than 27.

The digram table is reasonably small (8KB) and it is useful if the keystrokes to enter a character depend on the preceding character. However, it falls short if the keystrokes depend on more than one preceding character, as demonstrated later.

4.3 Words and Word Frequencies

In the word form, the corpus is reduced to a list of unique words. The result is a file of about 1MB containing 64,566 words with frequencies totaling 90,563,946. The top five entries are

```
the 6187925
of  2941789
and 2682874
to  2560344
a   2150880
```

Not surprisingly, 'the' is the most common word with 6,187,925 occurrences. Our list extends down to words with as few as three occurrences. Weighted by frequency, the average word size is 4.59 characters. A variation on this list containing 9025 entries was used by Silfverberg et al. [9] in their study of text entry on mobile phones.

For each word, we determine the keystrokes to enter the word in the interaction technique of interest. With this information, *KSPC* is computed as follows:

$$KSPC = \frac{\sum(K_W \times F_W)}{\sum(C_W \times F_W)} \tag{3}$$

where K_w is the number of keystrokes required to enter a word, C_w is the number of characters in the word, and F_w is the frequency of the word in the corpus. Importantly, K_w and C_w are adjusted to include a terminating SPACE after each word.

We begin by examining text entry techniques requiring more than one keystroke per character.

5 Keypads (*KSPC* > 1)

In mobile computing, full-size Qwerty keyboards are generally not practical. Alternate text entry techniques are employed such as handwriting recognition, stylus tapping on a soft keyboard, or pressing keys on miniature keypads. Since the keypad devices often have fewer keys than symbols in the language, more than one keystroke is required for each character entered.

5.1 Date Stamp Method

The date stamp method can be implemented with three keys: LEFT and RIGHT cursor keys and a SELECT key. It is so named because of similarity to a teller's date stamp, where characters are found by rotating a wheel containing the entire character set. As a text entry method, we assume the arrow keys maneuver a cursor over a linear sequence of letters and a SELECT key enters a letter. Players of arcade games know this technique, as it is often used to add one's name to a list of high scorers.

The date stamp method is well characterised by *KSPC*. We considered four variations, combining two character arrangements with two cursor behaviours. A good start is just to arrange letters alphabetically with a SPACE at the left:

```
_abcdefghijklmnopqrstuvwxyz
```

Interaction proceeds by moving a cursor back and forth with the arrow keys and entering characters by pressing the SELECT key. We call this date stamp method #1.

With the LEFT, RIGHT and SELECT keys operating as just described, we can append the requisite keystrokes to each entry in the digram-frequency table. The first five entries are as follows:

```
e_ 18403847 LLLLLS
_t 14939007 RRRRRRRRRRRRRRRRRRRS
th 12254702 LLLLLLLLLLLS
he 11042724 LLLS
s_ 10860471 LLLLLLLLLLLLLLLLLS
```

So, entering SPACE after 'e' requires six keystrokes (`LLLLLS`), a very frequent act in English. With a full digram table as above, *KSPC* is computed using Equation 2:

$$KSPC = 10.6598 \tag{4}$$

In method #1, the cursor is *persistent*: It maintains its position after each character entered. Since SPACE occurs with the greatest frequency in text entry tasks, it is worth considering a *snap-to-home* mode, whereby the cursor jumps to the SPACE character after each character entered. We call this method #2. Thus, inputting a SPACE requires just one keystroke, regardless of the preceding character. The improvement is only slight, however:

$$KSPC = 10.6199 \tag{5}$$

Another possibility is to position the SPACE character in the middle of the alphabet:

```
abcdefghijklm_nopqrstuvwxyz
```

Thus, SPACE is well-situated for English text entry. This letter arrangement combined with a persistent cursor bears further improvement (method #3):

$$KSPC = 9.1788 \tag{6}$$

However, a good leap forward is produced by combining a central SPACE character with a snap-to-home cursor (method #4):

$$KSPC = 6.4458 \tag{7}$$

English text is produced with about 40% fewer keystrokes per character using method #4 than using method #1. Numerous variations of the date stamp methods are possible, such as multiple SPACE characters, clustering common letters (e.g., 'e', 'a', 't') near the home position, or using linguistic knowledge to dynamically rearrange letters after each input to minimize the cursor-key distance to the next letter. Space precludes further elaboration here.

This simple exercise with the date stamp method illustrates the value of *KSPC* for a priori analyses of prospective text entry techniques. While avoiding labour-intensive implementations or evaluations, the exercise provided a reasonable sense of the outcome. One assumption is that *KSPC* is related to text entry throughput, for example, in "words per minute", and this seems reasonable. This is discussed in more detail later.

5.2 5-Button Pager Keypad

There are no commercial examples of date stamp method in mobile computing. Conversely, 5-button text entry is a reality on low-end two-way pagers. An example is the *AccessLink II* by Glenayre (www.glenayre.com). This device has an LCD display and five buttons for text entry (see Figure 1).

The five buttons are for LEFT, RIGHT, UP, and DOWN cursor control and SELECT. The display contains four lines of 20 characters each. When entering a message, the top line displays the message, and the bottom three lines present a virtual keyboard. The portion of the virtual keyboard of interest here is shown in Figure 2.

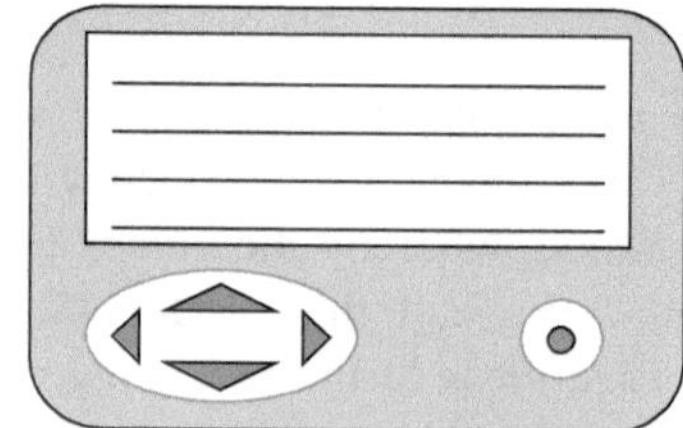

Fig. 1. Five-button pager keypad

a	b	c	d	e	f	g	h	i	j	
k	l	m	n	_	o	p	q	r		
			s	t	u	v	w	x	y	z

Fig. 2. Letter positions on the Glenayre pager virtual keyboard

Note the central position of the SPACE character ('_'), and its proximity to 'e', the most common letter in English. Text is entered by maneuvering the cursor with the arrow keys and entering letters with the SELECT key. On the Glenayre pager, the cursor snaps to a home position over the SPACE character after each entry. Since snap-to-home is used, keystrokes depend only on the current character, and so, the letter-frequency version of the corpus is sufficent to compute *KSPC*. A brief excerpt follows:

```
_ 90563946 S
a 32942557 ULLLLS
b 6444746  ULLLS
...
y 7929073  DRRRRRS
z 221512   DRRRRRRS
```

Entering text via the soft keyboard in Figure 1 as just described yields

$$KSPC = 3.1320 \tag{8}$$

This is about 50% fewer keystrokes per character than for date stamp method #4.

As with the date stamp methods, a variety of design alternatives are possible. See [1] for examples of techniques for linguistic optimization.

5.3 12-Key Mobile Phone Keypad

From three, to five, to twelve keys, mobile phones are better equipped for text entry than the devices just discussed; however, a significant challenge remains due to ambiguity in the keys (see Figure 1).

The letters a-z are spread across the keys 2-9 with 0 or # used for SPACE. The various methods for entering text using a 12-key keypad can be characterised by *KSPC*.

Multitap. Multitap is the conventional method for programming a mobile phone's address book. It is also widely used for entering text messages.

Fig. 3. Encoding of letters on a mobile phone keypad

With multitap, the user presses each key one or more times to specify the desired letter. For example, '2' is pressed once for 'a', twice for 'b', three times for 'c'. Multitap interaction can be accurately laid out in a digram-frequency file, however the examples are more interesting in the word form, so we'll use these here. The top five entries in the word-frequency file appear as follows, with multitap keystrokes appended:

```
the 6187925 84433S
of  2941789 666333S
and 2682874 2663S
to  2560344 8666S
a   2150880 2S
```

So, 'the' is entered as 8-4-4-3-3-S, where 'S' = SPACE.

Besides requiring multiple keystrokes for many letters, a mechanism is required to segment consecutive letters on the same key. The preferred technique is to press a special NEXT key between conflicting letters (see [9] for further details). As an example, our multitap word-frequency file includes the following entry for 'this':

```
this 463239 844N4447777S
```

Since 'h' and 'i' are both on the 4 key, NEXT is pressed to separate them. Note that 12 keystrokes produced a four-letter word. Assuming the word is followed by a space, we have $KSPC = 12 / (4 + 1) = 2.4$. Of course, a better measure is obtained by processing the entire file, as per Equation 3. Given this, we compute the following for English text entry using multitap mode on a mobile phone keypad:

$$KSPC = 2.0342 \tag{9}$$

Although a clear improvement over the pager method, this is still more than double our baseline of $KSPC = 1.0000$ for Qwerty. Not surprisingly, alternate text entry methods using the mobile phone keypad have emerged.

Dictionary-Based Disambiguation. Commercial telephone answering systems sometimes prompt the caller to enter the name of the person they wish to contact. The name is spelled by pressing the keys bearing the letters of the person's name. Each key is pressed just once. Thus, referring to Figure 1, 'smith' is entered as follows: (For clarity, letters are also shown.)

```
7 6 4 8 4
s m i t h
```

Although the key sequence 7-6-4-8-4 has $4 \times 3 \times 3 \times 3 \times 3 = 324$ renderings (see Figure 3), most are nonsense. The correct entry is found by searching a database for an entry matching the inputted key sequence. This technique is called *dictionary-based disambiguation.* Clearly, it also has potential for general-purpose text entry on a mobile phone. Examples include *T9* by Tegic (www.tegic.com), *eZiText* by Zi (www.zicorp.com), or *iTAP* by Motorola (www.motorola.com/lexicus). *T9* is the most widely used at the present time.

Of course the technique has limitations. If more than one word matches a key sequence, then the most probable is offered first. Alternatives appear by decreasing probability and are selected by successive presses of the NEXT key. A few examples from our word-frequency file illustrate this:

```
able     26890 2253S
cake     2256  2253NS
bald     569   2253NNS
calf     561   2253NNNS
calendar 1034  22536327S
```

The first four words above have the same key sequence: 2-2-5-3. 'able' – the most probable – is the default. If 'cake', 'bald', or 'calf' is intended, then one, two, or three presses of NEXT are required, respectively. Note that keystroke descriptions are not possible using letter- or digram-frequency files, because NEXT-key usage depends on the entire word, not just on the preceding letter.

Given a complete keystroke rendering of the word-frequency entries, as above, *KSPC* for dictionary-based disambiguation is thus calculated:

$$KSPC = 1.0072 \tag{10}$$

This is very close to 1.0000, suggesting presses of NEXT are relatively rare with dictionary-based disambiguation. The keystroke overhead above is only 0.72%, or about one additional keystroke every 1 / 0.0072 = 139 keystrokes. Assuming five characters per word, this is equivalent to about one additional keystroke every 139 / 5 = 27.8 words. Note that the most probable word in any ambiguous set (e.g., 'able' above) is the default and is entered directly, without pressing NEXT.

However impressive the *KSPC* figure above is, it is predicated on the rather generous assumption that users only enter dictionary words. It is well known that text messaging users employ a rich dialect of abbreviations, slang, etc. [4], and, in view of this, many systems allow users to enter new words in a dictionary. However, when confronted with non-dictionary words, dictionary-based disambiguation fails, and the user's only recourse is to switch to an alternate entry mode, such as multitap.

Prefix-Based Disambiguation. To avoid the problem just noted, Eatoni Ergonomics (www.eatoni.com) developed an alternative to dictionary-based disambiguation. Their method, called *LetterWise*, uses *prefix-based disambiguation* [7]. Instead of using a stored dictionary to guess the intended word, *LetterWise* uses probabilities of "prefixes" in the target language to guess the intended letter. A prefix is simply the letters preceding the current keystroke. Implementations currently use a prefix size of three. For example, if the user presses the 3 key with prefix '_th', the intended next letter is quite likely 'e' because '_the' in English is far more probable than '_thd' or '_thf'.

The distinguishing feature is that prefix-based disambiguation does not use a dictionary of stored words: it is based on the probabilities of letter sequences in a language. Thus, the technique degrades gracefully when confronted with unusual character sequences, as in abbreviations, slang, etc. Switching to an alternate entry mode is not needed.

Still, the wrong letter is occasionally produced, and in these cases the user presses the NEXT key to choose the next mostly likely letter for the given key and context.

The following is a brief excerpt from the word-frequency file, with *LetterWise* keystrokes appended:

```
the 6187925 843S
of  2941789 63S
and 2682874 263S
...
hockey  601 4N62NN539S
ecology 601 3NN2N65649S
```

Given a complete rendering of the word-frequency file, as above, the following *KSPC* measure results for prefix-based disambiguation, as typified by *LetterWise*:

$$KSPC = 1.1500 \tag{11}$$

Thus, the overhead is just 15% above KSPC = 1.0000 for Qwerty. This is well below the same figure for multitap (103%). Although the comparison with dictionary-based disambiguation is less impressive, the *KSPC* figure for *LetterWise* does not carry the same assumption with respect to dictionary words.

Other Keypad-Based Techniques. Other input techniques have been proposed for the 12-key keypad. One example is *MessagEase* from EXideas (www.exideas.com), which we include here to round out our *KSPC* measures for 12-key keypads. With *MessagEase*, the modified keypad shown in Figure 4 is used.

Digits are entered in the usual way, by pressing the corresponding key. For text entry, the alphabet is arranged by dividing letters into three sets. For the nine most-common letters (anihortes), the corresponding key is pressed twice (e.g., 4-4 for 'h'). For the next-most-common letters (upbjdgcq), the 5 key is pressed first followed by the key "pointed to" by the letter (e.g., 5-3 for 'p'). For the least-common letters (vlxkmywf) the corresponding key is pressed followed by the 5 key (e.g., 4-5 for 'k'). To enter a SPACE, 0 is pressed once.

Fig. 4. *MessagEase* keypad

MessagEase keystrokes can be accurately expressed using a letter- or digram-frequency form of the corpus; however, as before, the word-frequency form is more interesting. The following is a brief excerpt showing *MessagEase* keystrokes:

```
the 6187925 774488S
of  2941789 5595S
and 2682874 112258S
...
hockey  601 445554458875S
ecology 601 88545525555775S
```

Since each letter requires precisely two keystrokes, except SPACE which is pressed once, the *KSPC* measure for *MessagEase* is not surprising:

$$KSPC = 1.8210 \tag{12}$$

We now shift our focus to text entry techniques at the other end of the *KSPC* continuum — those requiring less than one keystroke per character of text entered.

6 Word Prediction (*KSPC* < 1)

With word prediction, there is the potential for *KSPC* < 1 because words can be entered without explicitly entering every letter. Given a sufficient portion of a word to identify it in a dictionary, an interaction technique is invoked to select the full word and deliver it to the application.

To explore *KSPC* for text entry techniques using word prediction, we must decide on interaction details, such as the input method, generation of the list of candidate words, and modeling the keystroke overhead in selecting the intended word from the candidate list.

For the analyses presented here, we assume an input method such as unistroke recognition, stylus tapping on a soft keyboard, or pressing keys on a miniature Qwerty keyboard. So, *KSPC* = 1 as entry progresses. Word prediction joins in as follows. A list of candidate words is produced as each letter is entered. The size of the list is a variable, n. If $n = 1$, only a single predicted word is offered, much like word completion in spreadsheets or URL completion in web browsers. If $n > 1$, the list contains the n most-frequent words beginning with the current word stem. The most-frequent word is first in the list, and so on. If the intended word appears, it is selected using a specified technique. This description is consistent with commercial examples, such as *WordComplete* by CIC (http://www.cic.com/). Given this, the number of keystrokes to enter the word is determined. This process is repeated for every word in the dictionary, and *KSPC* is computed as described earlier.

The number of keystrokes to enter a word has two components: (i) the number of characters in the word stem at the point where the intended word appears in the candidate list, and (ii) the keystroke overhead to select the intended word in the candidate list.

The keystroke overhead depends on the entry method. In our analyses, we consider two methods: stylus input and keypad input. With stylus input, the overhead is just 1 keystroke to tap on the intended word in the candidate list. The tap selects the word and adds a SPACE.

With keypad input, some effort is required to reach the intended word in the candidate list and to select it. The keystroke overhead is the lesser of (i) the number of characters remaining in the intended word, or (ii) the index of the intended word in the candidate list; plus one further keystroke to select the word and append a SPACE.

Two examples will help. If the user is entering 'vegetable' and $n = 5$, then the word stem and the candidate list after each keystroke are as follows:

Word Stem	Candidate List [a]
v	very voice view value various
ve	very version vehicle vehicles versions
veg	vegetables vegetation vegetable vegetarian vegetarians
[a] based on 64,566 word British National Corpus	

The intended word appears with the word stem 'veg' as the third entry in the candidate list. Although 'vegetable' followed by SPACE suggests 9 keystrokes, word prediction offers a savings. If stylus input is used, 4 keystrokes (viz. stylus taps) are required: 3 to enter the word stem, plus 1 to select the word from the candidate list and add a terminating SPACE.

If keypad input is used, 6 keystrokes are required: 3 to enter the word stem, plus 2 presses of NEXT to reach the intended word (because it is 3rd in the candidate list), plus 1 to select the word and add a SPACE. Thus,

```
vegetable 979 vegS        (stylus input, n = 5)
vegetable 979 vegNNS      (keypad input, n = 5)
```

One further example illustrates the "lesser of" proviso noted above for keypad input. Consider the word 'active' which normally requires 7 keystrokes (including SPACE). With word prediction, the following interaction ensues:

Word Stem	Candidate List [a]
a	and a as at are
ac	act actually across action account
act	act actually action activities activity
acti	action activities activity active actions

[a] based on 64,566 word British National Corpus

The intended word appears with the word stem 'acti' as the fourth entry in the candidate list. With stylus entry the word is entered with 5 keystrokes: 4 to enter the word stem, plus 1 to tap on the word in the candidate list.

However, the situation is less clear with keypad input. Potentially, 8 keystrokes are needed: 4 for the word stem, plus 3 to reach the intended word in the list, plus 1 to select the word. Or, the user could just finish entering the word in the usual manner: 7 keystrokes. Although intermediate strategies are possible, such as waiting for a word to rise in the list as more letters are entered, the impact on *KSPC* is minimal. For this analysis we assume the user acts when (or if) a word first appears in the candidate list, and that the user takes the best option — the lesser of the two tallies . Thus,

```
active 7290 actiS         (stylus input, n = 5)
active 7290 activeS       (keypad input, n = 5)
```

We calculated *KSPC* as just described for keypad and stylus input with four sizes of candidate lists:

$$\text{keypad} \quad n = 1 \quad KSPC = 0.7391 \tag{13}$$
$$\text{keypad} \quad n = 2 \quad KSPC = 0.7086 \tag{14}$$
$$\text{keypad} \quad n = 5 \quad KSPC = 0.7483 \tag{15}$$
$$\text{keypad} \quad n = 10 \quad KSPC = 0.8132 \tag{16}$$

$$\text{stylus} \quad n = 1 \quad KSPC = 0.7391 \tag{17}$$
$$\text{stylus} \quad n = 2 \quad KSPC = 0.6466 \tag{18}$$
$$\text{stylus} \quad n = 5 \quad KSPC = 0.5506 \tag{19}$$
$$\text{stylus} \quad n = 10 \quad KSPC = 0.5000 \tag{20}$$

With keypad or stylus input, *KSPC* = 0.7391 at *n* = 1, thus illustrating the potential benefit of word prediction text entry techniques. For keypad input, there is a slight gain at *n* = 2 (*KSPC* = 0.7086); however, there is a net loss at *n* > 2. In other words, the keystroke overhead in reaching and selecting candidate words more than offsets the gain afforded by word prediction.

For stylus input, just the opposite occurs: as *n* increases, *KSPC* decreases. The best result is *KSPC* = 0.5000, coincident with *n* = 10. Since the overhead is fixed at 1 tap with stylus input, increasing the size of the list always reduces *KSPC*, albeit with diminishing returns.

KSPC does not reveal the full picture with predictive interfaces or other interaction techniques for text entry. Some key performance issues are examined next.

7 Comparison of Text Entry Methods

The methods discussed above appear in Table 1, sorted by decreasing *KSPC*. *KSPC* varies by a factor of 20, from 0.5 for stylus input with word prediction using ten candidate words, to just over 10 for date stamp method #1. Midway is our Qwerty condition of *KSPC* = 1.0000.

Table 1. Comparison of Text Entry Methods

Interaction Technique	*KSPC*
Date Stamp (#1)	10.6598
Date Stamp (#2)	10.6199
Date Stamp (#3)	9.1788
Date Stamp (#4)	6.4458
Pager	3.1320
Multitap	2.0342
MessagEase	1.8210
LetterWise	1.1500
T9	1.0072
Qwerty	1.0000
Word Pred. (keypad, n = 10)	0.8132
Word Pred. (keypad, n = 5)	0.7483
Word Pred. (keypad, n = 1)	0.7391
Word Pred. (stylus, n = 1)	0.7391
Word Pred. (keypad, n = 2)	0.7086
Word Pred. (stylus, n = 2)	0.6466
Word Pred. (stylus, n = 5)	0.5506
Word Pred. (stylus, n = 10)	0.5000

7.1 Performance Issues

It is reasonable to expect an inverse relation between *KSPC* and throughput. In other words, the greater the number of keystrokes required to produce each character, the lower the text entry throughput in words per minute. However, numerous additional factors are at work that impact performance. A few seem particularly relevant.

Repeat Keystrokes. First, the date stamp, pager, multitap, and *MessagEase* techniques, by their very nature, include a preponderance of keystrokes repeated on the same key, either to advance the cursor position (date stamp, pager), to advance to the next letter on the same key (multitap), or to enter frequent letters (*MessagEase*). This behaviour is illustrated in Table 2, showing repeat keystrokes as a ratio of all keystrokes.

Not surprisingly, the ratios are quite high (> 74%) for the date stamp methods, since most keystrokes are cursor keystrokes. Auto-repeat, or *typamatic*, cursor strategies are clearly a desired interaction technique. Although lower for the pager (35%), multitap (47%), and *MessagEase* (33%), the values are significant, and should bear on performance. In key-based input the inter-key time is significantly less when successive entries are on the same key.[2] This mitigates the impact of *KSPC* on throughput for techniques with a high percentage of repeat keystrokes.

Table 2. Repeat Keystrokes as a Ratio of all Keystrokes

Interaction Technique	Repeat Keystrokes	Interaction Technique	Repeat Keystrokes
Date stamp #1	0.8156	Qwerty	0.0171
Date stamp #2	0.8454	Word pred. (k, 10)	0.0770
Date stamp #3	0.7858	Word pred. (k, 5)	0.0723
Date stamp #4	0.7453	Word pred. (k, 2)	0.0118
5-button pager	0.3501	Word pred. (k, 1)	0.0146
Multitap	0.4684	Word pred. (s, 1)	0.0146
MessagEase	0.3275	Word pred. (s, 2)	0.0122
LetterWise	0.0826	Word pred. (s, 5)	0.0085
T9	0.0817	Word pred. (s, 10)	0.0057

Attention Demands. The ratio of repeat keystrokes for the other techniques in Table 2 is either very low or of questionable impact compared with other aspects of interaction. With *LetterWise*, *T9*, and word prediction, for example, there is uncertainty on the outcome of keystrokes. And so, the user must attend to the on-going process of disambiguation ("*Did my keystroke produce the desired letter or word?*") or prediction ("*Is my intended word in the list?*"). Modeling these behaviours is complex as it depends on perceptual and cognitive processes, and on user strategies. See [9] for further discussion.

[2] Observed key repeat times are in the range of 128-176 ms [2, 9, 10, 12].

Visual processing times can be estimated, however. Keele and Posner [6] found that reacting to a visual stimulus (e.g., a letter or a word produced by a pressing a key), takes on the order of 190-260 ms. Interestingly, this range encompasses the "$n = 1$" point in the Hick-Hyman model for choice reaction time:

$$RT = k \times \log_2(n + 1) \tag{21}$$

with $k = 200$ ms/bit [11, p 68]. If the user is locating a word within a ten-word list, then $n = 10$ and reaction time is on the order of $200 \times \log_2(11) = 692$ ms! This is additive on the keystroke time, so the impact on throughput is considerable.

Many other issues impact performance, such as errors, error correction strategies, entering punctuation symbols, etc., but space precludes further examination.

8 Conclusion

We have demonstrated the calculation of *KSPC* over a variety of text entry methods, with values ranging by a factor of 20, from about 10 for date stamp methods to about 0.5 for word prediction methods. Notwithstanding other factors that influence performance, it is reasonable to expect an inverse relation between *KSPC* and text entry throughput, measured in words per minute.

We have shown the benefit of *KSPC* as a tool for a priori analyses of text entry techniques. Using *KSPC*, potential text entry techniques can undergo analysis, comparison, and redesign prior to labour-intensive implementations and evaluations.

References

1. Bellman, T., and MacKenzie, I. S. A probabilistic character layout strategy for mobile text entry, *Proc. Graphics Interface '98*. Toronto: Canadian Information Processing Society (1998) 168-176.
2. Card, S. K., Moran, T. P., and Newell, A. *The psychology of human-computer interaction*, Hillsdale, NJ: Lawrence Erlbaum (1983).
3. Francis, W. N., and Kucera, H. *Standard sample of present-day American English*, Providence, RI: Brown University (1964).
4. Grinter, R. E., and Eldridge, M. A. Y do tngrs luv 2 txt msg? *Proc. ECSCW 2001*. Amsterdam: Kluwer Academic Press (2001) 219-238.
5. Kaindl, H. Methods and modeling: Fiction or useful reality? *Extended Abstracts CHI 2001*. New York: ACM (2001) 213-214.
6. Keele, S. W., and Posner, M. I. Processing of visual feedback in rapid movements, *J. Exp. Psyc. 77* (1968) 155-158.
7. MacKenzie, I. S., Kober, H., Smith, D., Jones, T., and Skepner, E. LetterWise: Prefix-based disambiguation for mobile text input, *Proc. UIST 2001*. New York: ACM (2001) 111-120.
8. Rau, H., and Skiena, S. S. Dialing for documents: An experiment in information theory, *Proc. UIST '94*. New York: ACM (1994) 147-155.

9. Silfverberg, M., MacKenzie, I. S., and Korhonen, P. Predicting text entry speed on mobile phones, *Proc. CHI 2000*. New York: ACM (2000) 9-16.
10. Soukoreff, W., and MacKenzie, I. S. Theoretical upper and lower bounds on typing speeds using a stylus and soft keyboard, *Behaviour & Information Technology 14* (1995) 370-379.
11. Welford, A. T. *Fundamentals of skill*, London: Methuen, 1968.
12. Zhai, S., Hunter, M., and Smith, B. A. The Metropolis keyboard: An exploration of quantitative techniques for graphical keyboard design, *Proc. UIST 2000*. New York: ACM (2000) 119-128.

Lost or Found? A Usability Evaluation of a Mobile Navigation and Location-Based Service

Didier Chincholle, Mikael Goldstein, Marcus Nyberg, and Mikael Eriksson

Ericsson Research, Usability & Interaction Lab
Torshamnsgatan 23,
164 80 Kista, Sweden
{didier.chincholle,mikael.goldstein,marcus.nyberg,
mikael.x.eriksson}@era.ericsson.se

Abstract. Today's wireless devices have the capability to receive information that is tailored to fit customers' needs at a particular location. A location-sensitive prototype service, the Personal Navigation Tool (PNT), including user-solicited information, worldwide maps, route and location guidance, was created for the WAP-enabled Ericsson R380 Smartphone. Seven Smartphone-literate but PNT-naive users were given five typical tasks in an in-door evaluation, in order to evaluate the PNT when running over a circuit-switched GSM network. All users accomplished three tasks targeting retrieval of route directions successfully whereas two tasks targeting retrieval of location information were accomplished successfully by only two and one user, respectively. System speed was rated as inadequate. The download of traditional miniaturised maps contributed little to the mobile user's demands whereas route directions were considered valuable. User attitude towards location-based services was very positive and the potential usefulness is believed to be high.

1 Introduction

Most of the current wireless services have low take-up rates in the marketplace because users find them too difficult to use or of little real value when they are out and about. As found in recent studies [6, 10], this generation of wireless services almost always have low and miserable usability, slow connection speeds, and content which is often inappropriately designed for usage in a mobile environment. Often bad usability is attributable both to the mobile phone and the WAP (Wireless Application Protocol) service [4, 5]. In addition, the cost of accessing these services has also been found to be a major factor inhibiting usage [9]. The key to success in designing the next generation of services is actually to understand what content to develop and how to make it as simple and easy to use as possible. It is also to understand that exploiting the benefits of mobility is vital for building services of the future. Consequently, the next big wireless leap might be services that exploit the user's whereabouts, i.e.

F. Paternò (Ed.): Mobile HCI 2002, LNCS 2411, pp. 211–224, 2002.
© Springer-Verlag Berlin Heidelberg 2002

services that enable mobile users to receive personalized and lifestyle-oriented services relative to their geographic location.

Ericsson has designed a location-sensitive service, for use on WAP-enabled mobile phones, called PNT (Personal Navigation Tool). The service, as implemented on the Ericsson R380 Smartphone, will be explained in detail in the Section 3. A PC-demo, which differs slightly from the prototype, is also available [3]. The PNT prototype service has been examined in a usability evaluation and the outcome will be presented before moving into the discussion. The concluding parts include suggestions on how a location-based service like the PNT could be further improved in order to better match user requirements.

2 Usability and Design Issues

Usability is important for any device or service, but especially for small mobile devices whose constraints make them even harder for the user to interact with. Consequently and in order to design highly usable mobile services, it is crucial to first define the right audience, i.e. the potential user group of the service. As a matter of fact, mobile users will better understand the service to be designed, if first the designer understands the users, their needs and wants, and the way they will interact with the service when on the go.

2.1 Users on the Go

Mobile users have little patience for learning how to operate new services. They don't focus on their device in the same way as when they are sitting in front of their desktop computer [1, 7]. The demanding and/or stressing environment can also be very distracting. The mobile users have less "mental bandwidth" - capacity for absorbing and processing content - than a stationary user in front of a PC [10] since the interaction with the mobile phone often is reduced to a secondary task that must not interfere with their primary task (e.g., driving or walking). While on the go, time is also an important factor. The mobile users tend to use services that allow both quick manipulations of the interface and a reduction in number of steps to access information. They look for pieces of information on the fly, but do not need all the functions of a PC [1].

2.2 Design Limitations

Most mobile phones share two design limitations: small screens for displaying content and limited multifunctional keypads making data input tedious and frustrating. These limitations restrict both navigation and the amount of content that the device can display. Due to these constraints, the mobile service designer has to learn new ways of thinking about the presentation of information. In fact, designing an effective user interface for mobile services on a restricted screen space is hard and this is rapidly becoming a major issue for both the mobile service provider and the mobile phone manufacturer [4, 5, 9]. The key to success in designing mobile services is not only to understand what content to develop but also how to make it more appropriate for a mobile usage.

3 The Personal Navigation Tool

The PNT service consists of a set of location-based functions for users on the go. It offers them relevant maps, navigation tools and content depending on where they are or where they want to go. The current version includes:

- Route directions presented both as text and miniaturised maps with graphical representation of routes.
- User-solicited information, such as traffic news and what's on in the area.
- E-coupons providing special time-limited offer available from stores, restaurants, and pubs, etc.

This set of location-sensitive functions can fit many types of mobile phones, but the PNT service has been optimised for running on an Ericsson R380, a pen-based Smartphone.

3.1 Mobile Phone Description

The Ericsson R380 Smartphone combines the functions of a mobile phone with advanced communication features such as WAP, SMS and E-mail. It also includes a complete range of PDA-like tools including Address book, Calendar, Notes and support for synchronization. The Ericsson R380 has a grey scale touch-sensitive display (see Fig.1) with a resolution of 360 x 120 pixels (an active screen size of 83 x 28 mm). The device also features a full soft QWERTY keyboard.

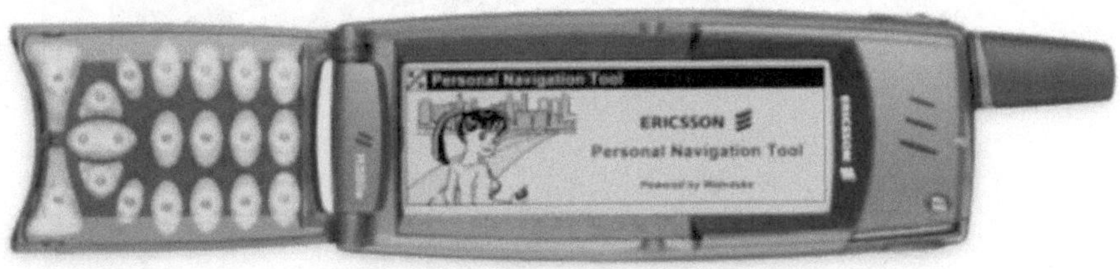

Fig. 1. The landscape-format screen of the R380 displaying the first splash screen of the PNT service.

3.2 Service Description

The core of the PNT service is the positioning technique that can be used for route planning and proximity search. The service uses the Ericsson Mobile Positioning System (MPS), a network-integrated positioning system that delivers location estimates for mobile phones [2]. Any Geographical Information System (GIS) can use the format. The Ericsson MPS enables the PNT service provider to know the whereabouts of mobile phones while protecting the end user's privacy.

In order to be obvious and make the first user experience comprehensible, the PNT service has been divided into four distinct sections:

- Route planning: Plan a route from start to end destination using positioning and fuzzy text input matching. Check "Traffic" state (see Fig. 4) of the route.
- Proximity search: Search for the location of a Point of Interest (POI) using positioning or a street address, a postcode or a recently visited location (Fig. 5). Retrieve location-related E-coupon advertisement offers (e.g. restaurants,etc., Fig. 6).
- Personalisation: Store personal details such as favourite routes, addresses, and time-limited E-coupons.
- Customisation: Customise the service according to personal preferences.

Route Planning

This service provides the mobile user with the best route (either by car or by walk) to a target address or a well-known place. It displays a miniaturised map (Fig. 2) by default that is stepwise zoomable. A table of route directions with turn-by-turn instructions can be obtained by tapping on the [Directions] tab (Fig. 3). Concurrently, it presents the distance and the time it takes to that destination. [Alternative Route] is also available. A list of Points Of Interest (POI), such as restaurants, shopping malls, museums and petrol stations, along the route is also supplied. Real-time [Traffic] information is also available when planning a route. Specific information about where the traffic is "HEAVY" or "Moderate" can be obtained (Fig. 4).

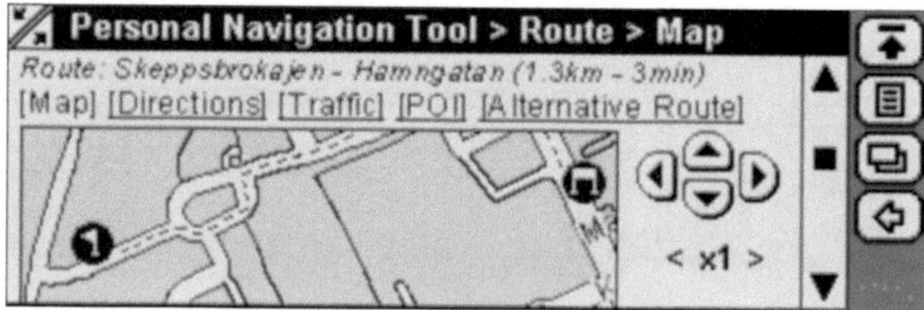

Fig. 2. Zoomable map displaying the requested route (dotted line) between "Skeppsbrokajen" and "Hamngatan". (The shortcut zoom feature (< x1 >) was not implemented in the PNT prototype).

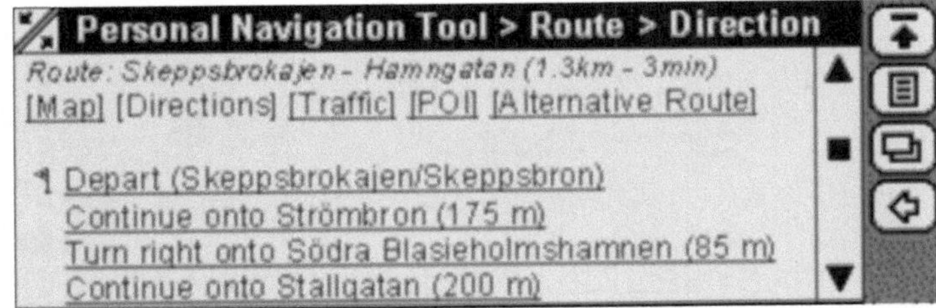

Fig. 3. By tapping on the [Directions] tab in Fig. 2, a table of turn-by-turn directions is displayed for the requested route.

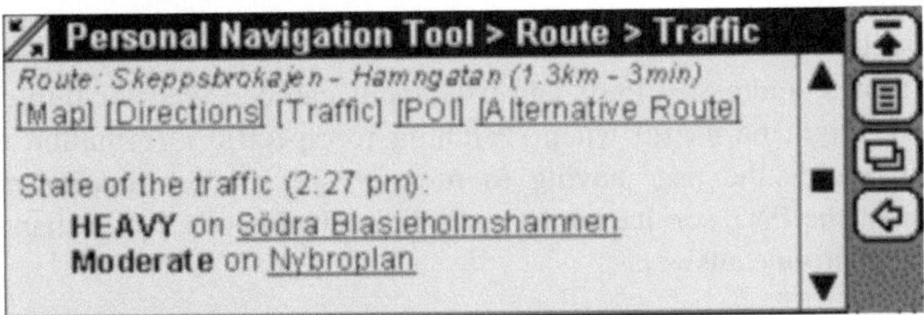

Fig. 4. By tapping on the [Traffic] tab in Fig. 3, information about traffic congestion for the requested route is displayed.

Proximity Search

This user can from a specific location (e.g. his current position) find a set of "nearest" requested points of interest (POI) and related maps (Fig. 5). Names, addresses and telephone numbers of these POIs can be saved or sent via either SMS or fax to another peer.

Fig. 5. Using the Find feature, a set of "nearest" requested POIs (represented by numbers) can be displayed on a map. Numbering was not implemented in the PNT prototype.

Remembering to bring paper coupons along to the store or restaurant is a major problem. Electronic (E)-coupons solve this problem since they can be accessed through the user's mobile phone. In the PNT service, the advertisements are only displayed when the user picks a category (e.g., restaurants) from the list of POIs. Then an E-coupon automatically pops up (Fig. 6) and disappears after 3 seconds (push service). During this time the user can save [Save as Favourite] or reject [Cancel] the E-coupon offer.

Fig. 6. Example of an E-coupon offering for a restaurant.

Personalisation

Favourite routes, addresses, POIs and E-coupons can be saved for future use (Fig. 7). This feature might be useful when searching for specific information or planning a route. This avoids the user having to re-enter specific information several times. Consequently, the PNT service conforms to individual needs, rather than attempting to employ the all-in-one answer.

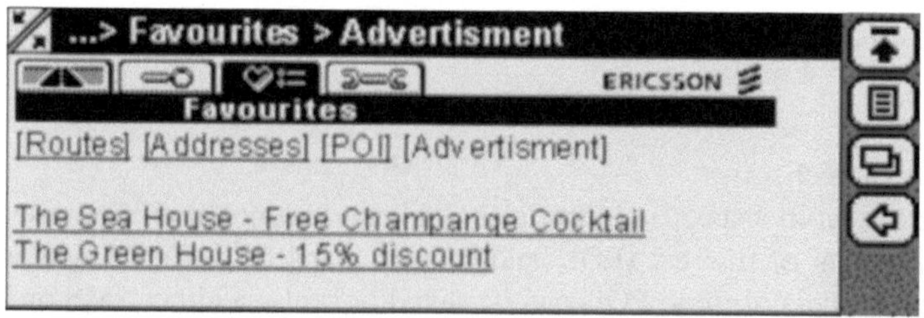

Fig. 7. List of previously saved E-coupons appearing in "Advertisement" under the "Favourites" tab.

Customisation

The user can customise the service in a way that restricts the geographical area of search and/or route planning (Fig. 8). As a matter of fact, when planning a route ("From:" "To:") or looking for information at specific places, the user can narrow down the search by excluding countries that are not of interest.

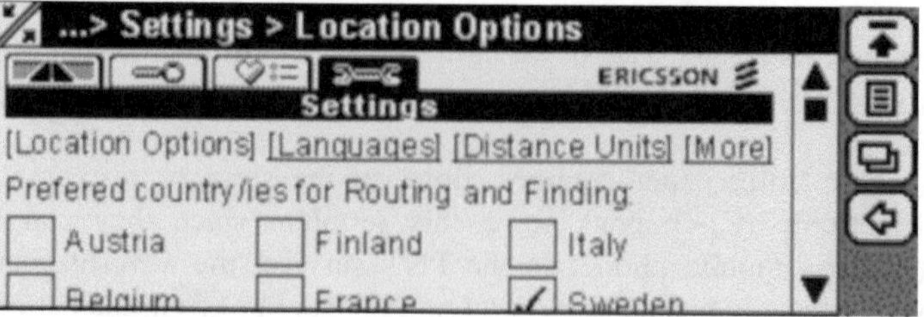

Fig. 8. List of countries to be selected when customising the search function.

4 Experiment

A usability evaluation of the PNT service was administered at Ericsson Research's Usability Lab in Kista, Stockholm. The lab features a test and a control room separated by a one-way mirror as well as several video cameras for recording (Fig. 9).

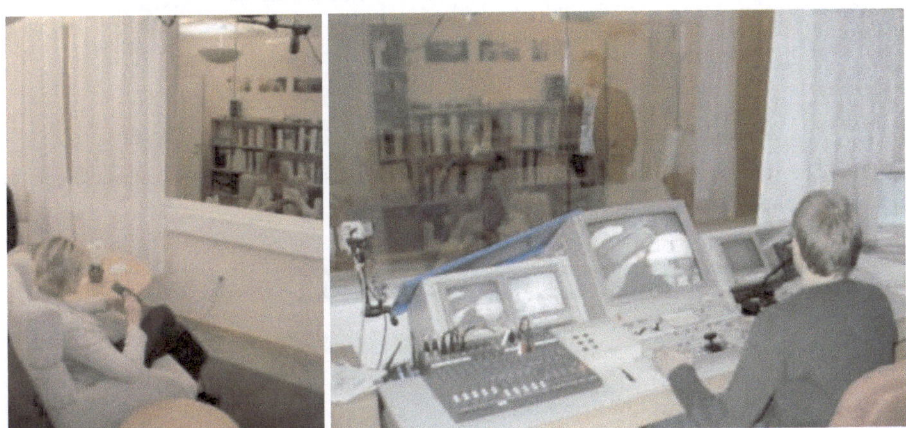

Fig. 9. Evaluation of the PNT service at Ericsson Research's Usability Laboratory in Kista. Left: The test room. Right: The control room.

4.1 Subjects

Seven (4 males and 3 females) Swedish-speaking PNT-naïve but Ericsson R380-literate users participated in the study. Their age varied between 26 and 59 years (M=38). Their previous Smartphone experience varied between 7- 24 months of active use. Four were familiar with other PDA (Personal Digital Assistant) devices and five used other mobile phones. All seven used other mobile WAP services whereas two had previous experience from using other mobile location-based services and six from using stationary navigation services.

4.2 PNT Prototype

The PNT prototype service was running on an Ericsson R380 Smartphone with flip open (Fig. 1) and screen backlight always on. The prototype lacked the ability to track the user's current location automatically (see Figs. 10 and 11). Thus, neither the <My Location> default nor the <Recent location> option could be employed. The user thus always had to enter both the "From:" and the "To:" destination when employing the Route and Find (proximity search) functionality. The WAP push functionality was not implemented and the users where thus not exposed to E-coupon offers (see Fig. 6). When searching for a predefined POI, the targets where *not* numbered on the displayed map (see Fig. 5). The maps in the prototype employed a step-by-step zoom implemented as a [-] and a [+] buttons where magnification size could not be selected by the user (Fig. 2)

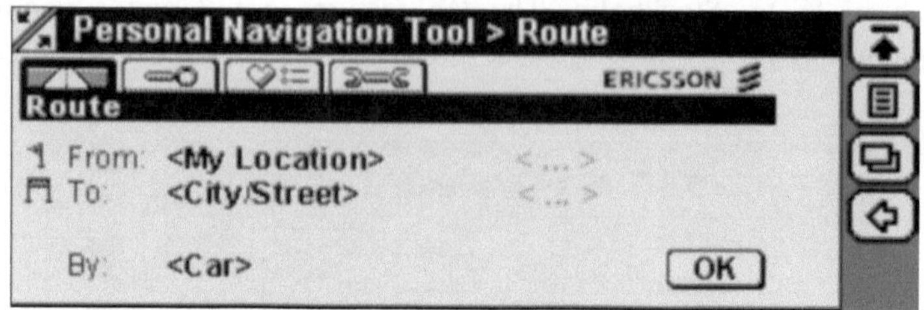

Fi.g 10. The start page of the PNT application, which is displayed once the splash screen (see Fig. 1) has disappeared.

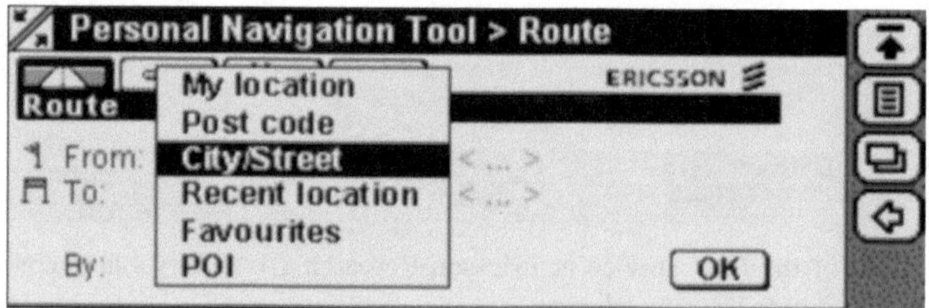

Fig. 11. Selection of various locations when tapping the "From:" or "To:" entry fields.

4.3 Tasks

Each user was exposed to five typical navigation tasks. The first three tasks (1-3) consisted of finding *car* route information when driving "From" a location "To" a destination:

- Task 1. Find the car route to a destination street in the same town (in Stockholm, Sweden).
- Task 2. Find the car route to a targeted destination street when travelling to a nearby town (in Nyköping, Sweden).
- Task 3. Plan a vacation trip by car between Stockholm, Sweden and Paris, France in a stationary/mobile context.

Task 4 and task 5 looked at the mobile user in an unfamiliar context when retrieving *walking* route directions by performing proximity search:

- Task 4. Select a nice restaurant within *walking* distance (in Paris, France). Find the directions of how to get there and save the restaurant in "Favourites".
- Task 5. Retrieve the previously saved restaurant Favourite in order to find the *walking* route to a new destination in Paris.

4.4 Subjective Inventories

After each task, the user was also questioned regarding what means (s)he would use to retrieve route directions along the way under the condition (s)he was travelling alone. Suggestions and remarks made by the users during and after the tasks were noted.

Upon completion of the five tasks, an inventory was administered consisting of totally 36 questions. They covered different areas such as: overall reactions to the PNT service, screen quality, terminology, learning, system speed and rating of the different features in the PNT service. The layout of the questionnaire was similar to the Questionnaire for User Interaction Satisfaction (QUIS) proposed by Shneiderman [11]. The users rated each attribute on a discrete nine-point scale (1-9) and at the anchor scale endpoints matching adjective opposites such as frustrating-satisfying, never-always, etc., were used.

4.5 Procedure

Each user was first asked to familiarise with the PNT service for approximately 5 minutes. No PNT manual was at hand. The experimenter stressed that it was a *prototype* service and that automatic location of the user's present location was not implemented.

The user was asked to accomplish each task while "talking aloud". Since clarifications regarding an opinion were seen as important, only accuracy was considered and completion times were not measured. If difficulties occurred, the user was first given a hint by the experimenter and if this information did not suffice, the user was guided through the task before starting with the next. A fully completed task scored as 1.0, a task that required hint(s) as 0.5 and a walkthrough guided by the experimenters scored naught 0.

5 Results

5.1 Objective Findings

The first three tasks (1-3) concerning retrieving car route information were accomplished successfully by all seven users, except for one user who needed a hint on task 2. The average Accuracy score was 1.0 on Task 1, 0.93 on Task 2 and 1.0 on Task 3.

Task 4, addressing the issue of finding a location (a restaurant) close to a target street, saving the restaurant as a bookmark as well as retrieving walking route directions to get to the restaurant, was accomplished successfully by only two out of seven users (Accuracy=0.43). Two out of seven users managed to accomplish the task with a hint, but the remaining three had to be guided through the task by the experimenter.

Task 5, addressing the issue of finding the walking route directions from the previously saved restaurant favourite to a new location was accomplished correctly by only one out of six (Accuracy=0.17) users (one user was not given this task due to lack of time).

5.2 Subjective Findings

Most users believed they would employ the directions information as it is presented on the Smartphone screen to retrieve route directions along the way (Fig. 3). In Task 1, four users selected this way, one wanted to print out the route directions and one would write them down on a paper. In Task 2, five users choose to have the Smartphone beside them while driving, whereas the other two were unsure. In Task 3, which was more a planning task, only one wanted to use the Smartphone to retrieve information from. Four would use a paper printout and two would try to find another (stationary) service than the PNT. In Tasks 4 and 5, when the walking route was retrieved, all users selected the Smartphone as their first choice to retrieve route directions.

The overall subjective ratings of the PNT service were relatively high (4.9-6.7). The ratings of the screen varied more (3.3-6.1). The lowest rating was given the "Quality of maps" (3.3), which were regarded as hard to read. The questions regarding terminology were rated between 3.7-6.9. The attribute "Length of delay between operations" was regarded as unacceptable (3.7). Four questions regarding learning were rated rather high (all above 6.6). System capabilities regarding speed, response time and system failure was rated low (3.7 - 4.1).

Ratings of features in the PNT service (11 questions regarding route map, directions, traffic information, etc.) varied between 5.1-8.7. The use of E-coupons was rated lowest (5.1) along with route maps (5.4). Rating of "Locations-sensitive options" (i.e. you do not have to enter your location) was given the highest rating (8.7) in spite of the fact this feature was not implemented in the PNT prototype. The users obviously deemed this feature to be absolutely mandatory in a mobile navigation service.

A question where two alternative presentations of turn-by-turn route directions (see Fig. 3) were displayed on paper yielded a mixed outcome. Three users preferred the first alternative where the current route directions in words (Turn left, Turn right, etc.) were substituted for directional arrows (⇐ ⇒) appearing in the left-hand column. Two preferred the second alternative where the displayed street names were centred and popped out to the left and right, accompanied by arrows on the edges. The remaining two preferred the current PNT layout for the table of directions.

6 Discussion

6.1 General Comments

The user's attitude towards this type of mobile service was very positive. All users foresaw several situations when they could benefit from location-based navigation. However, the service has its strength in mobile situations when the user does not have to enter his/her current position. The circuit-switched GSM connection proved to be far too slow for the PNT service, which indicates that a mobile location-based service has a time-critical component. This is important to remember even with future mobile systems when the bandwidth increases. The service design must still be designed for few input steps and short response times since easy and fast access to the service and the information is of highest importance. The functionality of the PNT service was seen as rather broad. The general opinion was that the service should not be further complicated.

6.2 Planning a Route vs. Finding a Location

For route planning, a stationary computer (PC) with its large screen and printing capabilities will definitely be the user's first choice. However, the mobile device is always with the user and can be used for planning in order to kill some time. It is then important to be able to send the information to e.g. a stationary computer so that it can be used at a later point.

Route directions (Fig. 3) were clear and concise and were regarded as essential for route navigation. However, they were rather detailed and required both scrolling as well as paging down. This is not a serious problem when walking, but might be when driving a car. Among the more advanced user suggestions voice prompt directions could be found. Another suggestion was to make continuous use of the location information and automatically update the actual location (and map image) as the user moved along the route. This would improve the route map functionality significantly.

Looking at more detailed interaction, the PNT service was seen as rather inflexible. It gives the user few options to skip an input field and only enter the street address for instance. Several users commented on the fact that most navigation would be performed within the same city. Therefore, the "To:" City name should be the same as the "From:" City name and appear as default (with the option to change it of course).

As mentioned in the Section 5, none of the users had any problem with finding a route from a location to a destination. However, it seems like the two parts of the service and the integration between them suffers from an unclear hierarchy. "Find", "POI", "Favourites", and "Map" are four expressions that several users did not fully understand in the context. The same terminology exists in several parts and on several hierarchical levels in the service. An interesting observation is that the "Route" section is about finding a route and the "Find" section is about finding a map or POI. However, the word Find is used only in the latter section. A user who is looking for a restaurant (as in Task 4) and does not have the initial goal to find a route can easily run into trouble.

6.3 Maps vs. Directions

The maps (see Figs. 2 and 5) were rated as less useful regarding navigation from the departure location to a destination. Still, some users commented that they might have a value after all. They can give the user an initial overview and a spatial understanding that the table of directions cannot provide. Although the maps can hardly be used for detailed navigation of any kind, users are familiar with maps and they are very common tools when travelling. By looking at the map for the full journey, it is possible to create a mental model of the route. The maps can also be useful if they are zoomed in on a street level and the user's position is clearly indicated on the map (this feature was not implemented in the prototype). It seems like the intermediary zoom levels were of less value and having to zoom in 3-5 times to reach the magnification when the street name is displayed is unacceptable. This must be possible to accomplish in one single tap. Distorted maps, only presenting the information needed for the navigation, is recommended to avoid cluttering the screen [7].

6.4 Push vs. Pull Information

E-coupons (Fig. 6) were a feature that could not be tested in the evaluation. Opposite to earlier investigations [8] where the users seemed very receptive to E-coupons and mobile marketing, user comments indicate that pushing this information can be a source of irritation. As shown by the HPI research group [8], being able to block and filter incoming (push) offers is very important. Perhaps the lack of this functionality in the PNT prototype is one explanation to the difference in user opinion. Several users stated that they would prefer to substitute the somewhat intrusive push alternative for initiating the viewing themselves. If we on the other hand look at the traffic information, that currently requires the user's active selection, we see that pushing of this information would be accepted. Instead of having to check traffic information (see Fig. 4), it is displayed (only) when there is congestion along the selected route.

6.5 Redesign Suggestions

Based on the experimental outcome it is suggested that the PNT service is simplified and designed with only one entrance, where the user can search for both "Routes" and POIs. A less advanced version of the "Favourites "section could also be integrated in this view if some of the functionality is traded for integration and one-window access to all functions. If the application is designed in two parts, there must be available shortcuts between them. E.g., a user who finds a restaurant using the "Find" section must be able to retrieve the route to the restaurant via a shortcut. Important to remember is that this problem is most significant when using the PNT service for location planning and less significant for retrieving a "Route" starting from <My Location> (see Figs. 5 and 10).

7 Suggested Improvements

- Route vs. Find location. Most users accomplished the (find) route tasks without any problems. However, when they had to find a location, several had difficulties. There should be only one "entrance" into the PNT service. Both (find) Route and Find (POI/map) should be accomplished in the same way and from the same page (starting page).
- The table of the route directions ([Directions] see Fig. 3) should be displayed as default instead of "Map". Most users selected the directions at once, since the cluttered map functionality did not provide the user with adequate information. Route directions should be continuously updated and the user should be informed when a turn is coming up.
- The map function should be improved, allowing the user to quickly zoom in from the overview image to a detailed street level. In this way, the advantage with the map overview is maintained and at the same time, the user can quickly get low-level context information on the immediate surroundings. Presenting distorted maps is also recommended. In this way, the mobile user can focus on the essential information only.
- Continuous update of location-sensitive information. When a user travels along a route, the user's position should be updated continuously. Both directions and map information should shadow the user's position. The user's position should be displayed in the centre of the map, since this was according to the user expectations.
- The [Traffic] tab is unnecessary. This information should be pushed automatically if traffic congestion occurs along the route. Details concerning congestion may then be further investigated if desired.
- Use more flexible search options and allow the user to enter information in a more flexible way. For instance, skipping the city name and entering the street address only or using the current city location as default.
- Improve feedback and inform the user of where he is in the search procedure and what s/he is supposed to do at all times.
- The use of acronyms should always be used with care. All users found the POI acronym difficult to understand.

8 Further Work

The next step of a location-sensitive study is to demonstrate that a service like PNT can fit any family of mobile phones (i.e. different screen size and format, different browsers, etc.). It is also to show how it can be customised to the service provider's needs regarding user interface, content, functionality and branding.

Investigating user behaviour when employing a location-based service in a mobile context is the next step. This type of service is mainly designed for the mobile environment and studying the applications there might reveal further important usability issues.

References

1. Chincholle, D., Designing Highly Usable Mobile Services for Display Devices. Tutorial notes, in Proceedings of IHM-HCI'2001, Lille, France, (September 2001) 225-226.
2. Ericsson corporate web site: Ericsson Mobile Internet. Available at: http://www.ericsson.com/mobileinternet/ (February 2002)
3. Ericsson corporate web site: Overview – Mobile Positioning System. Available at: http://www.ericsson.com/mps/ (February 2002)
4. Eriksson, T., Chincholle, D., Goldstein, M., Both the cellular phone and the service impact WAP usability, Short paper, Proceedings of IHM-HCI'2001, Volume II, HCI in Practice, by J. Vanderdonckt, A. Blanford and A. Derycke (eds.), 10-14 September, Lille, France, (2001) 79-8.
5. Ericsson, T., Goldstein, M. and Chincholle, D., Is poor initial WAP usability pertinent to the violation of the user's fixed Internet mental models? (paper submitted to NordiCHI'2002).
6. Helyar, V., Usability issues and user perceptions of a 1st generation WAP service. Available at: http://usability.serco.com/research/research.htm#research (January 2001)
7. Hjelm, J., Creating Location Services for the Wireless Web, John Wiley & Sons, Inc. (2002) 209-266
8. Nokia corporate web site: New Nokia research shows consumers ready for marketing via mobile handsets. Available at: http://press.nokia.com/PR/200201/846567_5.html (February 2002)
9. Ramsay, M. and Nielsen, J., WAP Usability Déjà vu: 1994 All Over Again. Report from a Field Study in London Fall 2000, Nielsen Norman Group. Available at: http://www.nngroup.com /reports/wap/ (December 2000)
10. Rischplater, R., The Wireless Web Development. Apress (2000)
11. Shneiderman, B., Designing the User Interface, 3rd edition. Addison-Wesley (1998).

Towards an Improved Readability on Mobile Devices: Evaluating Adaptive Rapid Serial Visual Presentation

Gustav Öquist[1] and Mikael Goldstein[2]

[1] Uppsala University, Department of Linguistics,
Box 527, 751 20 Uppsala, Sweden
Tel: +46-18-471-7006
Fax: +46-18-471-1416
gustav@stp.ling.uu.se

[2] Ericsson Research, Usability & Interaction Lab,
Torshamnsgatan 23, 164 80 Kista, Sweden
Tel: +46-8-757-3679
mikael.goldstein@era.ericsson.se

Abstract. Can readability on small screens be improved by using Rapid Serial Visual Presentation (RSVP) that adapts the presentation speed to the characteristics of the text? In this paper we introduce Adaptive RSVP and also report findings from a usability evaluation where the ability to read long and short texts on a mobile device was assessed. In a balanced repeated-measurement experiment employing 16 subjects two variants of Adaptive RSVP were benchmarked against Fixed RSVP and traditional text presentation. For short texts all RSVP formats increased reading speed by 33% with no significant differences in comprehension or task load. For long texts no differences were found in reading speed or comprehension but all RSVP formats increased task load significantly. Nevertheless, Adaptive RSVP improved task load ratings for most factors compared to Fixed RSVP. Causes, implications and effects of these findings are discussed.

1 Introduction

The mobile Internet has been widely predicted to cause a revolution in communications over the next few years and when the time is ripe it is likely to become an integral part of our everyday lives. However, before the revolution can start there must be devices, services and applications available that people really want to use. Enabling these new technologies, bringing them to market and making desirable content available to them is the challenge facing companies involved in the wireless business today. The limited screen space on most mobile devices is presently a bottleneck for efficient and usable retrieval of information [5]. Since it is the customers demand for small devices that sets the limit this constraint is also likely to remain tomorrow. Combined with the expected increase in texts to be read on mobile

F. Paternò (Ed.): Mobile HCI 2002, LNCS 2411, pp. 225–240, 2002.
© Springer-Verlag Berlin Heidelberg 2002

devices this quandary has made the issues concerning readability on small screens progressively more important.

Early research on screen reading showed that reading speed decreased by 20-30% when reading on large screens compared to reading on paper [17]. With time, as screen resolution improved and people got more used to them, readability on large screens became more or less equal to paper [18]. The evolution in readability on small screens is however not likely to follow the same pattern. The resolution will surely get better and thus improve legibility but decreased readability will still be intrinsic to limited screen space [4]. Readability may however still be increased by designing interfaces that display the text in a way more suitable for small screens.

Focus has thus shifted from how screens display static texts to how texts can be dynamically displayed on the screen. Leading and Rapid Serial Visual Presentation (RSVP) are the two major techniques that has been proposed for dynamic text presentation [17]. Leading, or the Times Square Format, scrolls the text on one line horizontally across the screen whereas RSVP presents the text as chunks of words or characters in rapid succession at a single visual location. Both formats offer a way of reading texts on a very limited screen space [17, 11, 21, 2]. Comparisons between the formats have so far been inconclusive [14, 10, 16] but since the eye processes information during fixed gazes it seems more *natural* to use RSVP, the reason for this is that the text then moves successively rather than continuously.

It is important to explore the possibilities of dynamic text presentation since improved readability on small screens also means improved usability of mobile devices. In this paper we look at the potential enhancement of the RSVP format by letting the presentation speed adapt to some linguistic characteristics of the text.

2 RSVP

RSVP originated as a tool for studying reading behavior [6, 10, 20] but has lately received more attention as a presentation technique with a promise of optimizing reading efficiency, especially when screen space is limited [11, 19, 21, 7]. The reason for the interest is that the process of reading works a little different when RSVP is used and that it requires much smaller screen space than traditional text presentation.

While you read this text on paper three distinct visual tasks are performed: Information is processed in fixed gazes or fixations, saccadic eye movements are executed to move between the fixations and return sweeps are used to move to the next line. Whereas saccadic eye movements and return sweeps are performed very quickly (~40 ms and ~55 ms respectively) the fixations take longer time (~230 ms for fast readers and ~330 ms for average readers) [22]. If you read this text using RSVP instead of paper, it would be successively displayed as small chunks within a small area. Each chunk would typically contain one or a few words depending on the width of the text presentation window. When reading in this fashion the text proceeds by itself and that reduces the need for saccadic eye movements and return sweeps [21].

The speed of the text presentation when using RSVP is usually measured in words per minute (wpm). The exposure time of each text chunk is calculated on basis of the set presentation speed and on how much that can be displayed in the text presentation window. Unfortunately there is little or no documentation in previous studies on exactly how the exposure times have been calculated [10, 15, 11, 19, 21]. What is known however is that the exposure times have generally been *fixed*. In this evaluation the following formula has been employed for calculating the fixed text chunk exposure times. (Eq. 1):

$$\text{time}_0 = \text{fchr}/(\text{wavg}*\text{wpm}/60)\ . \quad (1)$$

The average number of characters that can be displayed (fchr) is divided by the product of the average word length (wavg) for the current language and the presentation speed (wpm) divided by 60. The result is a fixed exposure time for each text chunk measured in seconds (time_0).

2.1 Previous Readability Evaluations with RSVP

Juola et al. [10] found comprehension between RSVP and traditional text presented on a screen equal whereas Masson [15] found that comprehension for text read using RSVP was poorer. A possible explanation for the different results may be the insertion of a blank screen for 200-300 ms between the sentences in the Juola et al. study. In a repeated-measurement experiment where long texts were read using RSVP with blank screens Goldstein et al. [7] found neither reading speed nor comprehension to differ from reading on paper. However, the NASA-TLX (Task Load Index) [9] revealed significantly higher task load when using RSVP for most factors [7]. One explanation to the high task load may be the fact that the exposure times in previous RSVP implementations have been fixed although the reading speed actually varies [12]. The relation between reading speed and exposure time for, what we from here on will refer to as, Fixed RSVP can be visualized in the following speed-exposure plot (Fig. 1).

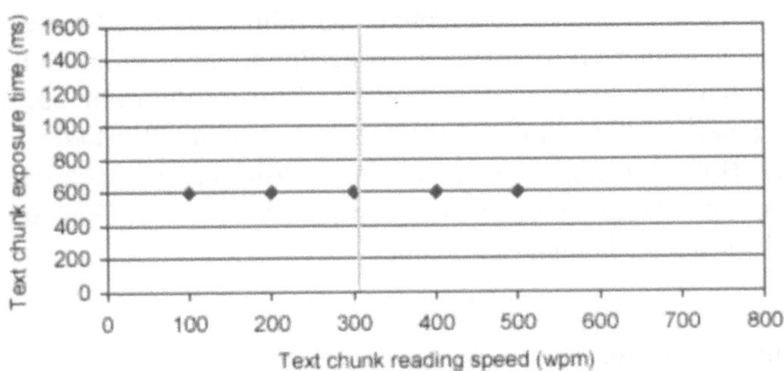

Fig. 1. Variations in reading speed for individual text chunks containing 1, 2, 3, 4 or 5 words presented at a speed of 300 wpm when using RSVP with fixed exposure times (data derived from the training text)

The plot is a result of presenting the training text used in the usability evaluation at a constant speed of 300 wpm using the fixed exposure time formula (Eq. 1). The width of the text presentation window was 25 characters (fchr=21) and the average word length was set to 7 characters. The obtained reading speed is a result of dividing each chunks exposure time by the number of words it contains (1-5). This explains why there is a variation in reading speed (100-500 wpm) although the exposure time is fixed (600 ms). Only the text chunks with three words match the selected reading speed (the vertical line in Fig. 1). If more words appear in a text chunk speed increases and if fewer words appear speed decreases. The reading speed is thus actually *inversely* related to the number of words that each text chunk contains which does not seem very natural.

3 Adaptive RSVP

Just and Carpenter found that "there is a large variation in the duration of individual fixations as well as the total gaze duration on individual words" when reading text from paper [12:330]. Adaptive RSVP [8, 25] attempts to mimic the reader's cognitive text processing pace more adequately by adjusting each text chunk exposure time in respect to the text appearing in the RSVP text presentation window. By assuming the *eye-mind hypothesis* [12], i.e., that the eye remains fixated on a text chunk as long as it is being processed, the needed exposure time of a text chunk can be assumed to be proportional to the predicted gaze duration of that text chunk. Since very common, known or short words are usually processed faster than infrequent, unknown or long words, the text chunk exposure times can be adjusted accordingly [12, 13]. Further, most new information tends to be introduced late in sentences and therefore ambiguity and references tends to be resolved there as well. A shorter sentence is also usually processed faster than a longer one since it conveys less information [12, 13]. Thus, processing time differs both within and between sentences and the text chunk exposure times can therefore be adjusted accordingly as well.

On basis of these findings two adaptive algorithms supposed to decrease task load were developed. Both were deliberately kept simple since mobile clients tend to be quite thin (i.e. have limited processing power). The first algorithm adapts the exposure time to the *content* of the text chunks whereas the second also looks to the *context* in the sentences. Both algorithms insert a blank window between each sentence if there is not enough space to begin on the next sentence in the same window, otherwise a delay is added to the sentence boundary instead.

3.1 Content Adaptation

In content adaptive mode the exposure time for each text chunk is based on the numbers of characters and words that are being exposed for the moment. Longer words are assumed to be more infrequent and take longer time to read than shorter words. A higher number of words are also assumed to take longer time to read and should thus receive more exposure time. The following formula is used to calculate the exposure time for content adaptation (Eq. 2):

$$time_1 = (nwrd+nchr)/(davg*wpm/60) . \quad (2)$$

The formula uses the number of words (nwrd) and the number of characters (nchr) as a basis for the results. Both arguments are added and divided by the product of the average word length including delimiters (davg) and the currently set speed in words per minute (wpm) divided by 60. The result is a variable exposure time ($time_1$) depending on the content the current text chunk.

The effect of using content adaptation compared to fixed exposure times can be visualized in a speed-exposure plot (Fig. 2).

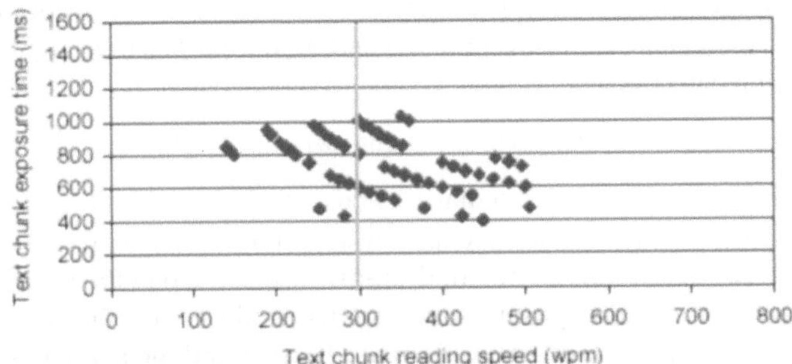

Fig. 2. Variations in reading speed for individual text chunks containing 1, 2, 3, 4 or 5 words (curved dotted lines) presented at a speed of 300 wpm when using content adaptation (data derived from the training text)

The plot is a result of presenting the training text at a constant speed of 300 wpm (vertical line in Fig. 2) but this time the formula for content adaptation is used instead (Eq. 2). The average word length including delimiters was set to 7,8 whereas the other variables were the same as those used for Fixed RSVP. Worth to notice is that even though the exposure times now varies the variation in reading speed is actually smaller for content adaptation than for Fixed RSVP. When using content adaptation the exposure time for each text chunk is also *directly* related to the number of words and characters it contains. This approach is assumed to decrease cognitive demand while reading since the relation between exposed information and time for exposure is more natural.

3.2 Context Adaptation

In context adaptive mode the exposure time for each text chunk is based on the following: The result of content adaptation, the word frequencies of the words in the chunk and the position of the chunk in sentence being exposed. To begin with each word in the chunk is looked up in a lexicon with word frequencies. If the word is common it receives a weight lower than one and if it is rare or not in the lexicon it receives a weight higher than one. The following formula is used to calculate how the exposure time is affected by the word frequencies (Eq. 3):

$$time_2 = time_1 * ((wfrq_1+\ldots+wfrq_{nwrd})/nwrd)\ . \quad (3)$$

The formula uses the exposure time for content adaptation ($time_1$) and the word frequency weights for the words in the chunk (wfrq) as a basis for the result. The word frequency weights are added and divided by the number of words in the text chunk (nwrd). The product is then multiplied with the content adaptive exposure time to get the weighted exposure time ($time_2$).

The next step is to give the chunk less exposure time if it appears in the beginning of a sentence and more if it appears in the end. The following formula is used to calculate the text chunk exposure time depending on the position in and the length of the current sentence (Eq. 4):

$$time_3 = (time_2+time_2 * \tanh(swrd/savg))/2\ . \quad (4)$$

The formula uses the intermediary exposure time reached earlier ($time_2$), the number of words in the sentence exposed so far (swrd) and the average sentence length (savg). In order to get a smooth drop-off in speed along the sentence, a mean of the previously calculated exposure time and its product with the hyperbolic tangent (tanh) of the division of the number of exposed words and the average sentence length is calculated. The result is a varying text chunk exposure time ($time_{3}$), the effect of using context adaptation is illustrated with a speed-exposure plot (Fig. 3).

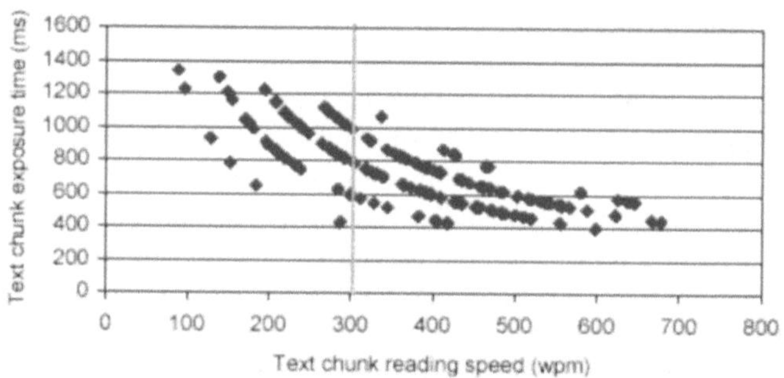

Fig. 3. Variations in reading speed for individual text chunks containing 1, 2, 3, 4 or 5 words (curved dotted lines) presented at a speed of 300 wpm when using context adaptation (data derived from the training text)

The plot is a result of presenting the training text at 300 wpm (vertical line in Fig. 3) when using context adaptation (Eq. 4). The average sentence length was set to 11,5 words and the word frequency weights ranged between 0,6-1,2. A lexicon with frequencies for the 10.000 most common words in a corpus of 11,9 million words (Press 97) was used to assign the weights according to a lognormal distribution [25]. Context adaptation causes larger variations in exposure time but is still assumed to decrease task load since the variations are supposed to better match the actual cognitive demand while reading.

In a recent experiment with an approach similar to adaptive RSVP Castelhano and Muter [3] found that the introduction of punctuation pauses within sentences was significantly favored compared to reducing common word exposure times. These findings are not necessarily pertinent to this evaluation since a combination of varying exposure times and punctuation pauses were used here and the number of words with reduced time was very small in the Castelhano and Muter evaluation (11 words).

4 BAILANDO: A Mobile Reader with Adaptive RSVP

Bailando is a prototype incorporating Adaptive RSVP capabilities that has been developed at Ericsson Research's Usability & Interaction Lab [8, 25]. Bailando is an acronym for: Better Access to Information through Linguistic Analysis and New Display Organization. The software was implemented on a Compaq iPAQ 3630 Pocket PC, a small Personal Digital Assistant (PDA) with a touch sensitive high-resolution color display. The size and weight was similar to a pocket book of approximately 250 pages (Fig. 4).

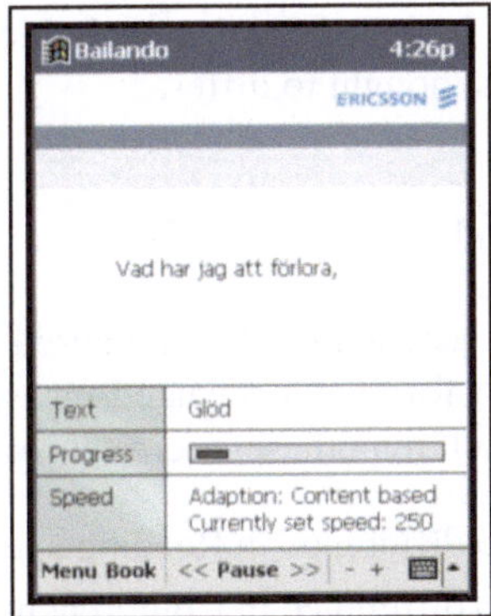

Fig. 4. The Bailando prototype running in reading mode on a Compaq iPAQ 3630 Pocket PC

Since ample screen space was available all the application controls were implemented in the graphical user interface (GUI). This also makes the Bailando software easier to run on other PDA's since button assignments differ between devices. The GUI contains controls to start, pause and resume the presentation. The presentation can also be paused and resumed by touching anywhere on the text presentation area of the screen. If the user feels he missed or did not understand a presented text window there is the possibility to go backwards (<<). It is also possible to skip text by going forward (>>). Reading speed is decreased (-) or increased (+) with the speed control buttons in steps of 10 wpm. In order to support memory of spatial location while reading there is a progress bar, the inclusion of a completion meter has previously been found to increase the user preference for the RSVP format [21].

The window width of the RSVP display was 25 characters wide with the text presented left justified in a 10-pt. sans-serif typeface; no legibility enhancing techniques were used. The text presentation window width was chosen on basis of the findings from an earlier evaluation [8]. Bailando supports all the three forms of RSVP that have been described in this paper. In all RSVP modes the text chunks that contained punctuation marks received an addition of 250 ms to their exposure times. In Fixed RSVP mode a blank screen was inserted for 250 ms between each sentence. In Content and Context adaptive RSVP mode a blank screen was inserted for 250 ms only if necessary, otherwise a delay of 250 ms was added to the exposure time of the text chunks exposing the sentence boundary.

5 The Usability Evaluation

The aim with the evaluation was to see how traditional text presentation, Fixed RSVP, Content adaptive RSVP and Context adaptive RSVP affected the ability to read on a mobile device. It was important that the same device was used for all conditions since the look and feel of the hardware was likely to bias the assessment. Long and short texts were included in the evaluation since the experience of extended and brief reading was thought to differ.

5.1 Method

In order to assess the effects caused by reading long and short texts using the four presentation formats a repeated-measurement experimental layout was adopted. The following null hypotheses were set for reading long and short texts:

- No difference in Reading speed
- No difference in Comprehension
- No difference in Task load

The hypotheses were tested in the SPSS V10.0 software using the repeated-measurement General Linear Model (GLM). The significance level was set to 5% and the level of multiple comparisons was Bonferroni adjusted.

Design. A within-subject Latin-square design was employed. Four experimental conditions were formed where each subject read one long and one short text using each presentation format. The combinations of long (A-D) and short (a-d) texts were fixed creating four text pairs Aa, Bb, Cc and Dd (Table 1). The text pairs were balanced against condition and order generating sixteen combinations. Each subject was randomly assigned to one of the sixteen combinations.

Subjects. Sixteen paid subjects participated in the experiment. They were all enrolled with the criteria that they were fluent in Swedish and had a self-reported interest in reading. The subjects had a mean age of 25 and half of them were male. All were computer literate and seven had some previous experience of using a PDA. Nine of the subjects had corrected vision and two were left-handed.

Apparatus. For all conditions a Compaq iPAQ 3630 was used, although the iPAQ offers far more screen space than the average mobile device it was chosen since it is easier to do prototyping on and it still shares many of the properties typical for smaller devices. Bailando was used for all RSVP conditions and the initial speed of the text presentation was always set to 250 wpm but the subjects were allowed to alter the speed at any time. Two commercial programs were chosen for traditional text presentation, Microsoft Reader for long texts and Microsoft Internet Explorer for short texts. It would probably have been more experimentally sound to use a single program for all traditional text presentation but it wouldn't have been realistic.

The reason for including two different programs was their intended *context of use*; the MS Reader is custom-made to present longer texts such as e-books whereas the MS Explorer is designed to present shorter web-content such as news articles. The MS Reader uses page-turn buttons to move between the pages whereas the MS Explorer uses *both* page-turn buttons and a scroll-bar, in addition to this the MS Reader also utilize a legibility enhancing technique called ClearType (Fig. 5).

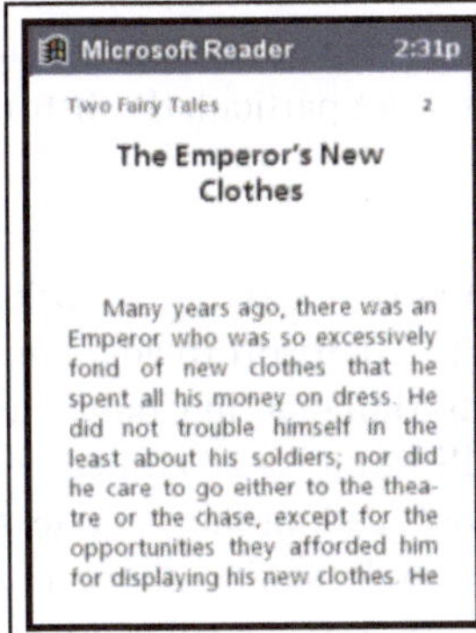

Fig. 5. The interface of the Microsoft Reader for long texts (left) and the Microsoft Internet Explorer for short texts (right)

Texts. Four long fiction texts and four short news articles where chosen to be included in the experiment. One shorter fiction text was also used as a training text. Texts in Swedish with different readability ratings were chosen. The readability rating was measured with LIX [1], a readability rating developed for Swedish texts that is comparable to the Flesh index [23] for English (Table 1).

Setting. The experiment took place in a dedicated usability lab outfitted with audio and video-recording facilities. While reading the subject was seated in a comfortable chair in a room separated from the experimenter by a one-way mirror. Before the experiment started each subject had some time to get acquainted with the facilities in order to create a relaxed, and consequently controlled, setting.

Table 1. Texts used in the experiment

Training text	Glöd	Annette Kullenberg, chapter 1.	705	25
Long text A	Röda rummet	August Strindberg, chapter 3.	4272	37
Long text B	Nils Holgersson	Selma Lagerlöf, chapter 1 and 2.	4230	27
Long text C	Valarnas sång	Wally Lamb, chapter 1.	4326	31
Long text D	Bara Alice	Maggie O´Farrel, chapter 1.	4170	29
Short text a	Makedonien	Dagens Nyheter 2001-06-03	430	44
Short text b	Mellanöstern	Dagens Nyheter 2001-05-27	384	54
Short text c	Alcala	Dagens Nyheter 2001-06-02	628	40
Short text d	Lundin Oil	Dagens Nyheter 2001-05-18	365	49

Instructions. Each subject received instructions before the experiment that pointed out that it was the applications and not the individual performance that were being tested. All were encouraged to ask questions whenever they wanted and also told that they could terminate the experiment at any time if they felt uncomfortable. Written instructions were administered before each session that described the principal features of the current user interface, what kind of text they were going to read and how long time it was likely to take. The subjects were particularly instructed to read at a pace as *comfortable* to them as possible.

Training. To begin with each subject participated in two training sessions. In the first session the subject read the training text using the Microsoft Reader and in the second session the subject read the same text again using Bailando in Content adaptive RSVP mode. The subjects did not train on using the MS Explorer as all had prior experience of using it on desktop computers. The idea behind reading the same text twice was to give the subjects an early success experience and making them more willing to experiment with the interface. After the training sessions the subject was introduced to the questions in the inventories and filled them in.

Procedure. Four experimental conditions were administered and each condition was divided into two sessions. In the first session the subject read a long text and filled in the inventories, in the second session the subject read a short text and filled in the inventories. Between the first and second condition the subject had a 15-minute break and between the second and third condition the subject had a 45-minute lunch break. Between the third and fourth condition the subject had a 15-minute break again. The total participation time for each subject was around five hours.

Inventories. After each experimental session there were two inventories to fill in. The first inventory was a comprehension test made up of multiple-choice questions with three alternatives, for long texts there were ten questions and for short texts there were five. The second inventory was the NASA-TLX Task Load Index [9], which was administered to check Mental, Physical, and Temporal demands, as well as Performance, Effort and Frustration levels. The NASA-TLX Task Load Index inventory was chosen as a measure of cognitive demand since the results would then be comparable to a previous evaluation were the measure was rewardingly used [7].

5.2 Results

All subjects completed the experiment and there were few problems with understanding what to do or how to do it. The subjects that were left-handed had some minor problems using the MS Explorer as the hand sometimes obscured the screen while using the scroll-bar. However, these subjects quickly resorted to using the page-turn buttons instead. The presentation of the results is divided into three sections: Reading speed, Comprehension and Task load. Under each section the null hypotheses set for long and short texts is tested.

Reading Speed. Reading speed was calculated as words read per minute based on the *total* time it took for the subjects to read a text including all kind of interruptions like pauses, regressions, speed changes etc.

Long texts. Adaptive RSVP improved reading speed some but the null hypothesis regarding no difference in reading speed between the conditions when reading long texts was kept (Table 2).

Short texts. The null hypothesis regarding no difference in reading speed between the conditions when reading short texts was rejected since the main factor for reading speed was significant ($F[3,45]=8.4$, $p=0.04$). Pair-wise comparisons revealed that all RSVP conditions increased reading speed significantly ($p \leq 0.002$) compared to using traditional text presentation with the MS Explorer (Table 2).

Table 2. Reading speed in words per minute (wpm) for long and short texts

Condition	Long texts		Short texts	
	Avg.	*Std. Dev.*	*Avg.*	*Std. Dev.*
MS Reader / MS Explorer	242	80,4	157	53,2
Fixed RSVP	249	58,5	212	46,5
Content adaptive	260	51,2	213	36,8
Context adaptive	258	79,5	203	43,9

Comprehension. Comprehension was computed as percent of correctly answered multiple-choice questions. For long texts there were ten questions and for short texts there were five.

Long texts. The null hypothesis regarding no difference in comprehension between the conditions when reading long texts was kept. Content adaptive RSVP showed the best results but the differences between the conditions were small (Table 3).

Short texts. The null hypothesis regarding no difference in comprehension between the conditions when reading short texts was kept. MS Explorer gave the best result but the differences between the conditions were small (Table 3).

Table 3. Comprehension scores in percent correct (%) for long and short texts

Condition	Long texts		Short texts	
	Avg.	*Std. Dev.*	*Avg.*	*Std. Dev.*
MS Reader / MS Explorer	73	19,6	70	21,9
Fixed RSVP	75	17,9	66	26,0
Content adaptive	76	17,5	59	19,9
Context adaptive	71	21,9	66	18,9

Task Load. Task load was calculated as percent of millimeters to the left of the tick mark on a 120-mm scale. The factors were not rated within each other.

Long texts. The null hypothesis regarding no difference in task load between the conditions when reading long texts was rejected as all main factors except Physical demand became significant ($F[3,45]\geq 5.2$, $p\leq 0.014$). Pair-wise comparisons revealed that the use of RSVP resulted in significantly higher ($p\leq 0.014$) task loads compared to using traditional text presentation with the MS Reader. Content adaptive RSVP decreased task load ratings and the only factor that was rated significantly higher compared to the MS Reader was Frustration level ($p=0.002$). Context adaptive RSVP also decreased task load, but in a different way. The only significantly higher factor compared to the MS Reader was Temporal demand ($p=0.001$) (Fig. 6).

Short texts. The null hypothesis regarding no difference in task load between the conditions when reading short texts was kept (Fig. 6).

6 Discussion

That no significant differences were found within the RSVP formats indicate that the effects caused by adaptation were quite small. Nevertheless, when the results obtained for RSVP were compared to those for traditional text presentation some significant differences were found. The discussion will primarily be based on these findings for reading speed and task load.

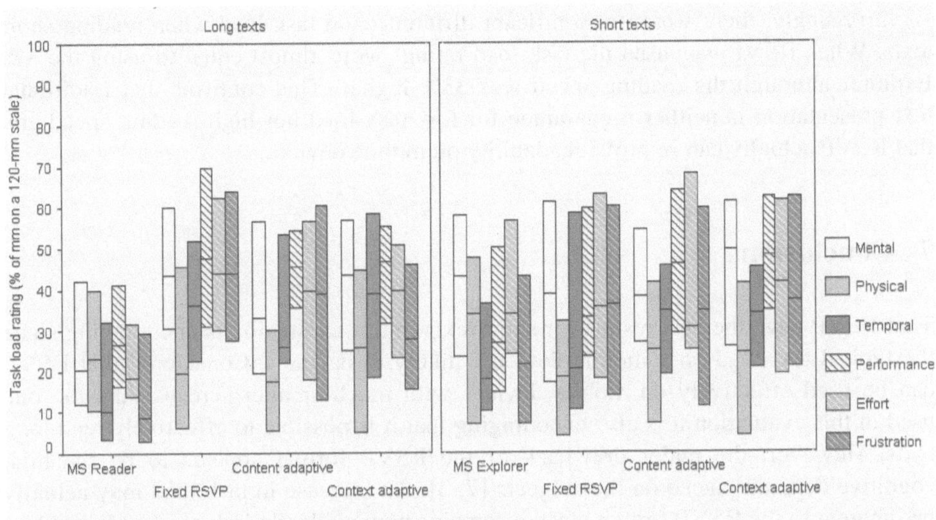

Fig. 6. NASA-TLX Task Load Index ratings for long and short texts with median, 25- and 75-percentile represented. Lower ratings are better

Since no significant differences in reading speed were found for long texts RSVP appears to be just as fast as traditional text presentation with the MS Reader. The lower reading speeds obtained for short texts is not very surprising as news articles are generally harder to read [1, 23]. However, the significant differences between using RSVP and the MS Explorer is very surprising since RSVP increased reading speed by 33%. RSVP has in previous studies not been much faster than traditional text presentation but these findings indicate that RSVP really can offer a significant increase in reading speed on mobile devices.

The task load ratings obtained for Fixed RSVP and the MS Reader were close to identical to those obtained for Fixed RSVP and paper-book in the Goldstein et al. evaluation [7]. This is surprising since the subjects now selected a comfortable reading speed. This may imply that the size of the assumed trade-off between reading speed and cognitive demand is small for the RSVP format, quite contrary to the size of the well-established speed-accuracy trade-off [24].

Adaptive RSVP was supposed to decrease task load and it seems to have worked as expected for long texts. Compared to the MS Reader the only factor significantly higher for Content adaptation was Frustration level. Probably some words were not exposed for a duration that matched the time needed for cognitive processing, it is however encouraging that even the most straightforward form of adaptation actually decreased task load. In Context adaptive mode the only significant factor compared to the MS Reader was Temporal demand. A probable cause for this is that the variations in exposure time were too large. However, the relation between what was exposed and the time for exposure was probably sound since the Frustration level decreased compared to Content adaptation. It seems that although the variations were too large they probably occurred at the right places.

Surprisingly, there were no significant differences in task load when reading short texts. When RSVP was used the task load ratings were almost equal to using the MS Explorer although the reading speed was 33% higher. This confirms that traditional text presentation is neither a guarantee for low task load nor high reading speed and that RSVP actually can improve readability on mobile devices.

7 Conclusions

That RSVP gave the best results for short texts in this evaluation is encouraging since the typical text read on a mobile device is likely to be short. Considered that RSVP can be used effectively on mobile devices with much smaller screens than the one used in this evaluation it is also encouraging that it is possible to effectively read long texts. However, the major drawback of the RSVP format appears to be the high cognitive demand placed on the subjects [7, 3]. An increase in task load may actually be inherent to the RSVP format since it remains high evidently independent of reading speed. Therefore, the most important finding in this evaluation is that task load can be decreased by using adaptation. Both adaptive algorithms were found to decrease task load for most factors, this means that better adaptation may decrease task load even further.

There is really no reason to use RSVP when traditional text presentation can be used efficiently. In this evaluation RSVP was found to be just as effective as the MS Reader but also significantly more demanding. However, when traditional text presentation becomes ineffective it seems to become more demanding as well. The MS Explorer was found to be just as demanding to use as RSVP but much slower. It is probably when traditional text presentation becomes ineffective, like it is on most mobile devices of today, that RSVP can offer a real improvement in readability. A slight increase in task load may then also be acceptable if it is compensated by an increase in efficiency, particularly since time often equals money in the mobile context. Therefore, since adaptation evidently decreases task load, Adaptive RSVP can be seen as one small step towards an improved readability on mobile devices.

References

1. Björnsson, C.H. (1968). *Läsbarhet*, Stockholm, Liber.

2. Bruijn, O. and Spence, R. (2000). Rapid Serial Visual Presentation: A space-time trade-off in information presentation. *In Proceedings of Advanced Visual Interfaces*, AVI'2000.

3. Castelhano, M.S. and Muter, P. (2001). Optimizing the reading of electronic text using rapid serial visual presentation. *Behaviour & Information Technology*, 20(4), 237-247.

4. Duchnicky, R.L. and Kolers, P.A. (1983). Readability of text scrolled on visual display terminals as a function of window size. *Human Factors*, 25, 683-692.

5. Ericsson, T., Chincholle, D. and Goldstein, M. (2001). Both the device and the service influence WAP usability, *IHM-HCI2001,* Volume II, Usability in Practice by J. Vanderdonckt, A. Blandford and A. Derycke (Eds.), Short paper, 10-14 September, Lille, France, 79-85.

6. Forster, K.I. (1970). Visual perception of rapidly presented word sequences of varying complexity. *Percep. Psychophys. 8,* 215-221.

7. Goldstein, M., Sicheritz, K. and Anneroth, M. (2001). Reading from a small display using the RSVP technique. *Nordic Radio Symposium,* NRS01.

8. Goldstein, M., Öqvist, G., Bayat-M, M., Ljungstrand, P. and Björk, S. (2001). Enhancing the reading experience: Using adaptive and sonified RSVP for reading on small displays. *Proceedings of Workshop on Mobile Devices* at IHM-HCI 2001, Lille, September 2001.

9. Hart, S.G. and Staveland, L.E. (1988). Development of Nasa-TLX (Task Load Index): Results of empirical and theoretical research. *Human Mental Workload,* by P.A. Hancock and N. Meshkati (eds.). Elsevier Science Publishers, B.V.: North-Holland.

10. Joula, J.F., Ward, N.J. and MacNamara, T. (1982). Visual search and reading of rapid serial presentations of letter strings, words and text. *J. Exper. Psychol.: General*, 111, 208-227.

11. Juola, J.F., Tiritoglu, A., and Pleunis, J. (1995). Reading text presented on a small display. *Applied Ergonomics*, 26, 227-229.

12. Just, M. A., and Carpenter, P. A. (1980). A theory of reading: From eye fixations to comprehension. *Psychological Review*, Vol. 87, No. 4, pp. 329-354.

13. Just, M.A., Carpenter, P.A. and Masson, M.E.J. (1982). *What eye fixations tell us about speed-reading and skimming.* (Eye-lab Technical Report) Carnegie-Mellon University.

14. Kang, T.J., and Muter, P. (1989). Reading Dynamically Displayed Text. *Behaviour & Information Technology*, 1989, Vol. 8, No. 1, 33-42.

15. Masson, MEJ. (1983). Conceptual processing of text during skimming and rapid sequential reading. *Memory and Cognition*, 11: 262-274.

16. McCrickard, D.S., Catrambone R. and Stasko, J.T. (2001). Evaluating Animation in the Periphery as a Mechanism for Maintaining Awareness. *In proceedings of INTERACT 2001*, 148-156.

17. Mills, C.B. and Weldon, L.J. (1987). Reading text from computer screens. *ACM Computing Surveys,* Vol. 19, No. 4, ACM Press.

18. Muter, P. and Maurutto, P. (1991). Reading and skimming from computer screens and books: The paperless office revisited? *Behavior & Information Technology*, 10, 257–266.

19. Muter, P. (1996). Interface Design and Optimization of Reading of Continuous Text. In *Cognitive aspects of electronic text processing*. H. van Oostendorp and S. de Mul (Eds.). Norwood, N.J.:Ablex.

20. Potter, M.C. (1984). Rapid Serial Visual Presentation (RSVP): A method for studying language processing. In *New Methods in Reading Comprehension Research.* Kieras, D.E. and Just, M.A. (Eds.). Hillsdale, N.J.:Erlbaum.

21. Rahman, T. and Muter, P. (1999). Designing an interface to optimize reading with small display windows. *Human Factors,* Vol. 1, No. 1, 106-117, Human Factors and Ergonomics Society.

22. Robeck, M.C. and Wallace, R.R. (1990). *The Psychology of Reading: An Interdisciplinary Approach,* Lawrence Erlbaum Associates, Hillsdale, New Jersey, Second Edition.

23. Tekfi, C. (1987). Readability Formulas: An Overview, *Journal of Documentation* 43(3) 261-73.

24. Wickens, C. D. (1992). Engineering psychology and human performance, 2nd edition, Chapter 8, Harper Collins Publishers Inc., New York.

25. Öqvist, G. (2001). *Adaptive Rapid Serial Visual Presentation.* Master's Thesis, Department of Linguistics, Uppsala University.

Mobile and Collaborative Augmented Reality: A Scenario Based Design Approach

L. Nigay[1,*], P. Salembier[2], T. Marchand[2], P. Renevier[3], and L. Pasqualetti[4]

[1] University of Glasgow, Department of Computer Science,
17 Lilybank Gardens, Glasgow G12 8QQ,
laurence@dcs.gla.ac.uk
[2] GRIC-IRIT, Université Paul Sabatier,
31062 Toulouse cedex 7,
salembier@irit.fr, marchand@irit.fr
[3] CLIPS-IMAG, Université de Grenoble 1,
38041 Grenoble cedex 9,
philippe.renevier@imag.fr
[4] FT R&D-DIH/UCE,
38-40 rue G. Leclerc, 92794 Issy-lesMoulineaux,
laurence.pasqualetti@francetelecom.fr

Abstract. In this paper we address the combination of the physical and digital worlds in the context of a mobile collaborative activity. Our work seeks to accommodate the needs of professional users by joining their physical and digital operational worlds in a seamless way. Our application domain is archaeological prospecting. We present our approach of the design process, based on field studies and on the design of scenarios of actual and expected activities. We then describe the conceived and developed interaction techniques via the MAGIC platform.

1 Introduction

Augmented Reality (AR) seeks to smoothly link the physical and data processing environments. This is also the objective of other innovative interaction paradigms such as Ubiquitous Computing, Tangible Bits, Pervasive Computing and Traversable Interfaces. These examples of interaction paradigms are all based on the manipulation of objects of the physical environment [10]. Typically, objects are functionally limited but contextually relevant [18]. The challenge thus lies in the design and realisation of the fusion of the physical and data processing environments (hereafter called physical and digital worlds). The object of our study is to address this issue in the context of a situation of unconstrained mobility (archaeological prospecting). Context detection and augmented reality are then combined in order to create a personalised augmented environment.

* On sabbatical from the University of Grenoble, CLIPS Laboratory, BP 53, 38041 Grenoble Cedex 9, France

F. Paternò (Ed.): Mobile HCI 2002, LNCS 2411, pp. 241–255, 2002.
© Springer-Verlag Berlin Heidelberg 2002

In this paper, we explain a design approach for mobile AR systems. We illustrate our approach, by presenting the outcomes of the design steps of our MAGIC platform. MAGIC is both a hardware and software platform, which carries out the fusion of the two worlds, the physical and digital worlds. Although our goal is to design and develop generic interaction techniques, we base our study on a specific mobile fieldwork: archaeological prospecting, whose main characteristics are presented in the following section.

2 Archaeology and Computer Support

In the domain of archaeology the recent progress in 3D technology have made it possible to develop advanced modelling tools for constructing detailed virtual representations of archaeological sites. Augmented Reality (AR) and Virtual Reality (VR) technologies have also allowed the design of tools to enhance the overall experience of a visitor to a site by providing an AR reconstruction of ancient monuments [19]. With respect to the tasks performed by the archaeologists themselves, computing resources have mainly been used in post-excavation tasks (for example referencing the shards and the objects in a database, reconstructive modelling of the site after excavation, etc.). But, as a fieldworker, the activity of the archaeologist has rarely been considered, and the problem of designing tools in order to support on-site activities has barely been addressed[2].

Within archaeological procedures, prospecting is very important because it principally influences whether excavation at a site will take place. It provides a global overview of the environment, including a systematic census of the archaeological clues. The goal is to check the state of the archaeological archives and the potential of the sites [7]. The archaeological evaluation must fulfil the following requirements: establish the location of the deposit, find the boundaries of the site, define their nature (habitat, necropolis, etc.), evaluate the density of the structures, and date the site [4]. Prospecting is done by a group of archaeologists and consists initially of a ground analysis based on a systematic division of the zone. If necessary, the archaeologists consult a specialist whose opinion will determine whether prospecting will continue. Currently, this consultation with the expert is inefficient, because it requires repeated trips to and from the site by the archaeologists. The prospecting requires long journeys between sites, where the topographic characteristics are often poorly known. The long distances of the journeys cause problems since they result in asynchronous interaction with distant specialists who possess specialised knowledge whose nature cannot be anticipated a priori. The characteristics of the prospecting activities (e.g., the co-operative process, the nature of shared information, the type of interaction), appear to be representative of the co-operative activities found in mobile situations.

[2] For example Ryan & Pascoe [22] in the context of the design of mobile systems for fieldworkers.

The objective of our study is to understand the use of mobile supports and services required in a collaborative situation for a user's task in the real world, justifying the fusion of the physical and digital worlds. In this context, two properties are fundamental: transparency of the interaction and ubiquity.

- *Transparency* enables the users to focus their attention on the task to be concluded and not on the usage of the tool. It is thus advisable not to separate the human from his physical environment while using the tool. The aim is to create a seamless operational field between the physical and digital worlds.
- *Ubiquity* stems from usage of the mobile supports. Users wish to access services and to collaborate with their colleagues at any moment and place. The objective is thus to design groupware on mobile supports. The user is no longer a prisoner of the workstation on his desk for collaboration and communication.

Transparency and ubiquity contribute to the current effort of HCI to Universality, i.e. the computer tool is integrated into the physical environment, and must be accessible from everywhere and by all.

3 Design Approach

3.1 Design Methods for Mobile Augmented Reality Systems

Because Augmented Reality seeks to smoothly link the physical and digital worlds it has expanded our view of the nature of interactive systems. But it must be emphasised that these new technologies also introduce a change in the conventional HCI design methods. Building effective AR applications for instance is not simply a matter of technical wizardry. Beyond the HCI classical design approach, mobile AR makes it compulsory to use a multidisciplinary design approach that embeds complementary methods and techniques for the design and evaluation phases [14]. This is due to the fact that physical objects of the user's environment take an increasing role in the design. We will focus here on two aspects that we have found relevant when designing AR applications: field study and scenario based design.

- *Field study*: So far mostly naturalistic analysis of activities, inspired by ethnomethodology [15] and francophone ergonomics, have been applied in designing AR applications. These approaches place a strong emphasis on the necessity for field studies and participatory design [9, 14]. According to these situated design approaches, accounting for the context in which the users are involved means that the design methodology cannot be limited to a description of the task at hand. The methodology has to consider the whole environment (physical, technical and social) in which the task is performed. This requires an in depth analysis of users' activities in order to understand their successful work practices and to identify the limitations of the current way of working.
- *Scenario*: As a way to concretely embody a view of users' actual and future activities, scenarios have proven very useful in AR design projects [16]. In particular scenarios enable the description of how AR devices would affect the way professional users carry out their individual and collective activities. In addition

various scenarios show the importance of supporting collaboration between the design team and the users.

3.2 Functional Role of Scenarios in Designing Interactive Systems

Scenarios are widely used in different disciplines, including HCI, Software Engineering, Information Systems, Requirements Engineering. Scenarios can account for the use of various resources, and the context of the current or projected activity of identified or potential users [5, 11, 12, 21]. Scenarios can occur in varied forms and fulfil several functions during the design process:

- Scenarios are simple and accessible to a variety of actors who participate in the development of a new technology [5].
- Scenarios provide a common language to the set of participants engaged in the design process. In particular, they are supposed to facilitate the exchanges with the future users who in return will contribute to enrichment of the design options [13].
- Scenario based design processes may boost the participation of the members of the project, and consequently may extend the scope of what is realisable and increase creativity [1].
- Scenarios provide concrete descriptions of design solution, which can be considered at various levels of detail. The initial scenarios are in general very brief. They provide a description of the system by indicating the tasks that the users can or must carry out, but without explaining in detail the way in which the tasks are performed [6].

3.3 Applying a Scenario Based Design Approach

The method we adopt follows some principles of scenario based-design methods, but is also further enriched by the methodological contributions of the francophone ergonomics tradition ("real task analysis" vs. "activity analysis"). Figure 1 presents the design and realisation steps of our method. As shown in Figure 1, the empirical analysis stage includes two steps:

- First we design scenarios based on an analysis of the real task (explanation by the end-users of her/his relevant work phases, away from the work setting).
- Second we design scenarios based on an analysis of the activity (observation and video recording of the activity of the end-users on site).

From these two steps, we derive a set of requirements. These requirements serve as the basis for the specification of the future system: We first specify the functions that address the requirements. Such functional specifications of the system are then evaluated on the basis of scenarios called "projected scenarios": To do so the specified functions are integrated in existing task and activity scenarios. This step is a still in the planning stage since the system is not developed. The interaction techniques are then designed based on the final functional specifications. After a developmental step, tests are conducted in the work setting in order to assess the functional properties of the system as well as its usability. The outcome may lead to modification of the functional specifications and the designed interaction techniques, as part of an iterative design approach.

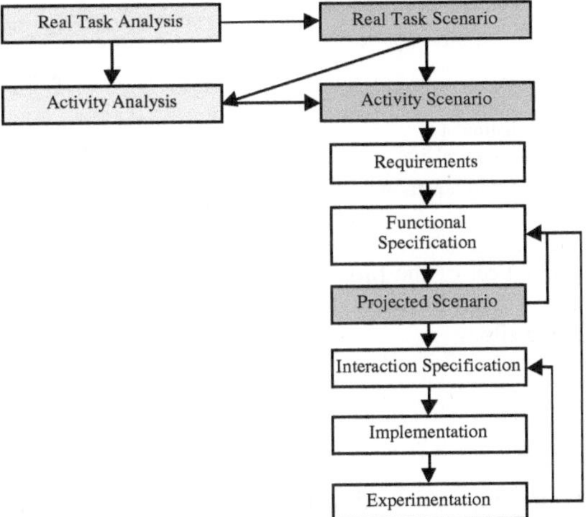

Fig. 1. Design and realisation steps

4 Design and Realisation of the MAGIC Platform

Having explained the design method, we now illustrate it by presenting the main results of each step of the design and realisation of our MAGIC platform (Mobile, Augmented reality, Group Interaction, in Context). We base our study on a specific mobile fieldwork: archaeological prospecting.

4.1 Task Analysis

Several levels of representation of the data resulting from the task analysis were used: global representation, narrative format of description, sequential chart. For reasons of brevity, only the global representation will be presented in this paper. The task analysis is completely described in [16]. At a general level of description, prospecting is a set of analysis operations of the ground, which will allow a final archaeological evaluation of a site. The site analysis consists of a detailed exploration of the ground, allowing specialists including material scientists, and geologists to evaluate the archaeological value of a site. As shown in Figure 2, this evaluation requires repeated journeys between the site and the research centre by archaeologists in order to provide relevant data to various specialists.

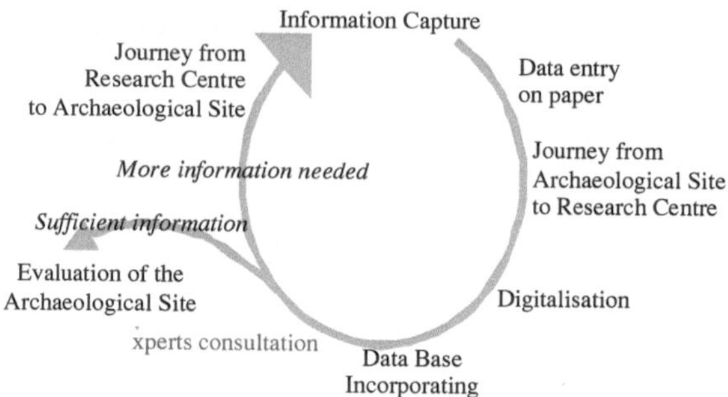

Fig. 2. Global representation of the prospecting task. This representation highlights the cycle (information capture - digitalisation - expert consultation) necessary to evaluate the archaeological value of a site.

4.2 Activity Analysis

An activity analysis was carried out during real (as compared to simulated) prospecting; the individual and collective activities were video recorded and commented upon by the archaeologists. As in task analysis step, several levels of representation of the collected data were used: a narrative description of activities, a graph representation of activities, a dynamic representation of displacements of the archaeologists and finally an integrated representation. We present here the narrative description, the graph of activities and the integrated representation.

4.2.1 Narrative Description of Activities

The narrative mode of representation is based on the concept of "meaningful event" for the archaeologists and/or analysts, i.e. a significant unit of action or a significant event occurring in the environment [24]. Table 1 gives an example.

4.2.2 Graph of Activities

As shown in Figure 3, this mode of description of the archaeologists' activities shows graphically the fragmentation of their activities into sub-tasks, the repetitions, the redundancies and highlights the prevalence of some tasks as compared to others. At a structural level this mode of description also makes it possible to reveal underlying regularities: sequences of actions, the communication patterns and collaboration between the users.

4.2.3 Integrated Representation of Activities

This representation enables the integration of the set of data collected during the **activity analysis (meaningful events expressed narratively, group members' displacements, video, snapshots, etc.). The representation is dynamic (Flash**[3] animation) and easy to handle. Figure 4 presents snapshots of such an animation[4].

[3] Flash Macromedia software.

[4] Available at: http://www.irit.fr/GRIC/public/scenario.htm

Table 1. Narrative description

Sequence K.	
Topics	Diffusion of contextual information, Geo-positioning, Data Entering, Collective Evaluation
Actor(s)	Three Archaeologists represented by the letters: V., C. & M.
Artefact(s)	Map
Output	Discovery of highly significant element
11:04:30 C finds a metal piece. She brings it to V who stops her activity immediately. They return towards the place where the metal piece was discovered and C tries to find the exact place where it was found, but the exact location remains approximate. Everyone gathers around this discovery. 11:06:45 the element is approximately located on a map and analysed by V while M and C seek other clues at the area of discovery. A first analysis is conducted of the element position to direct the search of other clues near this element (C: "Well, it must have gone down anyway")	
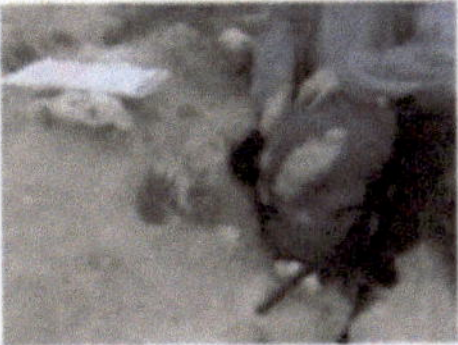  	

4.3 Requirements and Functional Specification: Projection of Existing Scenarios

The field analysis highlights the inherent limitations of the current situation for several types of activity, in particular for collaborative activities involving mobility:

- data capture and data entry of a significant element with its context, its mode of collection and its storage in a mobile and time-constrained context,
- contextual evaluation of an element, which requires communication between archaeologists in the field as well as between an archaeologist in the field and a distant expert,
- consultation or diffusion of previously analysed elements or situations in order to carry out an individual evaluation or to share information.

We define a set of functions, which may potentially overcome the problems encountered by the archaeologists in the current situation. These functions are then integrated in the scenarios of task and activity ("envisioning" phase) in order to project a new form of the activity as a solution to the present limitations. Different projection phases can be performed: alternative models of the future can then be analysed, debated and compared with the users. At this stage, interaction issues are not taken into account. The goal of the projection of the scenarios of activity, such as the scenario presented in Table 2, is to inform and to guide the functional specification step.

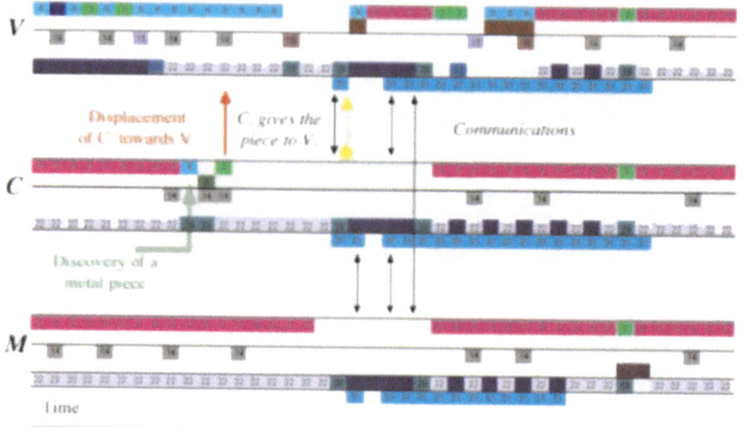

Fig. 3. Analysis of a part of the sequence of events associated with "the discovery of a metal piece by C". The data collected covers continuous observations of the actions and communications of the three archaeologists. This figure shows the graphical depiction of the activities of three archaeologists (V, C and M) in the context of the discovery of a metal piece. The three series of temporally ordered lines represent the sequences of activities performed by the archaeologists. The codes (number and colours) represent the categories of significant actions (hand-written notes, metric statement, photo, reading a map, etc.). The green arrow indicates the moment when the metal piece was discovered. The red arrow shows the move of archaeologist C. towards his colleague V. The yellow arrow expresses exchange of an element between archaeologists, and the black arrows show the communication between two archaeologists.

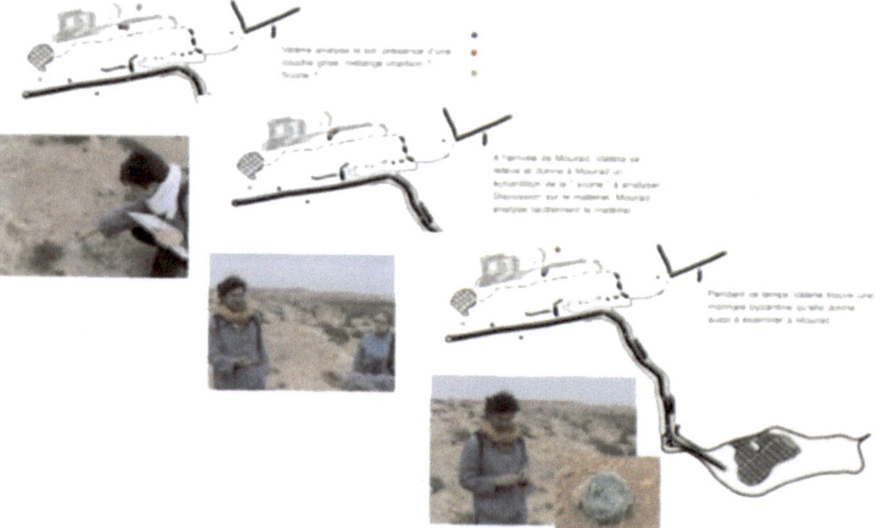

Fig. 4. Snapshots of the dynamic presentation of a scenario. These three snapshots show the animation at three different times in the sequence G of the scenario. It shows both the narrative scenario and some video snapshots of the activity, and also the positions of the archaeologists on the site as well as the dynamics of the displacements of each archaeologist (represented by the coloured points).

Table 2. Projection of a scenario of activity: envisioning phase

Sequence R. Projected	
Topics	Collective evaluation, Distribution of information about a previously discovered object.
Actor(s)	Three archaeologists represented by letters C., V. & M.
Artefact(s)	Editing system and location dependent information
Output	Sharing asynchronous knowledge
Case of location dependent information and database consultation. C arrives in front of a small rock structure already quoted but she does not know its significance. The system shows her that the rock structure on the ground was already described by V. She can thus read the notes, access the drawings and the pictures of this structure.	

4.4 Interaction Specification

Based on the functions integrated in the so-called "projected scenarios", different interaction techniques can be designed. The described interaction techniques are those that we have developed: the MAGIC platform. As part of a user-centred iterative process, experimental evaluation of MAGIC may lead to a redesigning of the interaction techniques.

MAGIC is both a hardware and software platform. In order to explain the interaction techniques, we first describe the hardware we used for those techniques. MAGIC is an assembly of commercial pieces of hardware. The MAGIC platform includes a Fujitsu Stylistic pen computer. This pen computer runs under the Windows operating system, with a Pentium III (450 MHz) and 196 Mb of RAM. The resolution of the tactile screen is 1024x768 pixels. In order to establish remote mobile connections, a WaveLan network by Lucent (11 Mb/s) was added. Connections from the pen computer are possible at about 200 feet around the network base. Moreover the network is compatible with the TCP/IP protocol. The hardware platform also contains a Head-Mounted Display (HMD), a SONY LDI D100 BE: its semi-transparency enables the fusion of computer data (opaque pixels) with the real environment (visible via transparent pixels). Secondly, a GPS is used to locate the users. It has an update rate of one image per second. The GPS at the University of Grenoble (France) has an accuracy one metre and the one at the Alexandria archaeological site 5 centimetres (Egypt, International Terrestrial Reference Frame ITRF). The GPS is also useful for computing the position of a newly found object and removed objects. Finally, capture of the real environment by the computer is achieved by the coupling of a camera and a magnetometer. The magnetometer is the HMR3000 by Honeywell that provides fast response time, up to 20 Hertz and high heading accuracy of 0.5° with 0.1° resolution. The camera orientation is therefore known by the system. Indeed the magnetometer and the camera are fixed on the HMD, in between the two eyes of the user. The system is then able to know the position (GPS) and orientation (magnetometer) of both the user and the camera. Figure 5-a shows a

MAGIC user, fully equipped: the equipment is quite invasive and suffers from a lack of power autonomy. Our goal is to demonstrate the feasibility of our interaction techniques by assembling existing commercial pieces of hardware and not by designing specific hardware out of the context of our expertise. For a real and long use of the platform in an archaeological site, a dedicated hardware platform must clearly be designed. For example we envision solar energy power, using retractable solar energy panels attached to the pen computer.

Based on the above described hardware platform, we designed and developed interaction techniques that enable the users to perform the functions associated with the "projected scenarios". The functions of the "projected scenarios" manipulate objects that are either digital or physical. Examples of digital objects are those contained in an online materials catalogue while examples of physical objects are materials such as metals or even a block-note that the archaeologist is using. As highlighted by our notation ASUR [8], some physical objects are tools while others are the focus of the activity. The block-note is a tool while the discovered items are the focus of the archaeological activity. During the design, we can only change the physical objects used as tools. In our design, we decided to use a pen computer instead of the block-note, and based the design of the pen computer display on the paper metaphor. Such a design solution will skip the digitalisation phase of Figure 2.

Having made the design choice of a pen computer, interaction techniques must be designed in order to let the users manipulate the two types of objects: physical and digital. For the flexibility and fluidity of interaction, such manipulation is either in the physical world or in the digital world. We therefore obtain the four cases of Table 3, by combining the two types of objects and the two worlds: the physical world (i.e., the archaeological field) and the digital world (i.e., the screen of the pen computer). The MAGIC interaction techniques cover graphical interaction on the pen computer (case (3) in Table 3) as well as the two cases of mixed interaction (cases (1) and (2) in Table 3).

Table 3. Four cases of interaction techniques

	Interaction with a physical object	Interaction with a digital object
In the physical world	Interaction purely in the real world	Mixed interaction (2)
In the digital world	Mixed interaction (1)	Interaction in the digital world (Graphical HCI) (3)

Figure 5-b presents the graphical user interface displayed on the tactile screen of the pen computer. The graphical user interface is fully described in [20]. Interaction on the pen computer (case (3) in Table 3) is related to communication, coordination and production functions as described by the CLOVER functional model of groupware [23]. For example, coordination between users relies on the map of the archaeological site, displayed in a dedicated window (at the bottom left of Figure 5-b). It represents the archaeological field: site topology, found objects and archaeologists current locations known by the GPS.

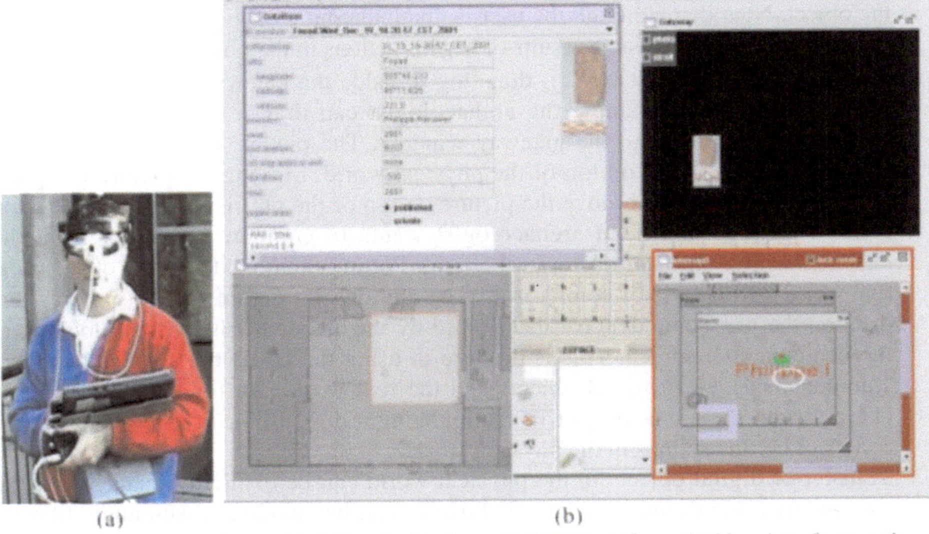

(a) (b)

Fig. 5. (a) A user wearing and holding the hardware MAGIC platform (b) User interface on the pen computer

The two cases of mixed interaction (cases (1) and (2) in Table 3) imply (i) that physical objects must be manageable in the digital world (case (1)) (ii) that digital objects must be manageable in the physical world (case (2)). Transfer of objects in between the two worlds is therefore necessary. To do so we designed a generic interaction technique, a gateway that plays the role of a door between the physical and digital worlds. As a door belongs to two rooms, the gateway exists in both worlds:

- the gateway is an area of the physical world, delimited by a rectangle displayed in a semi-transparency Head-Mounted Display (HMD),
- the gateway is a rectangular area in the digital world, on the pen computer screen.

Concretely the gateway is simply a window both displayed on the HMD (Java JFrame) on top of the physical world and on the pen computer screen (Java JInternalFrame). The gateway on the pen computer screen is the window at the top right of Figure 5-b.

Objects in the gateway are visible on the HMD (i.e., in the physical world) as well as on the pen computer screen (i.e., in the digital world):

- If the object is physical (case (1)), the object is transferred to the digital world thanks to the camera (fixed on the HMD, between the two eyes of the user). The real environment captured by the camera is displayed in the gateway window on the pen computer screen as a background. We allow the user to select or click on physical objects: we therefore call this technique "the clickable reality". Before taking a picture, the camera must be calibrated according to the user's visual field. Using the stylus on screen, the user then specifies a rectangular zone thanks to a magic lens [3] (kind of camera lens). The cursor displayed on the pen computer screen is also displayed on top of the physical world. The corresponding specified zone (magic lens), displayed in the gateway window on screen and on the HMD, corresponds to the physical object to be captured. The picture is then stored in the shared database along with the description of the object as well as the location of the object. Note that although the user is manipulating a magic lens using the stylus on screen, s/he perceives the results of her/his actions in the physical world.

- If the object is digital (case (2)) dragging it inside the gateway makes it visible in the real world. For example the archaeologist can drag a drawing or a picture stored in the database to the gateway window. The picture will automatically be displayed on the HMD on top of the physical world. Moving the picture using the stylus on the screen will move the picture on top of the physical world. This action is for example used if an archaeologist wants to compare an object from the database with a physical object in the field. Putting them next to each other in the real world will help their comparison. The motion of a digital object (ex: drag and drop on the pen computer) can be viewed by the archaeologist without looking at the pen computer screen. This is because in using the HMD the archaeologist can simultaneously view digital objects and the real world. As for the previous case (1), although the archaeologist is manipulating a digital object, s/he perceives the results of her/his actions in the physical world.
Transfer of digital objects to the physical world can be explicitly managed by the user by drag and drop as explained above or can be automatic. Automatic transfer is performed by the system based on the current location of the user. This technique is called "augmented field" ("augmented stroll" in [20]). Because a picture or a drawing is stored along with the location of the object, we can restore the picture/drawing in its original real context (2D location). When an archaeologist walks in the site, s/he can see discovered objects removed from the site and specified in the database by colleagues. The "augmented field" is an example of asynchronous collaboration and belongs to the mobile collaborative Augmented Reality (AR) class of systems as defined in our taxonomy of AR systems [20]. The "augmented field" technique is directly derived from the functions of the "projected scenario" presented in Table 2. The technique is fully described in [20].

In this section we have described the interaction techniques that are developed in the MAGIC platform. For the same functions of the "projected scenarios", we could derive different interaction techniques. For example in the case of the "clickable reality" technique, instead of selecting the area using the stylus on the screen, the user could perform a gesture captured by a 3D localizer. For the "augmented field" technique ("projected scenario" of Table 2), the approach adopted is to assist the user by providing extra information about the physical field via a device carried by the user. Another approach [15] would be to augment the physical environment by directly projecting extra information on top of the physical world. This approach would be possible in a confined environment and is not possible in the case of archaeological prospecting. Finally we would like to point out that without changing the designed interaction technique, some technical solutions adopted in MAGIC could be modified. For example instead of using a GPS and a magnetometer in the case of the "augmented field", markers could be positioned in the site and would communicate with a receiver carried by a user.

4.5 Implementation: Software Design

Having described the interaction techniques of the MAGIC platform, we now address their implementation. Our software design solution draws upon our software architecture model PAC-Amodeus [17]. The software architecture model is not new but the MAGIC implementation shows that it can be applied for the software design of a mobile collaborative Augmented Reality (AR) system. In the context of the overall software architecture of MAGIC [20], one point we would like to place particular emphasis on is the application of PAC-Amodeus architectural patterns for designing the software architecture. In particular one pattern [17] is dedicated to the implementation of multiple views: Use of a software agent to maintain visual consistency between multiple views. The gateway displayed on the pen computer screen and on the HMD is one example of multiple views. We therefore applied the Multiple Views architectural pattern to implement the gateway: Figure 6 shows the resulting three software agents.

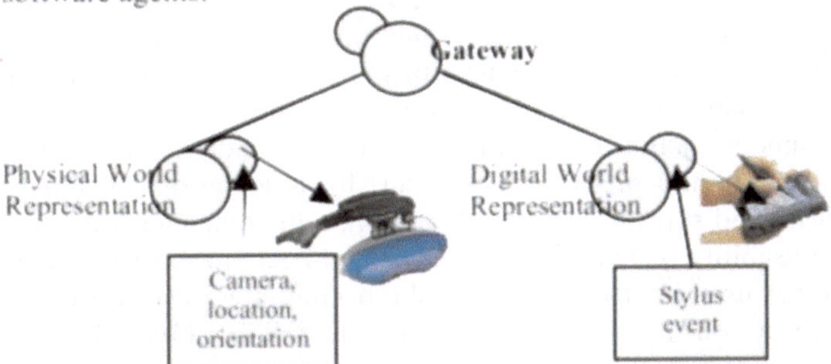

Fig. 6. Applying a PAC-Amodeus heuristic rule for the software design of the gateway: Three software agents implement the gateway.

5 Summary and Perspectives

The paper focuses on the design and implementation of mobile collaborative Augmented Reality AR systems. We have presented eight steps of a design method, depicted in Figure 1. In particular scenarios are shown to be very useful as a way to concretely embody a view of users' actual and future activities. Moreover an already established fact is that scenarios are very important in supporting collaboration between members of a multidisciplinary design team as well as between the designers and the final users. We applied our design method for the design and implementation of a mobile collaborative Augmented Reality (AR) system dedicated to a group of archaeologists prospecting fields. We have presented results of each design step including scenarios of present and future archaeological prospecting activities, designed interaction techniques and software architecture solutions of the system: the MAGIC platform. The next step is to experimentally test MAGIC in order to evaluate the usability of the developed interaction techniques (last step of Figure 1). To perform the experiments in the work environment, we must address several technical challenges, including the already mentioned problem of power autonomy, the ambient luminosity for on screen reading and the precision of the localizer (GPS).

As on-going work, we have two avenues:

- First we are currently developing a mobile collaborative game in order to show the generality of the interaction techniques of the MAGIC platform ("clickable reality" and "augmented field"). The game is based on the technique of barter, the mobile players discovering and exchanging between them physical objects that are augmented by magical power. Again scenarios of the game without the system and scenarios of how the game could be with the system ("projected scenarios") are being developed.
- Our second research avenue is methodological. During the design of the MAGIC platform, the applied design method underlined the central role of scenarios, including scenarios of actual activity as well as scenarios of future activity using the system. Additionally from the functions of the scenarios of future activity, we designed interaction techniques and their software design based on software patterns. We would like to extend this approach by establishing links between the scenarios and the software patterns. As shown in our previous work [17], ergonomic properties can be linked to software patterns. By tagging each scenario using ergonomic properties, we can, by association, establish links between scenarios and software patterns. Our work will complement the recent study by [2] where usability scenarios are linked to architectural mechanisms for the design of graphical user interfaces on a desktop computer. Our work is complementary because it considers the design of mobile collaborative AR systems. This study when completed will lead to a design method where scenarios are central and common materials for all the phases of the ergonomic design as well as the software design and implementation.

Acknowledgements
This work is supported by France Telecom R&D, under contract HOURIA I. We wish to thank our partners from the CEA (Centre d'Etude d'Alexandrie in Egypt) for welcoming us. Special thanks to G. Serghiou for reviewing the paper.

References

1. Ackoff, R.L.: Resurrecting the future of operations research. Journal of the Operations Research Society, Vol. 30, Issue 3. (1979) 189-199.
2. Bass, L., John, B.: Achieving Usability Through Software Architecture. In Conference Proceedings of ISCE 2001, IEEE Computer Society Publ., (2001) 684. CMU/SEI-2001-TR-005.
3. Bier, E. et al.: Toolglass and Magic Lenses: The See-Through Interface. In Conference Proceedings of Siggraph'93, Computer Graphics Annual Conference Series,. ACM Press (1993) 73-80.
4. Blouet, V.: Essais de comparaison de différentes méthodes d'étude archéologique préalable. Les nouvelles de l'archéologie, 58. (1994) 17-19.

5. Carroll, J. M.: The Scenario Perspective on System Development. In Carroll, J. M.: Scenario-Based Design: Envisioning Work and Technology in System Development. J. Wiley & Sons (1995).
6. Carroll, J.M.: Making use. Scenario-based design of computer interactions. MIT Press (2000).
7. Dabas, M., Deletang, H., Ferdière, A., Jung, C., Haio Zimmermann, W.: La Prospection. Paris : Errance (1999).
8. Dubois, E., Nigay, L., Troccaz, J.: Assessing Continuity and Compatibility in Augmented Reality Systems. To appear in International Journal on Universal Access in the Information Society, Special Issue on Continuous Interaction in Future Computing Systems. Springer-Verlag, Berlin Heidelberg New York (2002).
9. Geiger, C., Paelke, V., Reimann, C., Rosenbach, W.: Structured Design of Interactive Virtual and Augmented Reality Content. In Conference Proceedings of Web3D (2001).
10. Ishii, H., Ullmer, B.: Tangible Bits: Towards Seamless Interfaces between People, Bits and Atoms. In Conference Proceedings of CHI'97. ACM Press (1997) 234-241.
11. Jacobson, I.: The Use Case Construct in Object-Oriented Software Engineering. In Carroll, J. M.: Scenario-Based Design: Envisioning Work and Technology in System Development. John Wiley and Sons (1995) 309-336.
12. Kyng, M.: Creating Contexts for Design. In Carroll, J. M.: Scenario-Based Design: Envisioning Work and Technology in System Development. J. Wiley&Sons (1995) 85-107.
13. Kyng, M., Mathiassen, L.: Computers and Design in Context. MIT Press. (1997).
14. Mackay, W., Fayard, A.-L.: Designing Interactive Paper: Lessons from three Augmented Reality Projects. In Conference Proceedings of IWAR'98, International Workshop on Augmented Reality. Natick, MA: A K Peters, Ltd (1999).
15. Mackay, W., Fayard, A.-L., Frobert, L., Médini, L.: Reinventing the Familiar: an Augmented Reality Design Space for Air Traffic Control. In Conference Proceedings of CHI'98. ACM Press (1998) 558-565.
16. Marchand, T., Nigay, L., Pasqualetti, L., Renevier, P., Salembier, P.: Activités coopératives distribuées en situation de mobilité. Rapport Final Contrat FT R&D HOURIA Lot I (2001).
17. Nigay, L., Coutaz, 1997, J.: Software architecture modelling: Bridging Two Worlds using Ergonomics and Software Properties. In Palanque, P., Paterno, F. (eds.): Formal Methods in Human-Computer Interaction, ISBN 3-540-76158-6. Springer-Verlag, Berlin Heidelberg New York (1997) 49-73.
18. Norman, D. A.: The design of everyday things. London: MIT Press (1998)
19. Papageorgiou, D., Ioannidis, N., Christou, I., Papathomas, M., Diorinos, M.: (2000). ARCHEOGUIDE: An Augmented Reality based System for Personalized Tours in Cultural Heritage Sites. Cultivate Interactive, Issue 1, 3. (2000).
20. Renevier, P., Nigay L.: Mobile Collaborative Augmented Reality, the Augmente Stroll. In Little, R. Nigay, L. (eds): Proceedings of EHCI'2001, Revisited papers, LNCS 2254. Springer-Verlag, Berlin Heidelberg New York (2001) 315-334.
21. Rolland, C., Grosz, G.: De la modélisation conceptuelle à l'Ingénierie des besoins. In Encyclopédie d'informatique. Hermès: Paris (2000).
22. Ryan, N., Pascoe, J. Enhanced reality fieldwork: the context-aware archeological assistant. In Gaffney, V., et al. (eds.): Computer applications in Archeology (1997).
23. Salber, D.: De l'interaction homme-machine individuelle aux systèmes multi-utilisateurs. Phd dissertation, Grenoble University. (1995) 17-32.
24. Salembier, P., Zouinar, M.: A model of shared context for the analysis of cooperative activities. In Conference Proceedings of ESSLLI'98 (1998).

The Unigesture Approach

One-Handed Text Entry for Small Devices

Vibha Sazawal[1], Roy Want[2], and Gaetano Borriello[1,3]

[1] University of Washington,
Box 352350, Seattle, WA 98195-2350 USA,
+1-206-543-1695, +1-206-543-2969 (fax),
{vibha, gaetano}@cs.washington.edu
[2] Intel Corporation,
2200 Mission College Blvd, Santa Clara, CA 95052-8119 USA,
+1-408-765-9204,
roy.want@intel.com
[3] Intel Research Seattle,
1100 NE 45th St, Seattle, WA 98105-4615 USA,
+1-206-545-2530

Abstract. The rise of small modern handhelds mandates innovation in user input techniques. Researchers have turned to tilt-based interfaces as a possible alternative for the stylus and keyboard. We have implemented Unigesture, a tilt-to-write system that enables one-handed text entry. With Unigesture, the user can jot a note with one hand while leaving the other hand free to hold a phone or perform other tasks. In this paper, we describe a prototype implementation of Unigesture and outline results of a user study. Our results show that individual variations among users, such as hand position, have the greatest effect on tilt-to-write performance.

1 Introduction

Mark Weiser [1] defined a vision of ubiquitous computing that replaced the one-size-fits-all PC model with a variety of task-specific devices of different sizes. The "tab", a small credit-card sized device, the "pad", a tablet-sized device, and the "board," a whiteboard-sized device, would be peppered about a room, with tabs being the most plentiful.

The tab envisioned by Mark Weiser is now reality. The Hikari [2], designed at Xerox PARC, measures 80 × 64 × 14 mm. The Itsy [3], designed at Compaq, measures 118 × 64 × 16 mm. Both of these super-small computers have been exciting test beds for research, as well as useful devices in their own right.

As noted in [4] and [5], devices the size of the Hikari or smaller may be too small for traditional user input techniques. The Hikari has 4 buttons and smaller devices (so-called keychain computers [4]) are likely to have even less. There is no keyboard and little screen space on which to make use of a stylus. Entering text on these devices, despite their increasing computational power, is incredibly awkward.

F. Paternò (Ed.): Mobile HCI 2002, LNCS 2411, pp. 256–270, 2002.
© Springer-Verlag Berlin Heidelberg 2002

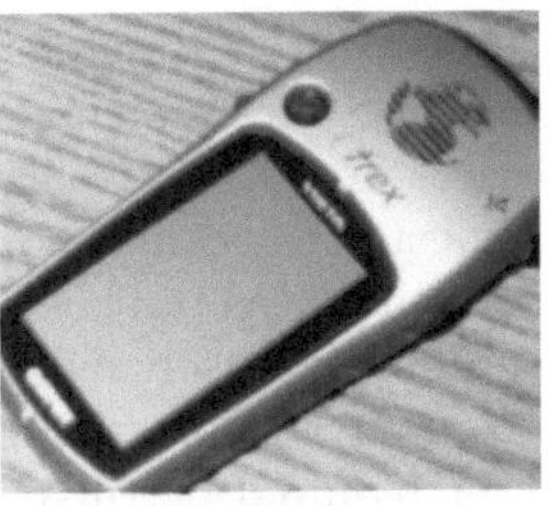

Fig. 1. Small devices with few or no buttons and varying amounts of screen space. User input on such devices can be very awkward. Pictured is the Hikari, a Timex Atlantis 100 TM watch (approx. 41mm in diameter), a Motorola TM BR850 pager (approx. 60 × 40 × 15 mm), and a Garmin eTrex Vista TM GPS receiver (112 × 51 × 30 mm).

Designers of the Hikari, the Itsy, and other small devices have turned to embodied user interfaces as a potential replacement for traditional techniques. Embodied user interfaces tightly couple the I/O of a device with the device itself. For example, a person may interact with a device by physically manipulating it – squeezing it, pushing it, and so forth. Devices such as the Hikari and the Itsy make use of an accelerometer that detects how much the device is tilted. Tilt-to-scroll and tilt-to-select have become useful input techniques for these devices.

In this paper, we tackle the problem of text entry for super-small devices. Text entry is a very important requirement for small device users. In December 2000, fifteen billion text messages were sent using phones alone [6]. As devices shrink in size, the challenges of enabling text entry grow. For example, users of the Garmin eTrex Vista pictured in Fig. 1 currently use the one button on the top left to enter text, which is incredibly cumbersome.

But what are designers to do with such small devices? There is no space for more than a couple buttons, and the small amount of screen space complicates use of a stylus. In addition, use of the stylus requires two hands, one to hold the small device and the other to hold the stylus. Two-handed use forces the user to discard whatever other objects the user was manipulating just to jot a note. A one-handed approach to text entry is an exciting alternative. With one-handed entry, the user's second hand is free to hold something else or perform other tasks.

Tilting has proved to be a successful mechanism to enable selection on tab-like devices, and so we decided to apply a similar approach to text entry. A tilt-based approach to writing offers an alternative to the stylus when screen space is small or nonexistent. Tilt-to-write also offers a one-handed method for text entry.

2 Unigesture: A Tilt-to-Write Method

Tilt-to-write is similar to tilt-to-select, but there are some important distinctions that stem from the differences between writing and selecting from a finite list.

The first is a difference between the number of commands entered in a single sequence. Even a short phrase can contain 8+ letters, whereas one is unlikely to select from 8+ finite lists all in a row. The second is the speed at which writing is expected to occur. People expect writing to occur at the speed of typing, whereas selecting need only occur at the speed of a mouse-based scroll bar. Thus, we have a problem – users will enter far more commands with tilt-to-write than with tilt-to-select, but users will expect their commands to be processed much faster than ever required by tilt-to-select.

Thus, we concluded that the most obvious implementation of tilt-to-write, where one had a list of 26 letters, and one tilted to scroll through that list, would be unacceptable. Even with word completion, such a system is likely to be cumbersome. Instead, we needed to produce a different scheme, one that could potentially offer speed, and would not strain the arm or wrist even after repeated commands.

For guidance we turned to successful methods for fast text entry in traditional systems. For example, the Quikwriting [7] method of stylus text entry abandons alphabet-like letters in favor of a zone-based text entry method. The letters are divided into zones. With the stylus, the user's first motion selects a zone, and the second selects a letter within the zone. Text entry on a cell phone is also a zoned input system. Here, each numeric button can be treated like a zone. There are 3 or 4 letters associated with each zone. One technique for writing with a phone involves pressing the associated button 1-4 times to reach the desired letter. For example, pressing the number "2" once writes "a", pressing the number 2 twice writes "b", and pressing the number 2 three times writes "c". Another approach, developed independently by Tegic Communications [8] and Dunlop and Crossan [9] [10], builds upon this simple phone text entry method. This new approach, commonly known as "T9" TM , requires only one press of each numeric button. When "2" is pressed, the phone has no idea whether "a", "b", or "c" is wanted. The phone must guess. As the user enters more letters, the phone can use a dictionary to help it make good guesses as to which word or partial word would best fit that pattern of button presses. By using inference, T9 manages to save the user from unnecessary button presses.

Unigesture applies these time-saving ideas to tilt-to-write. With Unigesture, the alphabet is divided into 7 zones. Space, or ' ', is allocated a zone of its own. To write a letter, the user tilts in the direction of a zone and then returns to center. An example is shown in Fig. 2. No additional gesture is needed to specify which letter is wanted within the zone. Rather, the Unigesture system will accept a sequence of tilt gestures and then attempt to infer a word.

Advantages of the Unigesture method include the variety of slight tilt gestures that can be used. Each tilt gesture returns the user to a neutral center position, which allows the arm and wrist to avoid strain. Slight tilts are also a rather quick motion, which suggests that Unigesture users should be able to write quickly once they have mastered the system. However, the inexperience we have with tilt-to-write systems results in a number of open design issues, each of which will have impact on Unigesture's possible success. Such design issues include the choice of

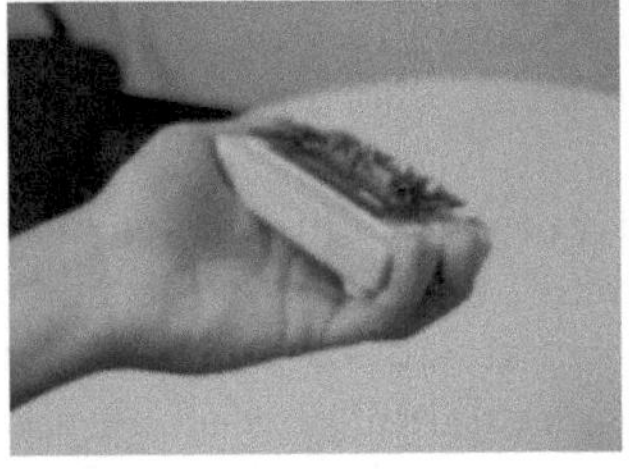

Fig. 2. Tilting a small device. Pictured is a user tilting up and then returning to center.

a letter layout and the minimum amount of tilt acknowledged by the system. It is also not known how users will respond to a tilt-to-write system. We decided to build a prototype of Unigesture in order to explore these issues.

3 The Unigesture Prototype

The Unigesture prototype is implemented using an Analog Devices ADXL202EB evaluation board. The evaluation board is mounted to a block of plastic with similar physical characteristics to the Hikari. The board converts 2-dimensional accelerometer data into an RS232-compatible format, which is then read by software on a desk-top PC. The device mockup is shown in Fig. 3.

Fig. 3. Device mockup. The Analog Devices ADXL202EB evaluation board is mounted to a block.

The software on the PC recognizes the gestures and provides basic features for user feedback. The recognition algorithm is quite simple, intentionally so in order to be easily implemented on a small computer, no matter how primitive. The PC prototype is implemented as a multi-threaded Windows application in C++.

3.1 Recognition Algorithm

We considered two approaches to recognizing accelerometer data. The first is a "rolling marble" approach: the user is presented with a letter map and a marble located in the center of it. Accelerometer data is mapped to a point on the screen. The result feels like a "game" – to write a letter, one guides the marble into a visually displayed zone.

The "marble" approach has many advantages. This system is very easy to implement, it gives the user a lot of control, and it also can make the act of writing feel like a game. However, we decided not to apply this approach for two reasons. One, this approach requires a visual display as an integral part of the recognition algorithm. We wanted our approach to be implementable on a small tab that may have a very limited display. In addition, we were worried that the system would be too slow, because the system required the user to carefully guide the marble into a zone.

We decided to go with a recognition approach that was entirely independent of a visual display or screen coordinates. We approach the problem by examining the change in acceleration over a set of contiguous data points (a "window") recorded by the accelerometer. The system will attempt to identify a gesture as the window changes with time. The idea is that the user will develop haptic memory, associating each zone with a tilt of specific degree, duration, and direction. Because we consider changes in acceleration, the user can hold the device in any comfortable position and measurements are made relative to it.

The "window" approach is dictated by the values of 3 variables. The data is first smoothed, and thus the first variable is the length, l_{smooth}, of the smoothing window. Then the change in acceleration is measured, in this case by taking the first difference of the smoothed data. The second variable is the threshold, δ, by which the first difference data must exceed in order to remove noisy data. The third variable is the length, $l_{\mathrm{recognize}}$, of the recognition window. The recognition window contains all of the first difference data that will be considered as a single unit. These three variables dictate the duration and degree of tilt that are required for recognition by the system.

For the prototype, we designed two versions of this algorithm. The first version, which we call the "deep tilt version", had a large l_{smooth}, a small δ, and a large $l_{\mathrm{recognize}}$. The result was a system that required a long, deep tilt. The system never misrecognized accidental motions as deliberate tilts. However, the long, deep tilts were somewhat slow to perform. The second version, which we call the "slight tilt version", had a small l_{smooth}, a high δ, and a small $l_{\mathrm{recognize}}$. The result was a system that required a short, fast slight tilt. The "slight-tilt" version required a good degree of coordination to operate; however, once learned, the user could write at a much faster speed.

Informal user studies caused us to refine our recognition algorithm. We found that users often made inadvertent motion directly after a deliberate motion. For example, a user would deliberately tilt up and then tilt back down to center, but then would tilt up and down inadvertently a few times before becoming stably still. To handle these unwanted oscillations, the system temporarily raises δ after every motion. The amount δ is increased is a function of the the degree of tilt formed when making the initial deliberate motion.

In the prototype, this algorithm is implemented using a pipe-and-filter architecture. Data from the accelerometer is first placed into a circular ring buffer. Each filter makes a transformation of the data and then places that transformed data into another circular buffer. The filters include a filter for smoothing the data, taking the first difference of the data, recognizing tilt motions from the data, and then recognizing zones from the tilt motions. A sequence of zones is then sent to the inference engine to produce a word.

3.2 Inference Engine

To reduce computational load, the system used a pre-computed dictionary tree that was keyed by zone sequence. In other words, the system would use a sequence of zones entered, which as a whole were intended to form a word, as a lookup key in the dictionary tree. The value returned for that key is an array of words that could all be formed given that same sequence of zones. The words and their ordering in each array were determined from sample texts. To reduce the search time, the array was ordered by the frequency with which each word occurred in the set of sample texts.

3.3 Letter Layout

For the prototype, we implemented two letter layouts. As stated above, the system had seven zones for characters. These zones can be named by direction: north, northwest, west, southwest, southeast, east, and northeast. The first letter layout placed commonly-used letters in zones that we believed were easier to reach. We had considered the cardinal directions: north, east, and west, to be the easier directions, with the diagonal directions somewhat harder. Thus, the first layout placed common letters in the north, east, and west directions, and uncommon letters in the diagonal directions. We refer to this layout as the "clustered layout", as the common letters were clustered together.

The second layout was designed to reduce the number of possible words that could be formed from each zone sequence. In this second layout, each vowel is placed in its own zone. For each pair of zones, there is only one likely pair of letters that would appear consecutively in an English word. The second layout reduces the number of words associated with every zone sequence, but perhaps at the price of making some common letters more difficult to enter. We refer to the second layout as the "spread-out" layout as the common letters are spread out across the zones.

The two letter layouts are shown in Fig. 4.

3.4 User Interface and Feedback

The user interface provides a small amount of visual feedback. This feedback is intentionally limited so that it would be transferable to a small screen. Users are presented with a visual letter map. This map is static and separate from feedback,

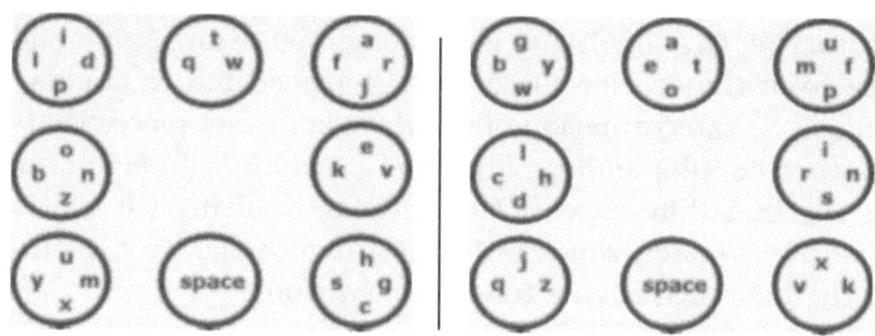

Fig. 4. Letter Layouts. Each layout consists of 7 zones for letters and the space zone. On the left side is the "spread-out" layout and on the right side is the "clustered" layout.

because the map is intended to be dropped as soon as the user memorizes it. Feedback from the system is implemented via a set of 8 radio buttons laid out in the same manner as the letter map and a set of check boxes that count the letters entered so far. A photo of the UI is shown in Fig. 5.

Users hold the mock handheld computer and tilt it. When a zone is recognized, the radio button associated with that zone darkens. A check box is also marked to show that a zone has been entered. As the user enters zones, the radio button of the current zone is darkened, while the number of checked boxes shows the number of zones that have been correctly entered so far.

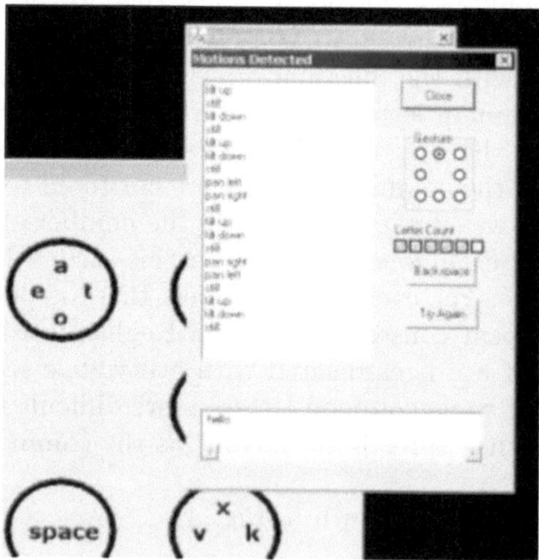

Fig. 5. UI for Unigesture prototype. The user is writing "hello there". In this example, the "clustered" layout was used. The system was photographed just after the last 'e' in "there" was written. The five checkmarks and darkened north zone can be seen. The word "there" will appear on the screen after the user tilts down to select the space zone.

After the user enters the zones for an entire word, the user inputs the "space" zone as a break. The zone sequence is then sent to the inference engine. The check boxes clear, and the user sees the word chosen by the inference engine. If the inference engine returns the word desired by the user, the user continues writing. Otherwise, the user clicks on the "Try Again" button. Clicking on the "Try Again" button repeatedly causes the system to traverse the array of words that match that zone sequence. One can easily imagine a more sophisticated system which remembered which words you write most often and would return them first. Other systems might allow you to manually set the frequency of words expected to be entered. For the purposes of our prototype, we did not need to implement these features, but one would want to implement them in a full-featured system.

The final piece of the UI worth noting is the "Backspace" button. Clicking on "Backspace" removes the last zone entered into the system. The last checkmark is removed as well, showing the user that the "Backspace" correctly took place.

On the PC prototype, the buttons are implemented as part of a Windows GUI and are selected via the mouse. In a full implementation, one would want these features to be buttons on the small device itself or implemented in some other fashion. For example, perhaps selecting the "space" zone multiple times could be used to tell the inference engine to "Try Again".

It is important to note that even though there are differences between our prototype and a full implementation, the prototype still forms an excellent base on which to explore the design space for tilt-to-write interfaces. For example, if all the users in our study hold the mockup a certain way, then we have excellent information with which to guide button placement on the real device. On the other hand, if users all hold the mockup differently, then it may be better to invest in non-button interfaces to "Backspace" and "Try Again". Because the prototype is clearly a mockup, users feel free to make suggestions, criticism, and comments, and feel freer with their use. This valuable interplay between user and device is often lost when users are given a "real device", which feels official and unquestionable [11].

4 User Study

4.1 Experiment Setup

Once the prototype was complete, we conducted a user study to learn more about how a tilt-to-write system would perform in practice. The first user test consisted of 12 users, all employees from Intel Corporation in Oregon. All 12 users used a desktop computer for 4 or more hours a day and owned a cell phone that they used regularly. 7 of the 12 users also owned a handheld computer that they used regularly. We were interested in participants who were already familiar with computers, and in particular, with technology "held in the hand".

The intent of the user study was to explore the design space of a possible tilt-to-write system. As described earlier, the Unigesture prototype has 2 letter map layouts ("clustered" and "spread-out") and 2 recognition versions ("slight-tilt" and "deep-tilt"). In our counterbalanced study, half the users received the "clustered" letter layout, while the other half received the "spread out" letter layout. Users were first trained in the Unigesture method. After their training, users were asked to write.

Because this method is intended for small devices, it was important to ask users to write messages of a realistic length. While users may appreciate the ability to jot down a quick note or the name of a GPS waypoint, they are unlikely to write their dissertation on a keychain computer. Thus, we defined 8 phrases, each containing between 10 and 12 characters. Users were asked to write these 8 phrases one at a time. Users were seated while they used the device.

Table 1. Phrases used in usability study

"on the web"	"thank you"	"be careful"	"try later"
"dont know"	"meet here"	"email jeff"	"how are you"

Half of these phrases were performed using the "deep tilt" recognition version and the other half with the "slight tilt" recognition version. Users would write the first 4 phrases using one version and then would write the latter 4 with the other version. They were permitted to write practice words to familiarize themselves with the new recognition algorithm if they chose.

Training consisted of the test-giver demonstrating how to use Unigesture, followed by the user trying out the device for a few practice words. Training was expedited if the user was already familiar with an inference system such as "T9". The user was allowed to practice as long as they wished, as long as the total training time was less than 20 minutes. The user was trained with either the "deep tilt" or the "slight tilt" recognition version, depending upon which version would be used for the first set of 4 phrases.

We tracked the user's mistakes and the time to complete each phrase. A mistake could have been one of two occurrences. The first is when the user tilts the device in a way that the system cannot recognize. When this happens, the user simply tries again. The second is when the system misrecognizes the user's tilts. When this happens, the user must press "Backspace" and then try again. Throughout the experiment, users were permitted to restart words if they chose. If the user restarted a word, only errors made during the last complete attempt of the word were counted.

As stated above, the purpose of the study was to explore the design space of a tilt-to-write system. We were interested in a number of questions:

1. Could all users write using the Unigesture method?
2. Would the performance of users vary with the level of tilt required by the recognition algorithm?
3. Would the performance of users vary with the choice of letter layout?
4. How would users react qualitatively to the Unigesture method?

The answers to these questions are explored in the next two subsections.

4.2 Quantitative Results

All 12 users could write phrases using the "deep-tilt" version of Unigesture. 10 of the 12 users could effectively write phrases using the "slight-tilt" version of Unigesture. Of these 10 users, 7 preferred the "slight-tilt" version, while 3 preferred the "deep-tilt" version. The 7 who preferred the "slight-tilt" version did so due to its speed and its ease on the wrist. The 3 who preferred the "deep-tilt" version preferred its accuracy.

Phrase-by-phrase, we can compare the number of errors made by users using the 2 layouts and the 2 tilt sizes. For the majority of phrases, there is not enough information to make statistically significant conclusions about the different versions. However, for one phrase, "try later", there was a significantly greater number of errors made using the "slight-tilt" version than the deep-tilt version. The table below shows the number of errors made by users when entering this phrase. Each number in the table refers to the quantity of errors made by a single user on the phrase; there were 3 users in each experimental condition. Statistical significance was determined by using a 2-factor analysis of variance (ANOVA) experiment [12] with the F-test at α=0.05.

Table 2. Number of errors made when writing phrase "try again"

	Slight-Tilt	Deep-Tilt
Clustered	0, 7, 8	1, 3, 0
Spread Out	4, 13, 4	1, 1, 2

For all phrases, the large bulk of variation in number of errors made per phrase occurs "within" treatments. In other words, variation due to the differences between each user had usually a greater effect on the number of errors made than the experimental setup that the user has been provided with.

We had initially believed that the diagonal zones would be more troublesome than the cardinal zones. However, this was not universally the case. Defining "trouble zones" as those zones that contained letters on which a user made 3 or more consecutive errors per phrase, 3 of the 12 users had problems with all 4 diagonal zones. But most users had trouble with a specific, customized set of zones, all based upon the way that the user held the device. For example, some right-handed users had trouble tilting to the right, and some left-handed users had trouble tilting to the left. 4 of the 12 users had troubles with the "space" zone.

4.3 Qualitative Results

Response to the Unigesture system was mostly positive. Users remarked that the system was "cool" and "clever". Only one user had something negative to say; this user found the system "annoying".

Users were asked which zones they found most troublesome. For the most part, the zones users identified as difficult were the ones where they had made their errors. However, users also reported that some zones felt awkward to select, but they were still able to enter a letter correctly.

Users with the "clustered" letter layout often had to click on the "Try Again" button in order to receive the word they had wanted. However, when asked, most users did not find the "Try Again" button annoying. The majority of users with the "clustered" layout felt that such a button was necessary, because they were giving the system incomplete information.

When finished with the experiment, we asked the user if they felt any physical discomfort. Of the 12 users, 2 users felt fatigued, and 1 user felt pain. These responses are definitely a concern. It is possible that users are exaggerating the tilt far beyond what they need to do to register a response with the system. Perhaps audio feedback or some other mechanism could be used to tell users to stop tilting. It is also possible that the device may need to be shaped in a more ergonomic form. Our study is too small to make any conclusive results, but it is clear that this concern is a primary issue for future work.

4.4 Numeric Map Experiment

We expected users in the experiment described above who performed well to still perform relatively slowly compared to an expert user, because the participants of the study were novices and had not memorized a letter map. In an attempt to measure the time spent looking at the letter map, we designed a second, smaller user study. In this study, we introduced a numeric layout that is not unlike a numeric keypad on a phone. (Numbers 5 and 9 are missing.) This numeric layout is quite easy to remember.

The experiment was designed as follows. The user was trained on a "slight-tilt" version of Unigesture and was given either the clustered layout or the spread-out layout. After training, users would write the first 6 phrases described in the earlier experiment, and they would also switch to the numeric layout and write two numbers. The experiment was counterbalanced.

Criteria for participation in this study was the same as the previous experiment. Of the 4 users who took part, 3 are employees at Intel Corporation in Hillsboro, Oregon and 1 is a computer science graduate student at the University of Washington in Seattle. The hypothesis was that the numeric layout would be familiar, and that users would perform faster when writing numbers. The numbers to be written were also intended to be familiar for the majority of the participants – the first number, 264, is a common phone exchange for Intel in Hillsboro, and the second, 97124, is the zip code for Hillsboro.

Table 3. Average time to completion per character (seconds)

	User 1	User 2	User 3	User 4
Phrases	3.6	8.9	5.8	5.4
Numbers	3.5	8.2	5.3	10.4

As in the earlier experiment, the letter layout used, either clustered or spread-out, did not affect performance very much. Users achieved times as low as 9 seconds for a 3-letter number (plus the space character), which implies a maximum data speed for novices as high as 2-3 seconds per character. However, fast users also obtained times as low as 33 seconds for a 10 character phrase, which can be considered equivalently fast as the speed achieved when writing numbers. In other words, taking away the letter map and using something simpler did not appear to speed up users. This suggests that the amount of time spent searching for letters on the letter map is not as high as we had previously expected. The speed per character only marginally increased with the simpler numeric map, if at all.

If looking for letters in a letter map is not the problem, what is taking up a user's time? Review of the user study suggests that users had zones that particularly troubled them, and selecting letters from these zones took up time. In addition, some users make mistakes and failed to notice this. The user would then end up repeating an entire attempt to enter the number or phrase.

4.5 Discussion of the User Studies

To summarize, we found the following results:

1. Users can tilt-to-write, and most of them have the dexterity to handle the "slight-tilt" version. However, a major concern involves whether the method results in unacceptable strain on the wrist. Future work is needed to resolve this concern.
2. There was not a significant difference between the performance of users given the "clustered" letter layout and the "spread out" letter layout. Rather, users each had different sets of trouble zones and easy zones, and while these sometimes coincided with the zone layouts in the "clustered" letter layout or the "spread out" letter layout, this was usually not the case.
3. Users did not spend much time looking at the letter layout. Time is spent instead trying to master tilting in trouble zones.

The usability tests show that everyone holds handheld devices in a slightly different way. Thus, the ideal system is one that can be customized to operate only in the non-trouble zones for the user. It is very easy to imagine a system that identifies which zones are troublesome and then arranges letters so that these zones are avoided. An automated trainer could also easily determine whether the user can write with the slight-tilt version or should be switched to the deep-tilt version. This customization approach could have its drawbacks, though – if a user borrows another's handheld computer, they may be unable to write with it!

The differences among how people hold handheld computers also complicates button placement. We propose that non-button techniques for entering "Try Again" and "Backspace" be used. Possible options include non-tilt motions, such as shaking or squeezing, which will be easily distinguishable from the tilt motions used for zone entry.

Potential augmentation options also include audio feedback. After the nth letter, the device could state "Zone [direction] selected. This is the nth letter in the word." Audio tones could also be used. Another potential option is force or tactile feedback, which could increase the haptic memorization of tilt motions.

In addition, it is possible that common phrases could be mapped to tilt motions instead of letters. These phrases would allow users to write faster at the cost of expressibility.

4.6 Issues Not Addressed in the User Studies

There are three issues that were not explored in the user studies. The first is the dictionary maintained by the system. All the words that users were asked to enter were in the dictionary. In practice, there would have to be a mechanism to enter words into the dictionary, and there would also have to be a second means for entry to "fall back to" in case one is writing a URL or other non-word. The obvious fall-back mechanism to Unigesture would be a Bigesture system. In a Bigesture system, the user would enter tilt gestures in pairs. The first tilt motion would select a zone, and the second would select a letter in the zone. However, we do not know if the obvious choice is the right one.

Another issue involves training. The user studies involved a human trainer. However, we believe that training would be easy to automate. In practice, people grasped the concept of Unigesture quite quickly. When they made mistakes, they tended to make similar types of mistakes, although the particular zones involved may be different.

A third issue is the need to support the entry of letters, numbers and punctuation by an integrated system. The obvious choice is to have a number of character layouts, not unlike the Quikwriting system [7]. A button or shake could potentially toggle the user between the layouts.

5 Conclusion

Tilt-to-write is ideally intended for "tabs", small credit-card sized computers with few buttons and small or no screen space for a stylus. Experiences with a tab-sized mockup show that users can learn a tilt-to-write system and can effectively enter text using only the hand that is holding the device.

In this paper, we explored a number of potential parameters of a tilt-to-write system. We offered users different letter layouts and provided systems that required varying degrees of tilt. We found that the natural variation among users had a greater effect on the number of mistakes made than any tweaking of a system parameter. Thus, an ideal system would customize itself to the user.

There are a great deal of open questions in this exciting area. What is the best way to customize a tilt-to-write system to the user? How can tilting be mixed with buttons or other input techniques to provide effective hybrid forms of text entry? How useful is letter-by-letter entry versus entry of entire phrases via one tilt? Can we teach users to tilt-to-write without causing too much wrist strain? These questions guide our future work.

6 Related Work

There are a number of stylus-based methods for text entry on handheld computers, such as Unistrokes [13] and Quikwriting [7]. All of these techniques require that one hand holds the device and the other hand holds the stylus. They also require ample screen space. Writing methods that could be implemented via tilt-based gestures include Dasher [14] and MDITIM [15]. Both of these methods seem quite promising. However, in their current forms, Dasher requires a great deal of visual attention and MDITIM requires a large number of tilts per character. It is possible that modifications to these systems could make them excellent tilt-to-write interfaces. Nonetheless, we propose that the results we found for Unigesture will also hold for any other non-trivial tilt-based method – the system must be customized to avoid troublesome tilts in order to be successful.

Although tilt-to-write is a new concept, the idea of tilt-based user interfaces is by no means new. Original work in this area was done by the University of Toronto's Chameleon project [16], Jun Rekimoto's tiltable screen project [17], the Extreme User Interface projects of Xerox PARC [2], and the Shakepad project from MIT [4]. More recent work in the area of tilt-based cursors was done by Weberg et al. [18], who coined the analogy "A Piece of Butter on the PDA Display".

New techniques for speeding up text entry on handheld devices is also an active research area. Work exists in word [19] [20] [10] and "partial word" prediction [6]. These techniques complement the Unigesture approach.

Acknowledgements

We thank Kurt Partridge, Jeff Hightower, and Steve Swanson for their helpful advice. We also thank all participants of our user study for their time and thoughtful comments. This research was supported by a DARPA Expeditions grant N66001-99-2-8924. This work was performed in part at Xerox PARC.

References

1. Mark Weiser. The computer for the 21st century. *Scientific American*, Sept. 1991.
2. Kenneth Fishkin, Anuj Gujar, Beverly Harrison, and Roy Want. Embodied user interfaces for really direct manipulation. *Communications of the ACM*, Sept. 2000.

3. William Hamburgen, Deborah Wallach, Marc Viredaz, Lawrence Brakmo, Carl Waldspurger, Joel Bartlett, Timothy Mann, and Keith Frakas. Itsy: Stretching the bounds of mobile computing. *IEEE Computer*, April 2001.
4. Golan Levin and Paul Yarin. Bringing sketching tools to keychain computers with an acceleration-based interface. In *Proceedings of ACM CHI '99 Extended Abstracts*, 1999.
5. Brad Myers, Scott Hudson, and Randy Pausch. Past, present, and future of user interface software tools. *ACM Transactions on Computer-Human Interaction*, March 2000.
6. I. Scott MacKenzie, Hedy Kober, Derek Smith, Terry Jones, and Eugene Skepner. Letterwise: Prefix-based disambiguation for mobile text input. In *Proceedings of ACM UIST '01*, 2001.
7. Ken Perlin. Quikwriting: Continuous stylus-based text entry. In *Proceedings of ACM UIST '98*, 1998.
8. Tegic Communications. T9 web page. URL: http://www.t9.com.
9. M.D. Dunlop and A. Crossan. Dictionary based text entry method for mobile phones. In *Proceedings of Mobile HCI 1999*, 1999.
10. M.D. Dunlop and A. Crossan. Predictive text entry methods for mobile phones. *Personal Technologies*, 4(2-3), 2000.
11. Yin Yin Wong. Rough and ready prototypes:lessons from graphic design. In *Proceedings of ACM CHI '92 Short Talks*, 1992.
12. Richard Hogg and Johannes Ledolter. *Applied Statistics for Engineers and Physical Scientists*. Macmillan Publishing Company, 1987.
13. David Goldberg and C Richardson. Touch-typing with a stylus. In *Proceedings of ACM INTERCHI '93*, 1993.
14. David Ward, Alan Blackwell, and David MacKay. Dasher – a data entry interface using continuous gestures and language models. In *Proceedings of ACM UIST '00*, 2000.
15. Poika Isokoski. A minimal device-independent text input method. Technical report, University of Tampere, 1999.
16. George Fitzmaurice, Shumin Zhai, and Mark Chignell. Virtual reality for palmtop computers. *ACM Transactions on Information Systems*, 11(3), July 1993.
17. Jun Rekimoto. Tilting operations for small screen interfaces. In *Proceedings of UIST '96*, pages 167–168, 1996.
18. Lars Weberg, Torbjörn Brange, and Åsa Wedelbo-Hannson. A piece of butter on the pda display. In *Proceedings of ACM CHI '01 Extended Abstracts*, 2001.
19. Fredrik Kronlid and Victoria Nilsson. Treepredict: Improving text entry on pda's. In *Proceedings of ACM CHI '01 Extended Abstracts*, 2001.
20. T Masui. Pobox: An efficient text input method for handheld and ubiquitous computers. In *Lecture Notes in Computer Science (1707), Handheld and Ubiquitous Computing, Springer Verlag*, 1999.

"Just-in-Place" Information for Mobile Device Interfaces

Jesper Kjeldskov

Department of Computer Science
Aalborg University, Denmark
jesper@cs.auc.dk

Abstract. This paper addresses the potentials of *context sensitivity* for making mobile device interfaces less complex and easier to interact with. Based on a semiotic approach to information representation, it is argued that the design of mobile device interfaces can benefit from spatial and temporal *indexicality*, reducing information complexity and interaction space of the device while focusing on information and functionality relevant *here and now*. Illustrating this approach, a series of design sketches show the possible redesign of an existing web and wap-based information service.

1 Introduction

With the launch of wap and browsers like AvantGo for PDAs, wireless information services targeted at mobile users are now beginning to emerge. Mobile technology is, however, still very much in its infancy regarding usability, network speed, display capabilities and computing performance. While the next generation of wireless devices (3G) promises a platform matching the performance of desktop computers, this does not in it self provide higher *usability* of mobile devices and applications.

Designing usable interfaces for tomorrow's mobile devices is not trivial but involves a series of challenges on human-computer interaction. Displays on mobile devices are small, means of input are limited and use-contexts are very dynamic. The usability of mobile information services consequently suffers from interfaces being very compact and cluttered with information, thus demanding the user's full attention. In mobile use-contexts (e.g. walking down the street or driving a car) requiring a change of focus away from activities in the real world can be problematic. If mobile devices are to have higher usability while users actually being mobile, the user interface must remain simple and the required interaction minimal. This call for new user interface design.

An example of a "minimal attention interface" is the use of audio output instead of graphical displays for GPS devices presented in [1]. Another example is the use of head-up displays in automobiles or augmented reality for mobile devices in general [2][3], allowing the users to stay focused on their physical surroundings. While these approaches focus on new means of *output* (sound, and transparent displays), new sources of *input* may contribute to minimal attention interfaces by supplying means for simplifying the user interface and reducing the demand for user interaction.

F. Paternò (Ed.): Mobile HCI 2002, LNCS 2411, pp. 271–275, 2002.
© Springer-Verlag Berlin Heidelberg 2002

2 "Just-in-Place" Information

One of the features of the next generation of wireless devices is the ability to access information about the user's *physical location* and combine this information with time. Besides facilitating spatial navigation, this feature more importantly enables the design of information services pushing information based on the continuously *changing* context of a mobile IT-user (see e.g. [4]). Instead of viewing the dynamic use context of mobile devices as a *problem*, context changes can be viewed as important means of *input* to mobile information services. Physical space becomes part of the interface, providing the user with information and functionality adapted to a specific location in space and time: "just-in-place" information.

It is, however, too simple to state that context sensitivity will automatically result in higher usability when this information becomes available in future 3G mobile devices. A number of questions have to be addressed. What are the potentials of context sensitivity in relation to HCI? How does the spatial and temporal context influence on information representation and how can general insight into this relation be transformed into specific interface design for mobile devices?

Some of these issues have already been addressed in recent literature on context awareness. See e.g. [5] and [6] for usability evaluations of context sensitive information service prototypes and [7] for a discussion on conceptualization and methodology for the design of such services based on field study analysis.

2.1 Representing "Just-in-Place" Information

The field of semiotics concerns the meaning and use of signs and symbols. A semiotic approach to the design of "just-in-place" information can contribute to a theoretical understanding of information representation and the design of minimal attention user interfaces, taking into account their spatial and temporal context.

From a semiotic perspective, information is viewed as representations of something else (their object). Faced with an interpreter, these representations cause a reaction or interpretation.

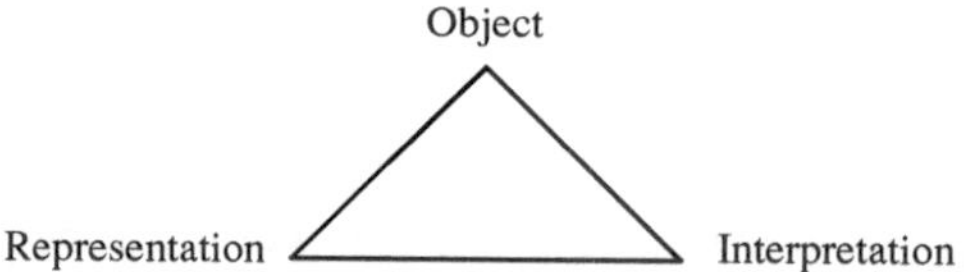

Fig. 1. Semiotics: relations between object, representation and interpretation.

The semiotics operates with three types of relationships between objects and representations: *symbolic* (conventional), *iconic* (similarity) and *indexical* (material/causal). Symbols and icons are ways of representing information *independent* of location in space and/or time like e.g. text and graphical illustrations in books or on web pages. Indexes, on the other hand, are ways of representing information with a *strong relation* to spatial and/or temporal location exploiting information present in the context. Indexical representations are e.g. used on signposts and information boards.

Locating information in time and space, symbolic and iconic representations can be converted into temporal and spatial indexical representations [8]. This is illustrated in fig. 2 and 3, showing two information representations of train departures.

The timetable shown in fig. 2 exemplifies symbolic and iconic information representation with no indexicality thus being valid independent of the user's location in space and time. The electronic timetable in fig. 3 exemplifies symbolic and indexical information representation being valid (and relevant) only at a specific location at a specific time. Increasing the indexicality from the traditional to the electronic timetable results in a significant reduction of information and need for interaction. Instead of having to look up departures from a specific location at a specific time (fig. 2), the user is presented with simple information adjusted to his physical location at the present time (fig. 3). Similar examples are numerous, but prohibited due to limited space.

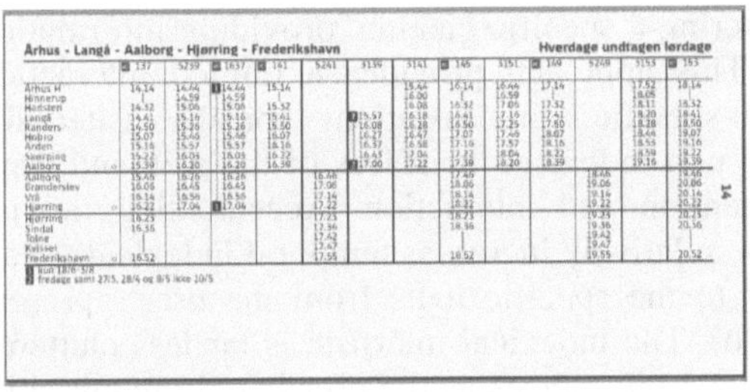

Fig. 2. Traditional timetable:
symbolic representation with no indexicality.

Fig. 3. Electronic timetable:
spatial and temporal indexicality

3 Designing an Indexical Interface for Mobile Devices

Developing prototypes that actually works and can be evaluated in real mobile use contexts is of huge importance to mobile HCI research. Discussing principles of context sensitive interface design, however, does not always require actual implementation. For the purpose of illustrating the principle of indexical interfaces for mobile devices, I have redesigned an existing information service using design sketches and scenarios. As subject for redesign I choose an information service for local cinemas as this service requires the user to browse movies in relation to time and place and supports mobile access via wap.

Using the website (fig. 4a), the user specifies day of week (temporal context) and is presented with a list of all movies on this day (even if in the past) at affiliated theatres. Accessing the service from a wap phone (fig. 4b, 4c), the user specifies a cinema (spatial context) and is presented with a list of movies. Selecting a title reveals playing times. While the website is straightforward to use, the wap-site requires a lot of clicking due to the division of information into a large number of sub-pages.

Spilletider for Lørdag d. 20. okt.		
Film titel	Spilletidspunkter	Biograf
A Knights Tale	16.00 18.40 21.20	Bio5
A.I.	15.00 18.30 21.20	SCALA
American pie 2	12.30 14.40 16.50 19.10 21.15	Bio5
American Sweethearts	16.15 18.50 21.10	SCALA
Anja og Victor (Kærlighed ved første hik 2)	14.20 16.20 19.00	SCALA
At klappe med en hånd	16.15 19.00 21.00	ASTORIA
Bridget Jones Diary	16.20 18.50 21.00	ASTORIA

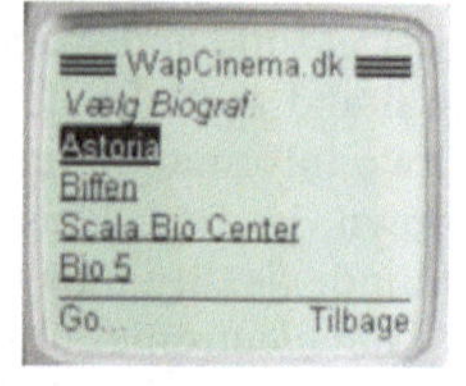

a) Web: "Movies playing tonight" b) Wap: "Choose Cinema" c) Wap: "Choose movie"

Fig. 4. An existing cinema information system for web (a) and wap (b and c)

3.1 An Indexical Mobile Device Interface

Figure 5 illustrates a *spatial and temporal indexical representation* of information designed to be available only when entering a specific cinema, providing information about movies playing *here, tonight*. This approach provides a limited interaction space of information meaningful only at specific spatial locations typically visited in a pre-defined sequence. Compared to the non-indexical interfaces of the web and wap-site, this design requires minimal attention and interaction. Accentuating movies playing within a limited window of time relatively increases temporal indexicality (a). Adding information about how to get to the specific halls from the user's present location increases spatial indexicality (b). The indexical interface is far less cluttered with information compared to web and wap interfaces as the user is only required be present at the cinema and select a movie from a limited list of choices.

a) Entering cinema b) Finding Hall A c) Outside Hall A d) Inside Hall A

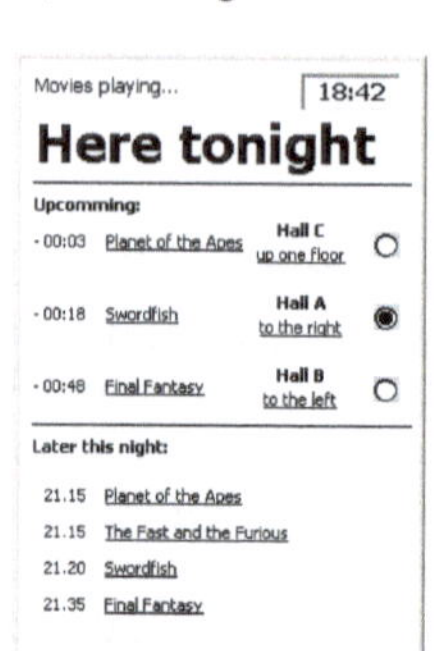

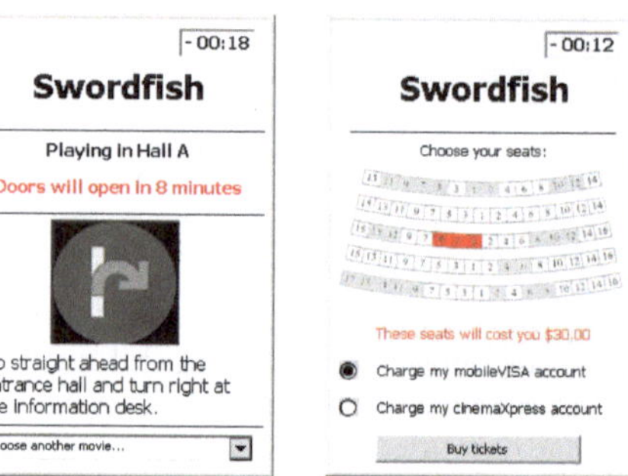

Fig. 5. A series of spatially and temporally indexical "just-in-place" interfaces for cinema information system.

One of the fundamental ideas of "just-in-place" information is that the content of the device changes when the context changes. A context sensitive mobile information service would thus typically consist of a *series* of indexical interfaces available at just the right places. The mobile cinema information service could involve the following use scenario and indexical interfaces: Having selected a movie (a), the device displays direction instructions (b). The representation of time changes from absolute to relative, increasing temporal indexicality. Outside Hall A, the user selects seats and buys tickets (c). When the movie begins, the user's mobile devices are silenced (d).

4 Conclusions and Further Work

Interface design for mobile devices can be simplified by increasing spatial and temporal indexicality, allowing information to be left out from the interface when present in the context. Simplifying the user interface on the basis of context reduces demands for user interaction and contributes to less required attention.

For further insight into the potentials of indexical interfaces, the presented design should be implemented and made subject for comparative usability evaluations. Such evaluations should preferably take place in context. A general test bed facilitating real world usability evaluations of information services based on location aware mechanisms embodying ideas such as the presented is currently being developed.

Finally, it should be noted that context is more than just space and time. Information about the user's tasks and activity etc. may contribute to further simplification of interfaces and required interaction. For interesting research on the use of contextual sensors for mobile interaction see [9].

References

1. Holland, S. and Morse, D. R.: Audio GPS: spatial audio in a minimal attention interface. Proceedings of MobileHCI, IHM-HCI, Lille, France (2001)
2. Feiner S., MacIntyre B., Höllerer T.: Wearing It Out: First Steps Towards Mobile Augmented Reality Systems in Onto, Y. and Tamura, H. (eds.): Mixed reality – Merging Real and Virtual Worlds, Springer-Verlag, London (1999)
3. Kjeldskov J.: Lessons From Being There: Interface Design For Mobile Augmented Reality. To appear in Qvortrup L. (ed.) Virtual Applications: Applications With Virtual Inhabited 3D Worlds, Springer-Verlag, London (2003)
4. Cheverst, K. et al.: Investigating Context-aware Information Push vs. Information Pull to Tourists. Proceedings of MobileHCI, IHM-HCI, Lille, France (2001)
5. Cheverst, K. et al.: Using Context as a Crystal Ball: Rewards and Pitfalls. In Personal and Ubiquitous Computing vol. 5:8-11, Springer-Verlag, London (2001)
6. Schmidt A. et al.: Context-Aware Telephony over WAP. In Personal and Ubiquitous Computing vol. 4:225-229, Springer-Verlag, London (2000)
7. Andersen, P. B. and Nowack, P.: The Space is the Computer. To appear in Qvortrup L. (ed.): Virtual Applications: Applications With Virtual Inhabited 3D Worlds, Springer-Verlag, London (2003)
8. Andersen, P. B.: Pervasive Computing and Space. Proceedings of IFIP WG8.1. Montreal, Quebec Concordia University, Pp. 106-125 (2001)
9. Hinckley K., Pierce J., Sinclair M., Horvitz, E.: Sensing Techniques for Mobile Interaction. In Symposium on User Interface Software and Technology, CHI Letters 2(2):91-100 (2000)

Pushing Web Pages into Personal Digital Assistants: Need, Tools and Solutions

E. Costa-Montenegro, F.J. González-Castaño, J. García-Reinoso, J. Vales-Alonso, L. Anido-Rifón

Departamento de Ingeniería Telemática, Universidad de Vigo, ETSI Telecomunicación, Campus, 36200 Vigo, Spain,
{javier,kike,reinoso,jvales,lanido}@det.uvigo.es

Abstract. In this paper, we analyze the problem of pushing Web contents into Personal Digital Assistants (PDAs), with minimum client participation. Typical HTTP or Javascript implementations require periodic client reloads, which impose an inefficient network utilization. We justify the existence of this problem, review existing alternatives, and propose a market-oriented implementation for wireless iPAQ Pocket PCs.

1 Introduction

Cell phone or PDA-based m-commerce [1,2] has a promising future. Wireless clients will allow customers to purchase items and services, and to receive (possibly profile-driven) store coupons, advertisements and guidance [3]. A m-commerce *service server* determines user position with the help of an *external location system*: 3G location [4], GPS or short-range beacons [3,5].

We can identify a number of situations in which a service server needs to push Web information into PDA clients in real-time. For instance, pushing location/profile-driven advertisements into user Web pages (when a customer arrives to a new mall store), sending alarms related to mall events (movie schedules, restaurant reservations), updating user location on a map or updating context information in a museum.

In the typical HTTP or Javascript approaches, the clients reload page contents periodically. In a mall with hundreds of users with wireless PDAs, this means a large network overload. To avoid it, reloads could be less frequent. However, the resulting elapsed time between service server events and customer notifications could not satisfy real-time constraints. Note that *a single push* at the right time would be enough. Also, note that, since users cannot be expected to be static, location information must be necessarily reloaded, and Web caching is useless.

For example, in the scenario in [5], if there is a new mall store each 10 m, there must be periodic updates each 12.5 s or less for a walking speed of 0.8 m/s. If a location update requires 10 KB, network load is 2,880 KB per user and hour. However, typical users are relatively static when shopping (say 50%

F. Paternò (Ed.): Mobile HCI 2002, LNCS 2411, pp. 276–280, 2002.
© Springer-Verlag Berlin Heidelberg 2002

of the time). Therefore, if the service server only updates information when user position changes, only 1,440 KB per user and hour would be necessary.

We are interested in developing a *connection server* for PDAs, to support server-side notifications. The target scenario is a m-commerce *intranet*, i.e. either external or internal malicious individuals cannot saturate the system, and all information is handheld-oriented (page transcoding [6] is not necessary).

This paper is organized as follows: In section 2 we describe two platforms for context-oriented mobile services and the technologies that are relevant to the purposes of the paper. Subsection*base software technologies* discusses PDA network programming capabilities. Section 3 presents our connection server architecture for Windows CE and its implementation. Finally, section 4 concludes.

2 Background

CoolTown [3] is a Hewlett-Packard distributed architecture for mobile PDAs with a WLAN interface and an infrared port. The user must aim the infrared port to location beacons, which push URLs into a PDA daemon that displays the corresponding Web pages.

In [5], the authors proposed an alternative system, m-Mall. M-Mall users carry a mobile PDA and a Bluetooth location *badge*. The badge interacts with an external Bluetooth location network, which provides the service server with real-time user position. Then, the service server may push context-oriented URLs into user PDAs via TCP/IP sockets. Thus, the main advantage over CoolTown is that no client action is required to generate user position (in CoolTown, the user must locate a location beacon and point his PDA). CoolTown authors argue that automatic detection of location information (without user participation) may have severe consecuencies in terms of nuisance (for example, if a Web page is suddenly supplanted by another one advertising frozen peas from a grocery store nearby). Obviously, we must expect the service server to be reasonable. Consider, for example, a museum, where the only updates are associated to new halls, and suppose that the current page is reloaded with a tiny icon at its bottom meaning "do you want to update context information"? We believe that asking the user to locate a beacon each time he enters a room full of visual distractions may be tiring, and signaling beacons with large red arrows unsightly. In any case, CoolTown may be more advantageous than m-Mall or viceversa depending on the specific application.

The philosophy in this paper is compatible with any of those two systems. In case of CoolTown beacons, GPS or 3G, location is obtained by some user interaction at the client side. The user would upload his position to the service server, which would push URLs back via TCP/IP sockets. In case of an user-independent location network, like [5], the client would simply receive URLs via TCP/IP sockets, when pushed by the service server.

Next, we will describe the technologies that are relevant to the purposes of this paper.

Hardware : Palm and Pocket PCs dominate the PDA market. Both systems support wireless TCP/IP socket programming: Palm Mobile Internet Kit [7] connects Palm handhelds with cell phones via cable or infrared ports, and there exist GPRS and WLAN Pocket PCs (like iPAQ) [8].

Base software technologies : We have identified some Java Virtual Machines (VMs) for PDAs, which would allow portable socket programming [9,10]. Java is relatively new in the PDA world. Pocket Internet Explorer does not include a VM, and the examples in [9,10] do not include all typical libraries. For instance, when this paper was written, Kada did not support swing libraries, which would be useful to let programs open Web pages themselves. Another reason to discard those VMs is the high PDA CPU load demanded, when compared to small binaries generated from high level languages.

Most PDA applications run on PalmOS or Windows CE. Since we are interested in mass-market m-commerce applications, we focused on those OS. Of course, there exist alternatives. For example, QNX is a real-time OS. Version 6.1 includes the IBM j9 Java VM, which is part of Visual Age Micro Edition 1.3 [9]. Pocket Linux is thoroughly described in [11]. Nowadays, despite of its potential, Pocket Linux is not an alternative for mass market applications. Finally, GENA is only a proposal for a Web notification system [12].

We finally selected Pocket PC + Windows CE for our prototype, because the programming tools are quite similar to Windows 9x ones, with a broad support [13]. Also, Windows CE includes Pocket Internet Explorer, and Palm OS has not included an Internet browser for a long time. Finally, for TCP/IP prototyping, there exist Pocket PC PCMCIA jackets and Ethernet cards.

3 Connection Server Implementation

Figure 1 shows our architecture. A PDA *connection server* process listens for socket connection requests at a given port. Once it accepts a request, it receives an URL (or any data) via a socket (A). Then, it submits the URL to the local browser (B), which requests and downloads the corresponding Web page (C, D).

We used a Compaq iPAQ H3630 PDA running Windows CE 3.0, with a PCMCIA Ethernet card. We wrote the connection server in C++ on a host PC running Microsoft Embedded Visual Tools 3.0. The host computer also acted as service server and, for that purpose, it ran a Java connection client. The host computer and the PDA had IP addresses in the same intranet.

The connection server code for Windows CE is freely available for research purposes (by sending e-mail to the authors). *The resulting binary only occupies 7 KB.*

4 Conclusions

We have adressed the need of a system to push URLs into PDAs via sockets, for state-of-the-art m-commerce platforms [3,5]. A prototype (figure 1) has been

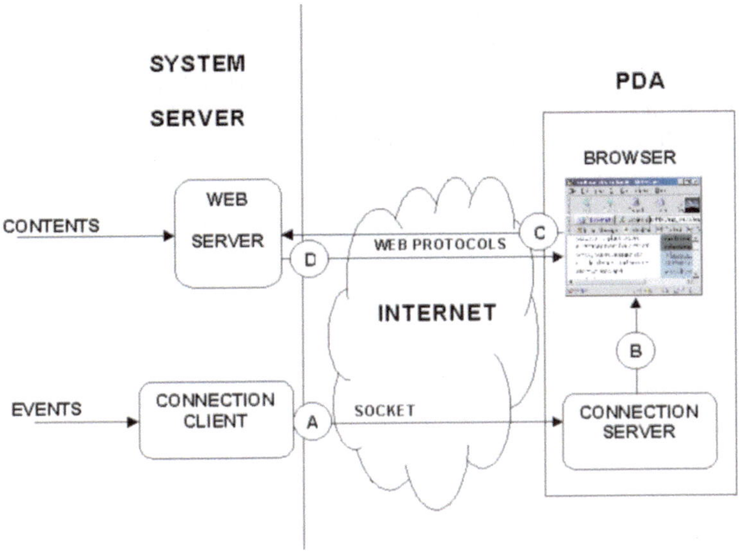

Fig. 1. Architecture prototype

developed. Future work is oriented towards developing PDA clients for meta-computers [14]. Also, we will follow the evolution of Java PDA VMs, which will allow individual frame upload, improving network efficiency.

References

1. Varshney, U., Vetter, R. J., Kalakota, R.: Mobile Commerce: A New Frontier. Computer. **10** (2000) 32–38
2. Darling, A.: Waiting for the m-commerce explosion. Telecommunications International. **3** (2001) 34–39
3. Kindberg, T., Barton, J.: A Web-based nomadic computing system. Computer Networks. **35(4)** (2001) 443–456
4. ETSI: ETSI document TS 122 071 V3.2.0
5. García-Reinoso, J., Vales-Alonso, J., González-Castaño, F. J., Anido-Rifón, L., Rodríguez-Hernández, P. S.: A New m-Commerce Concept: m-Mall. Lecture Notes in Computer Science. **2232** (2001) 14–25
6. Buyukkokten, O., García-Molina, H., Paepcke, A.: Accordion Summarization for End-Game Browsing on PDAs and Cellular Phones on the Road. Proc. of ACM CHI 2001 Conf. on Human Factors in Comp. Sys. 213–220
7. Palm Mobile Internet Kit. http://www.palm.com/software/mik/
8. A-Brand-New-World 4G jacket. http://www.abrandnewworld.se/
9. IBM Visual Age Micro Edition 1.3. http://www.embedded.oti.com/
10. Kada Systems. http://www.kadasystems.com
11. PocketLinux. http://www.pocketlinux.com
12. Cohen, J., Aggarwal, S.: General Event Notification Architecture Base. http://www.alternic.org/drafts/drafts-c-d/draft-cohen-gena-p-base-01.pdf.

13. Boling, D.: Programming Windows CE. Microsoft Press
14. González-Castaño, F. J., Anido-Rifón, L., Pousada-Carballo, J. M., Rodríguez-Hernández, P. S., López-Gómez, R.: A Java/CORBA Virtual Machine Architecture for Remote Execution of Optimization Solvers in Heterogeneous Networks. Software, Practice and Experience. **31** (2001) 1–16

Émigré: Metalevel Architecture and Migratory Work

Paul Dourish and André van der Hoek

Dept. of Information and Computer Science
University of California, Irvine
Irvine, CA 92697-3425 USA
{jpd,andre}@ics.uci.edu

Abstract. Migratory work extends traditional mobile work with an innate awareness of, and adaptability to, both technical and social surroundings. We are designing a technical framework, Émigré, that is based on the use of architectural meta-level representations to support rapid development and semi-automated run-time adaptation of migratory work applications.

1 Introduction

Not too long ago, for computer-based workers, moving outside the office meant leaving work behind—regardless of whether such a move involved getting coffee down the hall, attending a company meeting, visiting a customer in town, or traveling outside the country. Rapid and major technological advances, however, lead to the use of laptops, cell phones, pagers, Personal Digital Assistants, and many other electronic devices that allow users to bring work and stay connected with the office at all times and all places. Mobile computing has arrived; now the struggle is how to use it effectively.

Conventional approaches to mobile computing focus on *eliminating* boundaries by delivering seamless access to information and computation wherever a user goes. The motto of "any time, anywhere" is testament to this vision. Unfortunately, the vision of complete seamlessness is an illusion. In part, this is a technical matter, caused by variations in processor performance, storage architecture, and network throughput; in part, it is a social matter, reflecting the different forms of information needs and acceptable practice in environments as diverse as offices, meeting rooms, cars, homes, and restaurants. These technical and social variations create clearly visible boundaries that current mobile technology blithely ignores.

Rather than attempting to eliminate boundaries, our work focuses on *accommodating* boundaries. In contrast to mobile work, we focus on what we call *migratory work* – activities that move from place to place, device to device, and environment to environment, adjusting to suit changing circumstances. Migratory work applications, then, distinguish themselves in their ability to adapt to their surroundings—not only in terms of location, but also in terms of devices on which they operate, available infrastructure, additional devices and applications usable in the vicinity, and social and organizational settings. For example, consider the migration of an application from a person's office to a meeting room. The application

F. Paternò (Ed.): Mobile HCI 2002, LNCS 2411, pp. 281–285, 2002.
© Springer-Verlag Berlin Heidelberg 2002

must shift from an environment in which it executes on a powerful desktop with a range of input devices and multiple high-resolution displays, to an environment in which it executes on a simple PDA with a stylus as the single input device and a small text display as the only output device. Perhaps even more importantly, the application must be sensitive to its context of use: in the office it has the sole attention of the person and can behave rather intrusively, but in the meeting room it has limited attention and must be as unobtrusive as possible.

These are the challenges that our work aims to address. The broad research question we are tackling, then, is *how a migratory work application can relate its current and future behavior to the setting in which it is currently operating*—thereby allowing an application to effectively migrate from setting to setting and support its users in a manner appropriate to each of those settings. Our hypothesis is that *meta-level architectures provide a convenient and effective approach* for addressing this question. One common problem of meta-level architectures, however, is the complexity of developing both a base level and a meta-level representation, and maintaining the synchronization between them. We will address these problems by harnessing ongoing research into run-time architectural description languages. Specifically, software architecture promotes the use of explicit architectural models of components and connectors as part of the development process. We intend to exploit these capabilities in investigating the use of run-time architectural models as the basis for dynamic meta-level application control.

2 Background

2.1 Metalevel Representations

In traditional approaches to system design, each module offers a single interface to its client modules, creating an abstraction barrier between clients and the modules' implementations. This barrier confers many important benefits on the design, including portability, compositionality, and reuse. However, it is also a source of problems. Abstraction barriers hide implementation decisions that may, in fact, be important for specific clients (consider a virtual memory paging algorithm that is optimized for certain patterns of memory use but can lead to disastrous behavior with other patterns of use). To circumvent this problem, some systems now offer both a traditional interface through which they can be used and a meta-level interface through which they can be examined and controlled (Kiczales, 1992). In effect, through the meta-level interface, these systems offer a representation of their own behavior - a representation that can be manipulated in order to adapt the system to different needs and different circumstances.

These kinds of meta-level representations, and the principle of computational reflection on which they are based, were originally explored in the area of programming languages, although the same problems appear in many other areas of system design (e.g., Dourish, 1996). However, specific meta-level representations have typically been developed in an ad hoc fashion, in response to particular problems, but without a consistent frame of reference or grounding in software design practice.

2.2 Architectural Description Languages

Our starting point for infrastructure design is our current work on architectural modeling languages (Dashofy et al., 2001). These languages allow system developers and designers to construct models of system architecture - components, connectors, interactions, etc. xADL 2.0, the language with which we have been working thus far, provides us three important advantages over other ADLs. First, it models not just static architectures, but also aspects of dynamic run-time configurations, which makes it ideal as a basis for dynamic adaptability; second, it provides basic functionality for modeling architectural variants, making it ideal for modeling the alternative run-time incarnations of a migratory work application; and third, it is specified in XML, which makes it highly extensible and therefore suitable as a base modeling language to which we can add the necessary constructs that are unique to modeling migratory work applications.

One important component of the infrastructure is the means by which the system will sense aspects of its environment. A range of existing technologies provide support for local and remote service discovery (e.g., Sun's Jini), localized communication (e.g., BlueTooth), etc. Our intention is employ these existing solutions rather than to develop novel sensing techniques. However, the model-based approach allows us to decouple sensing from the use of this information, so that Émigré can incorporate novel mechanisms for sensing and discovery as they become available.

3 Approach

Figure 1 introduces the overall architecture of Émigré. The architecture separates the models, infrastructure, and application into three separate entities. The *models* are further subdivided in models of potential variants in which an application can reside at run-time, models of context, and models of the relationships and constraints that are present among application variants and contexts. Separating the models into these

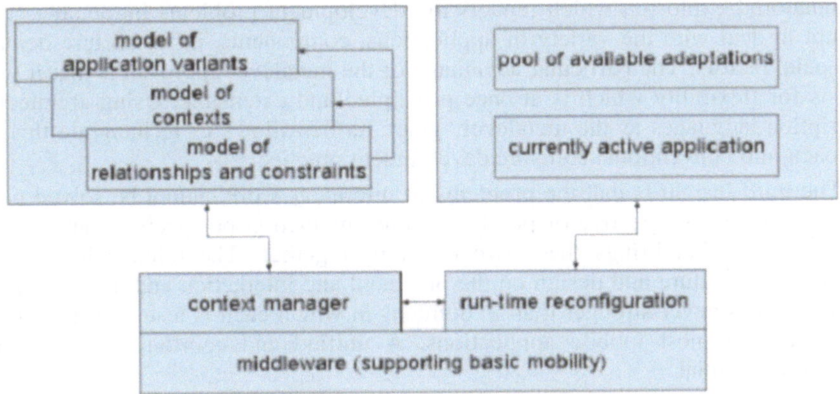

Fig. 1. Architectural outline.

three categories not only promotes a strong separation of concerns, but also promotes reuse of models among different migratory work applications.

The *infrastructure* is further subdivided in a basic middleware that supports mobility, a context manager, and a run-time reconfiguration component. The context manager, which is aware of different application contexts, combined with the run-time reconfiguration component, which is able to drive the reconfiguration of an application from one variant into another by drawing from the pool of available adaptations, extends the basic mobile middleware into a middleware that is suitable for migratory work.

4 Challenges and Directions

The work described here is at a very early stage. At the time of writing, we are engaged in initial experiments integrating various infrastructure components to demonstrate the use of xADL 2.0 as a metalevel representation. Further work will address the core challenges, which are to determine the effective bounds of the design space imposed by particular architectural models, and the appropriate design parameters that balance flexibility against cost of development. This work is ongoing and we look forward to being able to present it in the future.

However, even at this stage, we believe that some of the central insights that have arisen and which motivate our design are potentially valuable for others working in this area. Three of these are significant here.

The first is the separation of migratory from mobile work. Rather than respecting the detail of the different settings and situations in which work must be conducted, much research into mobility attempts to obscure that detail by fostering an attitude of seamlessness and ubiquity. In contrast, our approach recognizes the inherently heterogeneous nature of different social and technical settings and attempts to incorporate that heterogeneity into the interactive experience.

The second is the identification of the role for explicit metalevel representations. The diversity of settings for migratory applications potentially leads to a combinatorial explosion which renders the development problems intractable, as we attempt to deal with the variety of applications, components, infrastructure demands and social factors. The particular advantage of the metalevel approach is that it yields a basis for flexibility which is at once principled and extensible. Using architectural description languages as the metalevel "glue" further allows us to integrate this new approach into conventional software development practice.

The third insight is that the problems of migratory work cannot be solved purely from a software perspective or purely from an interaction perspective, but require a new approach that brings these two concerns together. The relationship between software architecture and design on the one hand and interaction and user experience on the other is never stronger than in domains in which each is tested to the limit, as is the case in most mobile applications. A unified and coordinated approach is absolutely essential.

Our current work with Émigré is an example of the integration of these three principles. Our early experiences are positive, and we hope to encourage others to adopt the principles and explore other parts of the design space.

References

[1] Dashofy, E., van der Hoek, A., and Taylor, R. 2001. A Highly-Extensible, XML-Based Architectural Description Language. *Proc.Working IEEE/IFIP Conf. Software Architectures WICSA 2001*. Amsterdam, Netherlands.

[2] Dourish, P. 1998. Using Metalevel Techniques in a Flexible Toolkit for CSCW Applications. *ACM Trans. Computer-Human Interaction*, 5(2), 109-155.

[3] Kiczales, G. 1992. Towards a New Model of Abstraction in Software Engineering. *Proc. IMSA Workshop on Reflection and Metalevel Architectures*. Tokyo, Japan.

Interactions Model and Code Generation for J2ME Applications

Davide Carboni[1], Stefano Sanna[1], Sylvain Giroux[2], and Gavino Paddeu[1]

[1] CRS4, Centro di Ricerca, Sviluppo e Studi Superiori in Sardegna, VI Strada OVEST Z.I. Macchiareddu,UTA, CA – Italy.
{dadaista,gerda,gavino}@crs4.it
[2] Dep. of Mathematics and Computer Science, University of Sherbrooke, Sherbrooke – Canada.
sylvain.giroux@dmi.usherb.ca

Abstract. The Java2 Micro Edition platform can really be considered an emerging standard for new generation embedded software. This article introduces a practical methodology aimed to automatically generate a software prototype starting from an abstract description which defines the dialogue between the user and the application by means of a device independent and abstract description. We will show how an agenda application for cellular phones can be described by means of a visual language called PLANES and present how the personal agenda prototype is implemented by an appropriate generation tool.

1 Introduction

With an army of 2.5 million programmers all around the world, ready to conceive and develop new intelligent applications for mobile devices, and the commitment of some sector giants (such as Motorola and Nokia), the Java2 Micro Edition (J2ME) platform [3] can really be considered an emerging standard for new generation embedded software. Developing new applications in J2ME is not as easy as developing for desktop computers or for the web. Some difficulties to overcome are:

- Integrated development environments (IDE) for J2ME provide less functionality than their counterparts for Java2 Standard Edition (J2SE).
- Application Programming Interfaces (API) for J2ME differ from their corresponding API for J2SE. Programmers need to learn them and adapt their code.

This paper describes a method and the implemented related tools for automatic code generation of the GUI and of the code that manages the sequences of forms and dialogs on the screen for small mobile devices as personal digital assistants, cellular phones etc. The dialogue between a user and the application is specified with a visual language. A device independent description is then produced. This description is used for code generation on the J2ME platform.

F. Paternò (Ed.): Mobile HCI 2002, LNCS 2411, pp. 286–290, 2002.
© Springer-Verlag Berlin Heidelberg 2002

2 Personal Agendas as Test Applications

The main task in a personal agenda is managing a set of appointments. From the viewpoint of GUI, this description is too abstract and needs to be refined towards more detailed and concrete steps. Concretely the agenda will initially provide a choice between two options:

- Create and insert a new appointment.
- Consult the whole list of appointments.

These choices then lead the user to other choices or sequences of more concrete tasks. For instance, the creation of a new appointment prompts the user with a form, then asks for a confirmation of the data and finally waits for data to be written in the database. We made two assumptions on the use of target devices:

1. Due to screen restrictions, it is not possible to work with more than one window at a time, thus the dialog is constrained in well defined pathways along a tree of choices and sequences of tasks.
2. The user should have the possibility to escape the current task at any time and return to the main menu by means of hot keys bound with physical buttons in the device.

With respect to these assumptions, the next section describes a practical approach to model the interaction between the user, the application, and the runtime environment.

3 Task Modeling with PLANES

Task modeling with PLANES involves a high-level description of choices and sequences of activities. Such a description is formalized by means of a visual language whose structure is basically a tree. The lexical elements of such a language are grouped into three classes: abstract tasks, concrete tasks and reification models. In the agenda example, creating a new appointment is a sequence of three tasks:

1. Form-filling: the user creates a new data object with a given structure; this object is copied into the current context.
2. Confirmation, the user is asked to provide his authorization to proceed to the next task. We coin such tasks as "shield" tasks.
3. Write operation, the data object is copied from the context to the system.

Tasks represent activities at different levels of abstraction. For instance, a task representing the insertion of a new appointment into the personal agenda must be considered an abstract task which will be reified by one or more concrete tasks.

The user is indeed an actor able either to create brand new data objects or modify existing objects by form-filling. A form is a tool which allows the user to create a data object with a given structure and write temporarily this object into the current context in order to make it available for the next task. The context acts as a container which stores the data during the interaction. Data objects have a structure composed by a number of primitive data types such as strings, numbers, booleans, enumerations and lists in a way similar to data structures used in procedural and object oriented programming languages.

To describe tasks and structure the dialogue, PLANES contains the following language elements:

Abstract Task: represents a high-level activity; for instance, “Inserting a new appointment” is an abstract task which must be reified by a sequence of concrete tasks.

Application Task: an application task represents an interaction with the user by means of a form. A task is completed once the user fills the form with required input.

Shield Task: a Shield Task represents an interaction in which the user must confirm his intention to proceed to the next task in a sequence. Typically, this task prevents the user from making accidental changes in the database.

Data Task: they represent an interaction between the application and the system in which a data object is read, written or deleted from/to the system and copied into/from the current context.

Reification Models: each abstract task must be reified by a set of detailed tasks. A Reification Model collects those tasks and defines the control flow between tasks. We have identified two types of reification models: Sequence and Choice.

4 The Visual Editor and J2ME Code Generator

PLANES is provided with a visual editor which allows the designer to build a new model and save it serialized in XML. In addition to these essential functionalities, the editor can perform a simulation of the application's behavior by means of dialogs and menus that follow the structure defined by the model. The generation process is depicted in figure below:

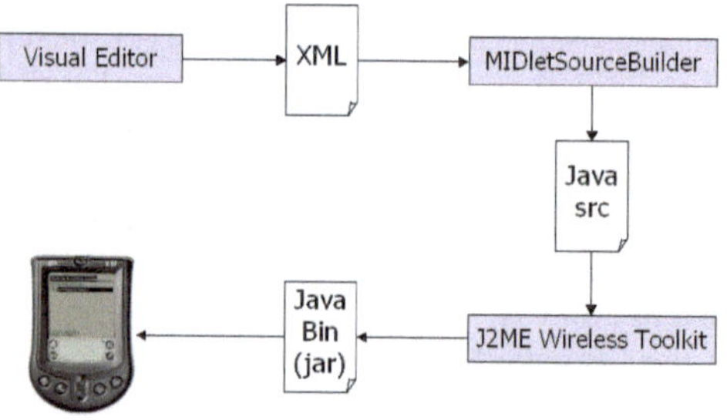

Fig. 1. The creation process of the prototype from the abstract model to the application.

The MIDletSourceBuilder reads the XML file and, starting form the root-level abstract task, it performs a depth-first search and generates for each node a corresponding Java source file. The main benefit of using an automatic tool for J2ME application development is for providing a complete composition and chaining of the GUI pieces. The tool described in this paper differs from commercial tools for application development which assist the programmer with visual composition of graphics widgets and project management. Visual Editor and MIDletSourceBuilder, instead, focus on the structure of interaction with end user.

5 Results for the Personal Agenda

This section depicts how we modeled the personal agenda in the visual editor. Figure 2 shows some screen shots (1) to (4) captured from the emulator and representing, in succession, the graphical components that lead the user along the creation and insertion of a new appointment in the personal agenda. Starting from the root node, the application shows a selection menu (1), from which the user selects the creation of a new appointment. Next form (2) contains the fields where the user can insert the details (3) of new appointment. When data have been inserted, the application asks the user (4) to confirm data writing on system database. The "write task" is the only concrete task the developer has to implement. In Figure 3, an excerpt of the XML code is shown.

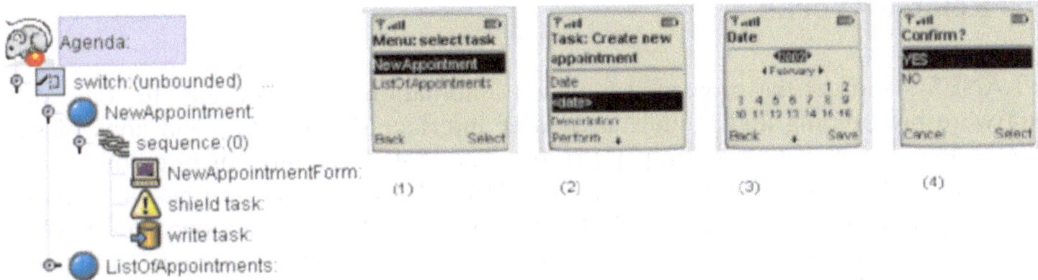

Fig. 2. Agenda model in the Visual Editor and screenshots of the prototype running on the J2ME emulator.

```
<?xml version="1.0"?>
<Interaction>

    ...snip...

    <AbstractTask ID="0_M_0" ancestorID="0_M" name="NewAppointment" type="abstract">
        <Model ID="0_M_0_M" ancestorID="0_M_0" type="sequence" description="A sequence model" repeat="0">
            <ModelItem taskID="0_M_0_M_0"/>
            <ModelItem taskID="0_M_0_M_1"/>
            <ModelItem taskID="0_M_0_M_2"/>
        </Model>
    </AbstractTask>
    <ConcreteTask ID="0_M_0_M_0" ancestorID="0_M_0_M" name="NewAppointmentForm" type="application" />
    <ConcreteTask ID="0_M_0_M_1" ancestorID="0_M_0_M" name="shield task" type="shield" />
    <ConcreteTask ID="0_M_0_M_2" ancestorID="0_M_0_M" name="write task" type="write" />
...snip....
</Interaction>
```

Fig. 3. An excerpt of XML code describing the sequence of tasks for the "Insertion of a new appointment".

6 Conclusion and Future Work

In this paper, we have sketched a visual language (PLANES) and an automatic code generator for prototyping new J2ME-MIDP applications. This process allows programmers to get a working prototype without the burden of writing all code lines for event handling and user interface creation and let them spend their efforts to customize the prototype and transform it in a real application. We have tested PLANES with a personal agenda application, which is a service largely available in many cellular phones. Application prototyping with a visual tool has demonstrated its efficacy both for showing and discussing application's functionalities with the customer, and for the quality of the generated source code which results untangled, easily readable, and customizable.

We believe that the adoption of model-based techniques is a valid support for software prototyping and application development not only for small devices but for web and desktop applications as well.

7 Related Works

PLANES stands at the crossroad between automation methodologies for embedded software prototyping and model based user interface design. Pros and cons of such methodologies are discussed in [2], [5], [6]. The tool which most inspired our work is ConcurTaskTree [4]. Another tool, MASTERMIND [1], [7], defines a technique based upon the definition of three distinct models, one for the functionalities available by means of the user interface, another defines the presentation structure in terms of widgets and a last model defines the dialog in terms of end-user interaction and how these affect the presentation and the application.

PLANES is not in competition with those methodologies, that represent milestones in HCI research, but rather, it is an attempt to experiment some of model-based design concepts and ascertain how they can be applied to the rapid prototyping of software for embedded systems and mobile phones.

References

1. T. P. Browne et al. Using declarative descriptions to model user interfaces with MASTERMIND. In F. Paterno and P. Palanque, editors, Formal Methods in Human Computer Interaction.
2. Janssen C., Weisbecker A., Ziegler J.: Generating user interfaces from data models and dialogue net specifications. In: Ashlund S., et al. (eds.): Bridges between Worlds. Proceedings InterCHI'93 (Amsterdam, April 1993). New York: ACM Press, 1993, 418-423.
3. Sun Microsystems, MIDP Specification (Final V1.0), 2001 (available at http://java.sun.com)
4. Paterno, F., Model-Based Design and Evaluation of Interactive Applications. Springer Verlag, ISBN 1-85233-155-0, 1999.
5. Puerta A.R. 1993. The Study of Models of Intelligent Interfaces. In Proceedings of the 1993 International Workshop on Intelligent User Interfaces. Orlando, Florida, January 1993, pp. 71–80.
6. Schlungbaum E., Elwert T.: Automatic user interface generation from declarative models. In: Vanderdonckt J. (ed.): Computer-Aided Design of User Interfaces. Namur: Presses Universitaires de Namur, 1996, 3-18.
7. Szekely, P. et.al. Declarative interface models for user interface construction tools: the MASTERMIND approach. In Proc. EHCI'95 (Grand Targhee Resort, August 1995).

An Interface with Weight: Taking Interaction by Tilt beyond Disembodied Metaphors

Daniel Fallman

Interaction Design Lab
Umea Center for Interaction Technology
Umea University Institute of Design
SE-901 87, Umea, Sweden
daniel.fallman@dh.umu.se
http://daniel.fallman.org

Abstract. We propose a novel way of using tilt as a style of interaction for palmtop computing, as well as a theoretical base for doing so. Beyond superficial and disembodied metaphors, the force of gravity has been chosen to guide navigation within a space larger than the palmtop's screen. Gravity allows the user to experience 'an interface with weight'; an understanding put forward as embodied which diverge from previous implementations of tilt interaction, as both the input style as well as the interface answer to the physical world.

1 Introduction

Mobile information technology, such as palmtop computers and cellular phones, is rapidly gaining ground as a new paradigm in computing. While their computational power increase dramatically, concerns remain regarding human—computer interaction issues, caused primarily by small form factors. In current research, inadequate input methods and small screens for output have been suggested as the two main barriers which confine usability in mobile computing [7]. Several suggestions on how these problems should be approached have been put forward. A growing body of research suggests that physical configurations of mobile devices should be determinant of their operation [2][3][7], following a trend in HCI away from dependency on superficial metaphors towards the phenomenological notion of embodiment, where meaning is created through engaged interaction with artifacts within the physical world [1]. This advance is also evident in exploration of tangible aspects of virtual phenomena [4].

1.1 Tilting as a Means of Interaction

Tilt has been previously suggested as an input method to ease interaction with palmtops. When utilized, tilting the device itself becomes the means of interaction, i.e. the way the user produces input to the system, as opposed to using a traditional stylus

F. Paternò (Ed.): Mobile HCI 2002, LNCS 2411, pp. 291–295, 2002.
© Springer-Verlag Berlin Heidelberg 2002

pen or a small-sized keyboard. Through tilting, computer devices could also be understood as embodying their interfaces, making the device as a whole an interface [2]. Previous work in the field has been largely concerned with scrolling and pointing on graphical user interfaces, and for various kinds of menu selection [2][3][7].

2 Beyond Disembodied Metaphors

Previous implementations of palmtop tilt interfaces tend paradoxically to rely on metaphors and interaction abstractions similar to those that their authors reject—e.g. where tilt is used to control a pointer on the screen—or they introduce new metaphors with equally superficial connections to the physical world—e.g. where tilt left is used to scroll the screen to the left. This latter is a metaphor suggested by for instance aircraft and motorcycle maneuvering, but with weak connections to computer interface control. We recognize and draw on such previous work concerning this style of interaction, but our contribution is to found embodiment not only in terms of the input method, but also in terms of the feedback given to the user. In doing so, we intend to deepen the user's experience and take a step further away from resorting to disembodied metaphors to guide HCI design, by connecting the virtual world with the physical in terms of both input and output, corporal intimacy, and multimodal interaction.

2.1 Gravity Embodied

The force of gravity was chosen as the common denominator between the human user and the computer system, a form of shared understanding substantially deeper than a traditional user interface metaphor. In HCI, metaphors have traditionally been understood as devices for conveying a complex of concepts by presenting one thing as if it were another [6]. In contrast, Lakoff & Johnson [5] provide a thorough understanding of metaphors, in which they suggest that our mind operates by them—i.e. that we always think metaphorically—and that our everyday experiences are shaped by different kind of metaphor. Through language, we often tend to structure experience in terms of spatial orientation, viable in expressions such as *'I am feeling down'* and *'things are looking up'* [5][6]. From the perspective of the metaphor used for interaction in this prototype, it should be noted that Lakoff & Johnson suggest that meaning is fundamentally rooted in basic, bodily experiences of us residing in physical bodies with certain configurations, located on a planet with certain characteristics. Among these characteristics, gravity takes a principal position [5]. Hence, according to Lakoff & Johnson, at the core of the human mind is an embodied understanding of gravity.

Gravity *per se* can be defined as the force that attracts a body to the center of the earth or to other physical body having mass. Implemented in software and used in the context of a palmtop graphical user interface, the law of gravity provides the user interface the property of having weight and as such affords a basic understanding of a device's mode of interaction and explains its behavior. Use of software gravitational models is also frequent in related fields of research, e.g. computer generated graphical effects, 3D animation, Virtual Reality (VR), and perhaps most prominently and frequently so in computer games.

3 Prototype Implementation

The research context for this prototype implementation is a large, ongoing multidisciplinary project funded by Swedish industrial company ABB; established to explore novel ways of interacting with mobile information technology suitable for assembly manufacturing settings. The work presented here transcends this particular use context, as the prototype has been designed and tested as a generic interaction technique with a graphical user interface free from assembly manufacturing specific information.

3.1 Interaction by Gravitational Tilt

A starting point and a working hypothesis for the project was that if we consider tilt as an interaction style, we will probably need to design the actual user interface in a way that answers to alleged benefits of the new interaction paradigm, rather than to force tilt to operate on a WIMP style interface originally designed for a 2D input device, such as a mouse. Our assumption was that this would do justice neither to traditional GUIs nor to tilt as an interaction style.

The basic setup of the prototype interface provides the user with a number of screens (320x240 pixels) aligned horizontally next to each other, each of which offers conceptually related functionality to its close neighbors. The number of available screens is dynamically determined by the executed application. Navigating the application's interface is hence a matter of horizontal scrolling of a flat surface, which is typically substantially larger than the palmtop screen. Selection is made by finger tapping the screen. A custom made tilt sensing device with two degrees of freedom, resembling a gyro sensor, is connected to the palmtop device. A model of gravity is implemented in software, controlling the user experience of the interaction where the interface slides one screen at a time in the direction the palmtop is tilted—hence in the direction opposite to which scrolling by tilt is typically implemented.

Acceleration of the sliding surface is based on a software model of gravity, and altering the 'friction' to the surface on which the interface is seemingly placed controls acceleration and sensitivity, adding to the user experience of having an interface with weight. Alternating the angle of the device hence controls the scrolling speed, which for instance allows for easy switching between two screens, quickly browsing a number of screens, returning rapidly to the leftmost screen, etc.

3.2 Technology Implementation

A custom made tilt device has been designed for this project, consisting of small, inexpensive, and standard electronic components. An AVR 2313 microcontroller is connected to a 2G accelerometer with PWM output (ADXL202). The tilt sensor's signal is sampled by the AVR and transmitted to a Compaq iPAQ through a RS232 serial communication link.

4 User Evaluation and Future Work

Preliminary user testing (n=6) suggests that the proposed 'interface with weight' style of interaction is useful for navigating through the prototype system's different screens, that subjects very quickly seem to learn to control the interface and predict its behavior, and also that it is rated highly on a subjective scale of acceptance and appeal. A more systematic evaluation is being designed. Earlier in-house tests suggest, however, that it is much less useful for also controlling traditional interface widgets on these screens. The working hypothesis is that this is because the embodiment relation that is established between the user and the system for navigating between different screens is broken when interface widgets instead become the focus of attention. Operating widgets by tilt seems not to bring about embodied activity, possibly so because there is no natural mapping between gravity and the operation of these widgets.

These early user assessments have had two impacts: first, we have iteratively redesigned the interface to better fit the tilt style of interaction; and second, a larger quantitative study on the relative merits of tilt as a style of interaction is being designed to examine different ways of navigating a space larger than the actual screen.

5 Conclusions

This paper has presented ongoing work on exploring ways of interacting with mobile information technology. Primarily, we have proposed, explained, and discussed a novel way of using tilt as an interaction style to navigate a space larger than a palmtop screen. The notion of gravity has been introduced as the metaphor that forms a shared means of understanding between the user and the system, with the purpose of making the act of virtual navigation physical. Theoretical support for the use of gravity in this context has too been pointed out, drawing primarily on the work of Lakoff & Johnson.

References

1. Dourish, P.: Where the Action Is: The Foundations of Embodied Interaction, MIT Press. Cambridge MA (2001)
2. Fishkin, K.P., Gujar, A., Harrison, B.L., Moran, T., Want, R: Embodied User Interfaces for Really Direct Manipulation. Communications of the ACM, Vol. 43, No. 9 (2000) 74–80

3. Harrison, B.L., Fishkin, K.P., Gujar, A., Mochon, C., & Want, R.: Squeeze me, Hold me, Tilt me! An Exploration of Manipulative User Interfaces. Proceedings of CHI'98. ACM, Los Angeles CA (1998) 17–24
4. Ishii, H. & Ullmer, B.: Tangible Bits: Towards Seamless Interfaces between People, Bits, and Atoms. Proceedings of CHI'97. ACM, Atlanta GA (1997) 234–241
5. Lakoff, G. & Johnson, M.: Metaphors We Live By. Chicago University Press (1980)
6. Lund, A. & Waterworth, J.A.: Experiential Design: Reflecting Embodiment at the Interface. Computation for Metaphors, Analogy and Agents. University of Aizu, Japan (1998)
7. Rekimoto, R.: Tilting operations for small screen interfaces. Proceedings of UIST'96. ACM, Seattle WT (1996) 167–168

UML Modelling of Device-Independent Interfaces and Services for a Home Environment Application

O. Mayora-Ibarra, E. Cambranes-Martínez, C. Miranda-Palma, A. Fuentes-Penna, and O. De la Paz-Arroyo

ITESM, Paseo de la Reforma 182-A, Cuernavaca, Morelos 62589, México
omayora@campus.mor.itesm.mx,
{00377869,00377865,00376431,00376595}@academ01.mor.itesm.mx

Abstract. In this work, we use the UML language to model and describe an approach for generating graphical (GUI) and speech-based (SUI) user interfaces from a single source in a home environment application. The proposed method introduces a generic dedicated widget vocabulary that aids in defining user interface descriptions written in the UIML language. Subsequently, this generic description may be converted to multiple UI implementation formats suitable for the specific client terminals. These targets include GUI-like formats (e.g. HTML, WML and Java), as well as voice-based formats(VoiceXML).

1 Introduction

The promise of information anytime, anywhere is becoming an everyday reality. In the upcoming years, specialized and easy to use information appliances, such as intelligent hand-held devices, will proliferate and practically dominate the future of human-machine interactions. This will lead to the need of adding this new devices with intuitive interfaces that should be used independently of the physical limitations of each specific device. Key-design aspects for achieving complete user satisfaction will include features like interface personalisation, augmented reality, multi-modal interaction, device-independence, context adaptation and non-obtrusive computing. In this way, it will be possible to perform the same tasks by using different appliances that may vary in physical characteristics (like size, screen type, input modality, etc).

In particular, the design of user interfaces for appliances at home will be going through a major restructuring. Instead of the traditional one device – one UI concept, control of devices will be done via other devices, which may even be in remote locations. Not only will be the UI front-end thus separated from the back-end, also the interaction techniques may vary with the properties of the control device. These properties include, for instance, the interaction modalities, display dimensions, and the different software platforms. Designing a user interface supporting this kind of foreign control mechanism sets new requirements for the format used to specify the UI. The uncertainty about the final features of the device rendering the UI calls for a very generic, flexible and device independent format.

F. Paternò (Ed.): Mobile HCI 2002, LNCS 2411, pp. 296–301, 2002.
© Springer-Verlag Berlin Heidelberg 2002

The task becomes even more complicated when inherently different modalities need to be supported, like graphical and speech-driven UIs. Traditional approaches would suggest to write different code versions for each possible type of interaction modality. The former solution will require to write as many versions as there were modalities, which will originate great maintenance and coherence problems. These considerations motivate the development of a more generic methodology that allows for the generation of user interfaces regardless of their physical characteristics. A language that fulfils all the previous requirements is UIML [1]. In this work, we present an approach for modelling UIML-based interactions and services in home environments with the standard Unified Modelling Language (UML) [2]

2 The Problem

The proliferation of intelligent devices in the home ambient, will pose the problem of choosing an approach for controlling them. A first solution would be to design a user interface for every single artefact at home. This approach will introduce other problems like designing exhaustively ad-hoc interfaces that may vary according to different physical constraints like size and shape. Besides, the user would have to learn to use multiple interfaces and to carry a huge quantity of remote controls. A better solution would be to add communication capabilities to home artefacts and to present their UI descriptions in a limited number of controlling devices. This would imply not to operate each device via its own separate user interface, but to operate them as part of a network of appliances. These appliances may be operated remotely via a set of control devices that may be enhanced with context awareness capabilities (like the user's identity or the other devices status. The technology for implementing the previous vision is clearly available and it is only a matter of time when standards are finalised and technologies are tailored to work smoothly together. Figure 1 shows a vision of a home device controlling application using device-independent interfaces. Every controllable device is associated with a UI description written in the UIML language and delivered through a UI broker to the different control devices. The UI broker consist on a series of target language transcoders that render the interfaces to the specific clients.

3 Modelling the Solution

3.1 UML Modelling of Interfaces and Services

In the past few years, the Unified Modelling Language (UML) has become the best candidate language for representing object oriented systems. The versatility of UML has motivated its use for software engineering as well as for describing systems in other knowledge fields. In particular, the use of UML for modelling human-machine interfaces has been successfully achieved in recent developments.

In this work, we used the UML to model the interactions and events among humans, devices and services within the home environment application described before. Our approach included 3 different views of the system corresponding to the controlling devices, the generic UI transcoding system and the controlled home appliances respectively. Each one of these views was associated to a UML class with specific attributes and methods. Figure 2 shows a UML collaboration diagram that displays in a general form the whole process of service request, UI transcoding and interface deployment. The control device class requests services from the server class in order to modify the status of the controllable devices. In this work, we choose to include the collaboration diagram as one that included the most general view of interaction information within this application. Other UML diagrams were realised (like class, sequence, activity, states and use case diagrams) that allowed more detail views of the system model.

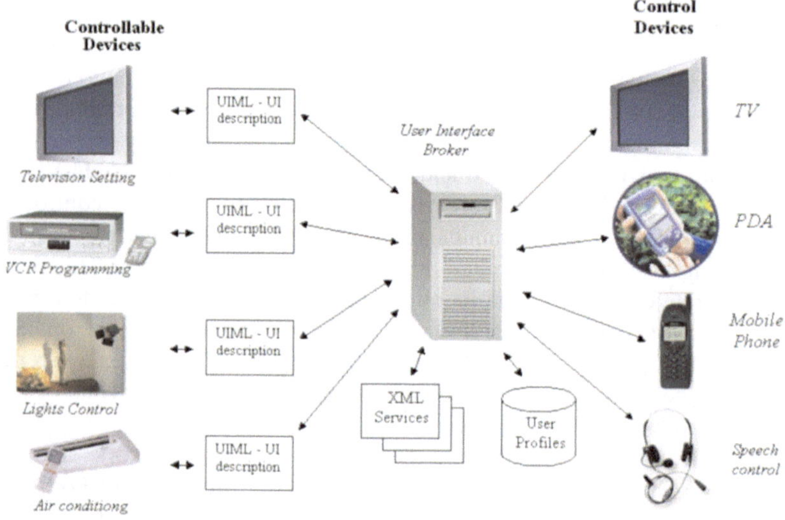

Fig. 1. Scenario for home-artifact controlling via device-independent interfaces

3.2 Generic User Interface Design

A generic interface is one whose aspect may vary in different devices while its functionality prevails in some of them. The design considerations involved in this kind of approach may permit to define interaction features that may be rendered according to the target device characteristics[3]. Devices with larger screens may display the information and allow interaction in a very different way than in smaller interfaces or screen-free devices (like the telephone). In any case, the functionality of the system may allow to perform the same task regardless the device that is used to do it. For this purpose, a set of interaction elements with different functionality must be identified and associated to different rendering formats.

A generic-class vocabulary can be defined as one vocabulary for all kind of platforms. The generic class vocabulary consists of a set of elements and combinations of them used to create generic user interface descriptions. A desired characteristic of such vocabulary would be its easy adaptation to graphic interfaces (parallel oriented interaction) and to voice-based interfaces (serial oriented interaction)[4].

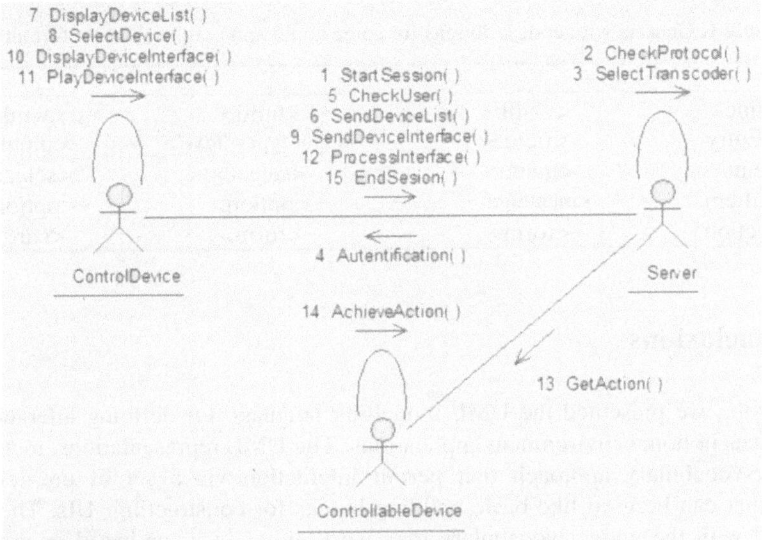

Fig. 2. UML Collaboration diagram for home environment application

However, not all the graphic elements may have a correspondent equivalence for voice-based interaction systems. This last consideration made us to identify and define the most adequate one-to-many mappings from a set of generic elements to a set of platform-specific elements. One example of this will be to make the attributes definition of generic elements the most flexible as possible. We defined properties that provided style flexibility to widgets like "label" and "extended-label". The first one was used for mapping concise information regarding a widget to a GUI (for example a text label on a button), while the second one provides extended information for a speech user interface (a text-to-speech prompt played in a telephone). For example, a generic button widget may be associated to a text-label and an on-click event in a graphic interface while its voice-based representation may include an extended-label played as a text-to-speech prompt and a dedicated grammar for shooting an acknowledge event. Other interface widgets may be defined considering the same design philosophy for allowing complete GUI and SUI interaction.

The definition of the complete set of widgets may be satisfied according to the different roles that they play within the UI. However a set of common attributes was found to apply to all the widgets in the vocabulary. These shared attributes were grouped into a general and common class (*Element* class) serving as a generic template or ancestor of all classes. Other classes of elements were grouped according to the particular role they play within the UI in three more groups: *Input class*, *Output* class, *Collection* class Each generic class contained a set of specific widgets for representing different interaction moments. Table 1 shows an example of some of the elements presented in each interaction group and their representation in different target languages.

Table 1. Generic widget descriptions for some audio and graphical target formats

Frame	<vxml>	<html>	<wml>
TextEntry	<field>	<input type="text">	<input>
Menu	<menu>	<select>	<select>
MenuItem	<menuitem>	<option>	<option>
Collection	<form>	<form>	<card>

4 Conclusions

In this work, we presented the UML modelling language for defining interactions and services in home environment applications. The UML representations, included a generic-vocabulary approach that permit interaction via a set of pre-defined widgets that can be used like basic building blocks for constructing UIs. The UIs generated with the widget vocabulary may work either in voice-based or graphic interacting modalities. This means that whether the interaction will be performed via a GUI or a voice-based device, the UI design should be robust enough to support either formats. The former is achieved associating a series of attributes that bring flexibility for handling either interaction formats with the generic widgets

Designing for both speech and graphical interfaces was reflected specifically in the properties attributed to the basic classes defined in the vocabulary. These properties represent a certain redundancy, since target formats will only need a subset of the provided properties. However, the resulting UI description will be more coherent and concise than a listing of target-specific UI descriptions.

References

1 Abrams, M., Phanouriou, C., Batongbacal, A.L., Williams, S.M., and Shuster J.E. "UIML: An appliance independent XML user interface language". *Computer Networks, vol. 31*, Amsterdam: Elsevier pp. 1695-1708.

2 Booch G. Rumbaugh H and Jacobson J. "The Unified Modeling Language User Guide", Addison Wesley,1999.

3 Paternò F., "Towards a UML for interactive systems", Eng. HCI 01, Lect. Notes in Comp. Sc., pp. 7-18, 2001

4 Plomp J. and Mayora-Ibarra O., "A Generic Widget Vocabulary for the Generation of Graphical and Speech–Driven User Interfaces". International Journal of Speech Technology. V. 5 2002, pp. 39–47.

iTutor: A Wireless and Multimodal Support to Industrial Maintenance Activities

Luka Tummolini, Andrea Lorenzon, Giancarlo Bo, and Roberto Vaccaro

Giunti Ricerca S.r.l., Via Portobello, Abbazia dell'Annunziata, 16039 Sestri Levante, Italy, {l.tummolini,a.lorenzon,g.bo,r.vaccaro}@giuntilabs.it, http://www.giuntilabs.com

Abstract. This paper is focused on the description of the iTutor application. iTutor addresses the increasing needs originating in contemporary industrial workspaces for distribution of electronic information within learning, knowledge transfer and information management processes. In order to achieve the goal of a flexible and adaptable information delivery in the working environment, iTutor integrates a mobile wearable device with a voice and gaze controlled, web-based graphical user interface based on the XML standard. The application is wirelessly integrated with the enterprise information system where the technical information is stored. iTutor is meant to support the working activity of maintainers needing to have hands free in order to complete their tasks.

1 Introduction

Mobile services support the emerging increasing need of leisure as well as of working sectors to receive information and training anytime and anywhere. More specifically, in the industrial maintenance sector, the organization of productive plants requires teams of spatially distributed working people. In order to improve productivity, these groups need to retrieve information as quickly as possible to complete repairing tasks. Moreover, due to downsizing of personnel, new ways of improving learning tasks are needed to complement the activity of experts providing on-line courses. Wearable and Mobile computers combined with wireless technology has been widely adopted to meet these general user requirements, [1], [2]. While mobility and remote access to enterprise informative system are generally pursued, only few applications have considered that traditional human-computer interaction modalities strongly impede the repairing task. The general goal of iTutor is to support and ease maintenance activity and to provide professional training in an industrial setting where a user must keep his/her hands free to work with the target of intervention. To this end it has been designed and implemented in order to run on a wearable device (Xybernaut MAIV) that is wirelessly integrated with the industrial information system. The user can browse multimedia technical information (drawings, schemas, procedures and manuals) in a multimodal way, by simply using eye movements and speech commands. In this way, iTutor offers a seamless mobile integration of information processing tools with the existing working environment. The role of gaze based interaction with the system has been deeply investigated and a prototype providing information as a response to natural eye fixations has been developed and integrated. In the following sections the main features and the underlying architecture

F. Paternò (Ed.): Mobile HCI 2002, LNCS 2411, pp. 302–305, 2002.
© Springer-Verlag Berlin Heidelberg 2002

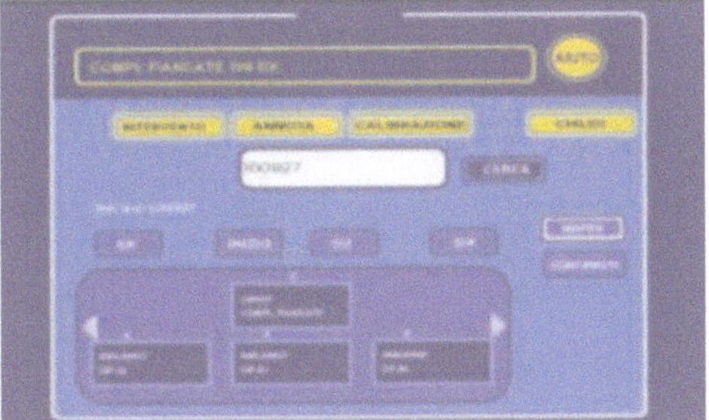

Fig. 1. Architecture of iTutor: a Web-based client-server application

of the system will be described. A brief discussion about the results obtained during a preliminary usability evaluation campaign will be provided too.

2 iTutor: A Gaze and Speech Controlled Application

Besides the extremely complex architecture that has been developed, the novelty and originality of the system mainly resides in the Graphical User Interface and the modalities of interaction with it. Essentially, the *iTutor User Interface* is a web-based application that allows the user to remotely navigate the information contained in the enterprise information system by exploiting a common Internet browser (Fig.1). The complete structure of an industrial plant is usually represented by a huge amount of hierarchically organized information. In iTutor a plant is represented in the *Map Area* : maintainers can browse through the different levels of a plant structure simply by "looking" at the different items in order to select them. Each rectangle in the map represents a different level in the actual plant structure (i.e lines, machines, groups and elements). Through few simple voice commands, a user can then access specific information and assets linked with a selected component (i.e. its position in the warehouse, available manuals and procedures, mechanical and electrical schemas, etc.). Speech input is acquired and managed by the IBM Via Voice software, which has been integrated in the application. To ease the specific task of electronic troubleshooting the concept of *interconnected schemas* has been introduced in iTutor in order to let the user access the same detail/component in different representations (*electrical schema*, *block diagram*, *serigraphy*) by simply selecting the element of interest through eye fixations (Fig. 2). An *Intention Extraction Module* (*IEM*) represents the core of the eye-based interaction. It is devoted to the interpretation of the user's fixations, which are acquired by a small USB eye tracker (the *EyeMouse* , produced by *SensoMotoric Instruments*), with respect to the active portions of the graphical interface. The problem of locating a gaze point on the user interface has been treated as a classification process (adopting a K-NN algorithm, see [3]) while the decision space is represented by an R-Tree, [4]. The IEM guarantees the direct interpretation of the user's intentions from the statistics of the eye gaze, therefore allowing a *non-command style of interaction* : the computer senses what the user wants to do without the user having to give any explicit command, [5]. From the point of view of

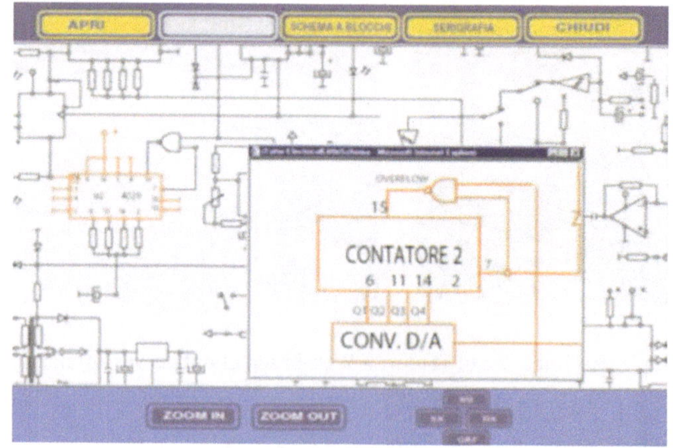

Fig. 2. Architecture of iTutor: a Web-based client-server application

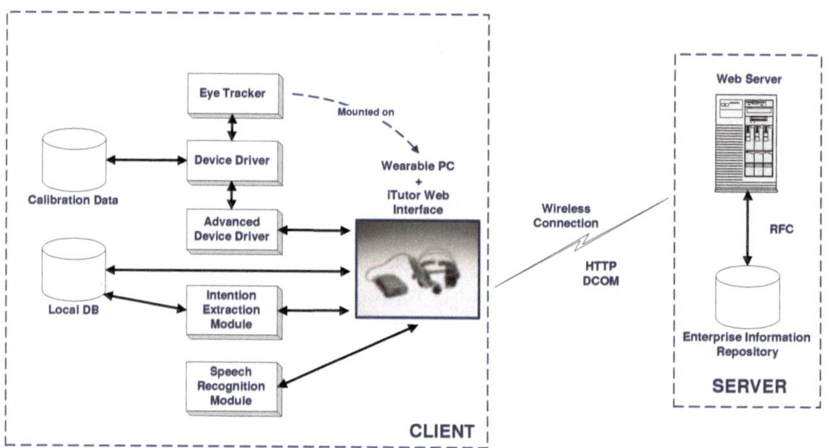

Fig. 3. Architecture of iTutor: a Web-based client-server application

the overall architecture, iTutor has been conceived and designed as a *Web-based client-server application* (Fig. 3). On the *Client Side* three main components of the architecture can be located: the *Eye Tracking Device* and the related *Advanced Device Driver* , the *Intention Extraction Module* and the *Speech Recognition Module* . The *Server Side* of the application is organized as a *two-layer* server architecture: a *Web Server* is mainly devoted to the generation of the ASP pages that set iTutor's interface, while an *Internal Server* supports the communication with the DBMS (Database Management System) of the Enterprise Information System. Finally, the integration in iTutor of an IEEE 802.11b compliant WLAN allows maintainers to freely move through the plant, while working, without loosing contact with the source of the needed information.

3 Conclusions

In this project we have designed and realized a mobile and multimodal computer system to support maintenance and repair processes. iTutor has been designed and developed in collaboration with COMAU Service (FIAT group) which provided also the industrial site for scenario based evaluation with real users (automotive plant maintainers). The positive feedback that has been obtained clearly underlines how the system is effective and fulfills the main users requirements, even if some modifications to the interface will be necessary for its improvement and specific functionalities will be added to the system.

Acknowledgments

This work has been undertaken in the framework of the iEye Project (IST-1999-11883) which is partly funded by the European Commission under the Information Society Technologies 5th Framework Programme. The Authors would like to acknowledge the contributions of their colleagues from University of Tampere, Conexor, SensoMotoric Instruments and University of Nottingham.

References

1. Siewiorek, D., Smailagic, A., Bass, L., Siegel, J., Martin, R., Bennington, B.: Adtranz: A Mobile Computing System for Maintenance and Collaboration. In: Proc. IEEE International Conference on Wearable Computers, Pittsburgh, PA, October 1998.
2. Ockerman, J. J., Najjar, L. J. & Thompson, J. C.: Factory automation support technology (FAST). In D. Adelson & E. Domeshek (Eds.) International Conference on the Learning Sciences, 1996 (p. 567). Charlottesville, VA: Association for the Advancement of Computing in Education.
3. Roussopoulos, N., Kelley, S., Vincent, F.: Nearest neighbor queries. In: Proceedings of the 1995 ACM-SIGMOD Intl.Conf. on Management of Data, San Jose, CA, June 1995.
4. Guttman, A.: R-trees: A dynamic index structure for spatial searching. In: Proc. ACM SIGMOD Int. Conf. on Management of Data, pages 45 - 57, 1984
5. Nielsen, J.: Non-command user interfaces. In: Communications of the ACM, 36(4), 1994, 83-99.

ArtemisExpress: A Case Study in Designing Handheld Interfaces for an Online Digital Library

Kathleen Luchini, Chris Quintana, and Elliot Soloway

University of Michigan, 3111 IST Building, Ann Arbor, MI 48109, USA
{kluchini, quintana, soloway}@umich.edu

Abstract. Learner-Centered Design (LCD) is an approach to building software that supports students as they engage in unfamiliar activities and learn about a new area. LCD has been successfully used to support students using desktop computers for a variety of learning activities, and in this paper we discuss new work to extend LCD to the design of educational software for handheld computers. We discuss some of the challenges of designing handheld software and present a case study of ArtemisExpress, a tool that supports learners using handheld computers for online research.

1 Introduction

Learner-Centered Design (LCD) is an emerging technique that extends traditional user-centered design (UCD) techniques to address the unique challenges of building technology that can support learning. While UCD methods typically focus on making more usable software to support the work of expert computer users, LCD techniques focus on developing software that provides learners with the supports they need to learn about the content, tasks, and activities of the new domain they are exploring.

LCD principles have been used to develop software tools that support learning activities as diverse as modeling complex systems [1], conducting online research [2] and using the scientific method to explore questions about the environment [3]. While LCD tools such as these have been successful in helping students learn about a variety of subjects and activities, in order for students to gain the full benefits of LCD software they must have access to computer tools wherever and whenever their learning activities take place. Mobile handheld computers, such as PalmOS and Pocket PC devices, offer new opportunities to provide learners with one-to-one computer-to-student access. These devices provide students with "ready-at-hand" access to the tools and information they need for a variety of learning activities [4]. By extending the principles of LCD to the design of handheld software, we hope to build learning tools that can support students wherever and whenever learning happens – whether students are outside studying the local stream or in the classroom using their personal handhelds to find information online to contribute to a group discussion. In this paper we discuss an initial effort to extend LCD techniques to the development of ArtemisExpress, a handheld tool to support learners in conducting online research.

F. Paternò (Ed.): Mobile HCI 2002, LNCS 2411, pp. 306–310, 2002.
© Springer-Verlag Berlin Heidelberg 2002

2 Learner-Centered Design

Traditional User-Centered Design (UCD) techniques focus on developing efficient, usable computer tools to support the work of experts. Since experts are by definition knowledgeable about their fields, UCD methods typically try to make computer tools that help experts complete familiar work more productively [5]. A tool designed for experts is not generally usable for learners, however, since learners do not have the background knowledge or experience necessary to use the experts' tool correctly or productively [6]. In addition to their lack of experience, learners are often unmotivated to learn or to work, and individual students have different learning styles and paces [6]. The Learner-Centered Design (LCD) approach has evolved to address these unique learner needs by developing educational software that is not only usable, but that can also help people learn new skills and information.

To address learners' unique needs for support, LCD approaches typically include a variety of *scaffolds*. Scaffolds are temporary supports that provide learners with the additional assistance they need in order to mindfully engage in unfamiliar work [7]. A variety of scaffolding techniques are commonly used in classrooms, such as when a teacher provides coaching to help students complete a task or models a new process for students before asking them to try the work on their own. Scaffolding techniques have also been successfully incorporated into desktop software tools such as Model-It [1], which uses a series of prompts to guide students through the process of building and testing dynamic models of complex environments such as local ecosystems. The scaffolds in Model-It are designed to fade away over time as learners gain knowledge and learn to complete the task without the scaffold's support; this fading allows the Model-It interface to adapt as the learners' needs change over time.

3 Usability and Visibility: The Challenges of LCD for Handhelds

In order to extend LCD principles to the development of educational software for handheld computers, we must understand both the UCD research on designing small handheld interfaces as well as the LCD research on the design of effective scaffolds. UCD research suggests that small screens do not make users more error-prone when using handheld devices for activities such as looking for online information. However, the layout and usability of the interface has a significant impact on users' experiences with handheld devices [8]. One common method for developing usable handheld tools is to streamline the interface so that users can easily access tools and information with a minimum of effort. This can be an effective technique for creating more usable software for experts, but is not necessarily appropriate for educational software. If the handheld software is too automated then the learner may complete the task without truly engaging in the work or developing an understanding of the underlying processes involved with the task. So one challenge of extending LCD to handheld learning tools is understanding how to build tools that are "just usable enough" – meaning that the interface is sufficiently usable to avoid frustrating learners, yet the tool is not so automated that learners can complete the task without truly engaging in the learning process.

In addition to balancing these usability issues, we must also consider how LCD guidelines for scaffolding desktop software can be adapted to design scaffolds within the constraints of smaller handheld screens. For example, LCD guidelines call for scaffolds to be highly visible within the interface, readily accessible to learners, and closely associated with the activity or task that the learner is currently engaged in [3]. In many desktop tools these guidelines are realized by placing most of the scaffolds within the main program interface so that they are always visible and accessible to the learner. An example of this is Artemis, a desktop tool that supports students engaged in online research projects [2]. The Artemis interface makes the search tools available in the center of the screen while keeping the driving question tools, used for collecting and organizing references, always visible on the left of the screen (Figure 1).

By arranging all of the tools and scaffolds on the same screen, Artemis follows the LCD guidelines for desktop scaffolds by allowing students to see the entire context of their research activity and providing easy access to the tools and information they need. However, the small screens and limited input methods available on handheld computers constrain how we design and implement scaffolds on mobile devices. In order to extend LCD design guidelines to the development of scaffolded handheld tools, we must make tradeoffs between the need to make scaffolds visible and accessible to the learner with the need to design a usable handheld interface that organizes tools and streamlines information to avoid frustrating or overwhelming the students. As an initial effort to explore these issues and tradeoffs, we developed ArtemisExpress as a handheld interface to the same digital library used by the desktop Artemis tool. While both programs support similar learning activities, the scaffolds and interface for ArtemisExpress have been re-designed to function more effectively on smaller handheld screens, and in the following section we describe some of the design decisions and tradeoffs involved with the development of ArtemisExpress.

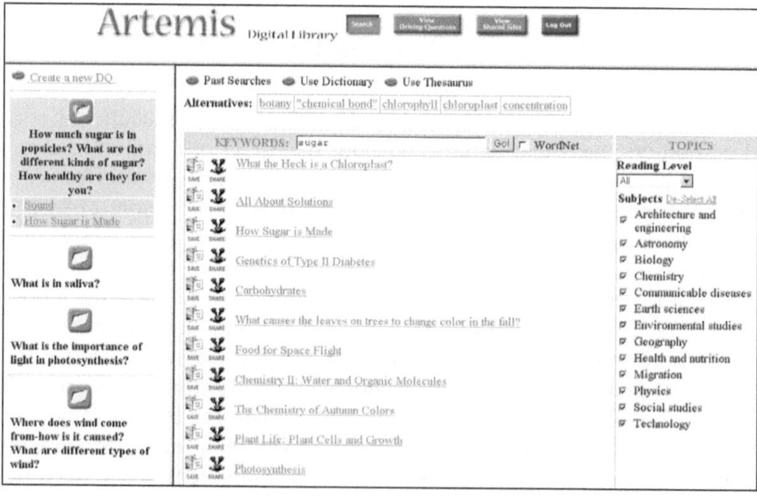

Fig. 1. Desktop Artemis Interface

4 ArtemisExpress: Making Tradeoffs in Scaffold Design

For learners, finding relevant, credible information online is a challenging task. In order to successfully conduct an online investigation, learners must be able to pose a question, utilize search tools to find relevant information, organize, analyze, and synthesize multiple resource, and reflect on the progress of their inquiry over time to determine when they have answered their question or if they need to revise their search [2]. To assist learners in these tasks, both Artemis and ArtemisExpress provide a number of scaffolds, the most prominent of which are the Driving Question (DQ) folders where students can record the questions they are trying to answer and store relevant information that they find during their searchs.

In desktop Artemis, the DQ folders are always onscreen, making these supports highly visible and closely linking the DQ folders to the learner's task of searching for information to answer the driving questions. The playing card-sized screens available on handheld computers made it impossible to achieve the same level of visible scaffolding in ArtemisExpress as in the desktop Artemis. Consequently, we had to make trade-off decisions between keeping scaffolds visible while not crowing so much information onto the screen that the interface becomes unusable. One decision we made was to design separate screens for the search and DQ functions. The search window in ArtemisExpress (Figure 2) allows learners to execute keyword searches and view the results, while the DQ window (Figure 3) provides access to the learners' driving questions and saved references. The benefit of creating separate search and DQ windows in ArtemisExpress is that within each area the individual scaffolds are visible and students can easily access related tools and information. However, one of the drawbacks of this design decision is that learners cannot see both the search and the driving questions simultaneously, and this separation may cause learners to lose sight of the overall context of their research activities.

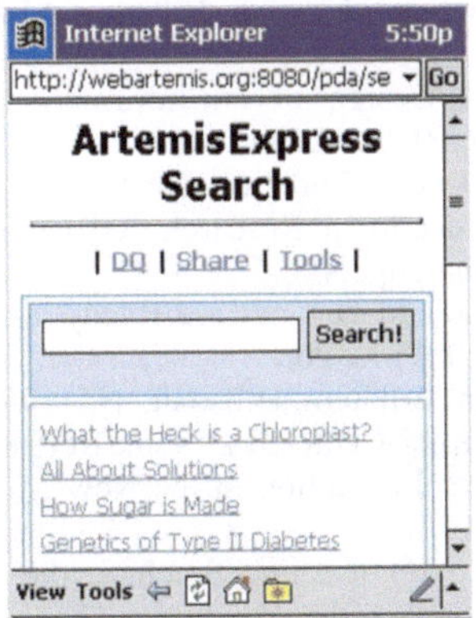

Fig. 2. ArtemisExpress Search Window

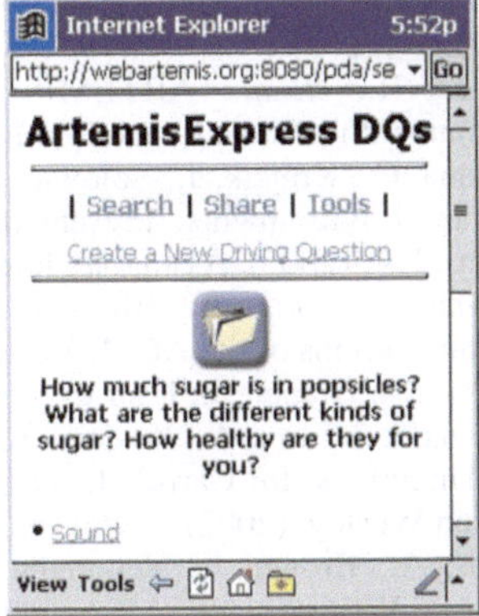

Fig. 3. ArtemisExpress DQ Window

5 Conclusions and Future Work

Our current efforts to extend LCD guidelines to the development of scaffolds for handheld learning tools are being tested as part of an ongoing classroom study. This study involves 33 eighth graders (students approximately 13 years old) who are using ArtemisExpress in science class. We are comparing students' performance using both the handheld ArtemisExpress (on iPAQ Pocket PC computers with wireless Internet access) and the desktop Artemis website for the same search activities. Through this comparison testing we hope to identify whether the handheld scaffolds we developed are effective, and how the trade-off decisions between scaffold visibility and interface usability in the handheld software impact students' work processes. We hope that this study will help identify additional methods for extending LCD guidelines to address the unique challenges of building handheld tools for learners.

Acknowledgements

The work described here is supported in part by Intel Research, Microsoft Research, and the National Science Foundation under grant number NSF ITR 0085946 and a Graduate Research Fellowship. Any opinions, findings and conclusions or recommendations expressed in this material are those of the authors and do not necessarily reflect those of the NSF.

References

1. Metcalf, S., Krajcik, J., Soloway, E.: Model-It: A Design Retrospective. In : Jacobson, M., Kozma, R (eds.): Innovations in Science and Mathematics Education. Erlbaum, Mahwah, NJ (2000) 77-115.
2. Wallace, R., *et. al.*: ARTEMIS: Learner-Centered Design of an Information Seeking Environment for K-12 Education. In: Proceedings of CHI '98. ACM (1998) 195-202.
3. Quintana, C., Krajcik, J., Soloway, E.: Scaffolding Design Guidelines for Learner-Centered Software Environments. Presented at American Educational Research Association, New Orleans, LA, (2002). Available: http://hice.org/quintana-papers
4. Soloway, E., Norris, C., Blumenfeld, P., *et. al*: Handheld Devices are Ready-at-Hand. In: Communications of the ACM, Vol. 44. ACM Press (2001) 15-20.
5. Norman, D., Draper, S.: User Centered System Design. Erlbaum, Hillsdale, NJ (1986).
6. Quintana, C., Carra, A., Krajcik, J., Soloway, E.: Learner-Centered Design: Reflections and New Directions. In: Carrol, J. (ed.): Human-Computer Interaction in the New Millennium. Addison Wesley, (2002).
7. Collins, A., Brown, J., Newmann, S.: Cognitive Apprenticeship: Teaching the Crafts of Reading, Writing, and Mathematics. In: Resnick, L. (ed.): Knowing, Learning, and Instruction: Essays in Honor of Robert Glaser. Erlbaum, Hillsdale, NJ (1989) 453-494.
8. Kim, L., Albers, M.: Web Design Issues When Searching for Information in a Small Screen Display. Presented at: SYSDOC '01, Santa Fe, NM (2001).

Addressing Mobile HCI Needs through Agents

Gregory O'Hare and Michael O'Grady

Practice & Research in Intelligent Systems & Media Laboratory (PRISM),
Department of Computer Science, University College Dublin (UCD),
Belfield, Dublin 4, Ireland,
{gregory.ohare,michael.j.ogrady}@ucd.ie

Abstract. Addressing the needs of the mobile computing community in all its guises is one of the critical challenges facing the research community in the coming decade. Given the constraints under which mobile computers must operate, significant effort must be expended to ensure that the end user's experience is satisfactory. In this paper, the selective use of intelligent agents as a means of augmenting the user experience through interfacing with the physical environment and anticipating user requirements is proposed.

1 Introduction

Mobile computing has dramatically changed the way users view and use computers. It has also raised serious technological challenges for the research community. From a potential user perspective, it has raised questions about security and privacy. Addressing these issues and concerns will be amongst the most import challenges facing researchers in the near future.

Context-aware computing [1] has been the subject of much research in the last decade and has been extensively described in the literature. One of the core premises of context-aware computing is that the computing device should be aware of the user's circumstances and should be able to interpret any interaction in a manner appropriate to these circumstances. It has already been demonstrated that context can form a key component in mobile HCI [2]. More recently, researchers have been exploring the concept of Ambient Intelligence, a convergence of ideas from ubiquitous computing and intelligent user interfaces amongst others, which seeks to provide a seamless and intuitive interface to computational resources in everyday life.

In this short paper, the use of agents for interpreting captured context and using this to anticipate user's needs is proposed. An example based on a mobile context-aware tourist guide is used to demonstrate feasibility.

2 Modeling Context via Agents

Though the benefits of incorporating a context-sensitive component into a mobile computer application may be immense, the actual practicalities of doing so

F. Paternò (Ed.): Mobile HCI 2002, LNCS 2411, pp. 311–314, 2002.
© Springer-Verlag Berlin Heidelberg 2002

present some difficult design and implementation problems. For example, identifying the user's context with a reasonable degree of certainty at any given time may prove difficult or even impossible. Even assuming that the user's context has been determined, identifying what action, if any, should be taken can be difficult. Presenting the user with information at an inappropriate time, or information that is based on false premises, will quickly lead to user dissatisfaction. The potential for such a scenario is high, and for this reason, context has tended to be used conservatively. However, the successful deployment of applications that seamlessly capture and interpret context will mark a milestone in mobile computing and offer significant scope for further research in mobile HCI.

Intelligent agents have been deployed in a multitude of settings with the goal of addressing some of those problems that traditional computer science find difficult to handle. Agents by definition are autonomous entities, can monitor and react to changes in their environment and are proactive in attempting to fulfill their objectives. If agents are considered from an AI perspective and endowed with mentalistic attributes such as beliefs and commitments, various scenarios can be modeled in a highly intuitive manner including the capture and interpretation of a user's context.

3 Deploying Agents on a PDA

One of the results of the massive strides that have been recently made in handheld technologies is that the deployment of sophisticated Multi-Agent Systems (MAS) on such devices is practical. A number of such systems have been successfully deployed on mobile devices including LEAP [3] and Agent Factory Lite[4], which was used to develop the WAY [5] system on a PDA. This environment was developed in Java and can be deployed on any PDA or laptop that supports a JVM that complies with the PersonalJava specification.

4 Agent-Enabled HCI: The Gulliver's Genie Story

Gulliver's Genie is a mobile context-sensitive handheld tourist guide. The modus operandi of the genie is quite simple: as the tourist wanders a city, they have online access to a large repository of information that is customized to their location and personalized to match their interest profile. While there are obviously a number of technological challenges that must be overcome before deploying such a system, for the purposes of this discussion, just two are focused on. The first concerns the problem of interpreting context. The second concerns using context in a meaningful way so as to augment the tourist's experience.

4.1 Interpreting Context

In this case, those elements of context that are of particular interest are the location and orientation of the user. Output from GPS devices usually adhere to some international standard so capturing it is not particularly difficult. However, interpreting it can be a more subtle process. For example, is a particular reading consistent with previous readings? Is the quality of the readings varying? And if so, what can be inferred from this? If the tourist's orientation is changing rapidly, is it safe to deduce that the tourist might be lost and would welcome some navigation hints? Or is he/she just avoiding obstacles? Dynamically reconciling the different objectives and priorities that a simple tourist based scenario may give rise to poses a number of difficult challenges for design and implementation.

4.2 Just-in-Time Information Delivery

One of the key issues that a mobile tourist information system must address is that of ensuring that information is delivered in an appropriate spatial and temporal setting. The use of GPS aids the resolution of the spatial element but ensuring that information is delivered at the appropriate time is a challenging task. In this case, one of the principle obstacles to be overcome is that of poor network bandwidth, which is a characteristic of most cellular wireless networks. For example, GSM supports a standard data rate of 9.6kb/s. Though the deployment of 3G technologies will improve the situation, bandwidth will remain an issue for the foreseeable future. Traditionally, one of the standard methods of overcoming this limitation has been through the implementation of a pre-caching/pre-fetching strategy when deploying mobile computing applications. However, given the limitations of standard PDA devices, particularly memory and disk space, the implementation of such a strategy requires an intelligent and adaptable software solution. Indeed, the nature of the data i.e. multimedia based, reinforces this requirement.

5 Deploying Intelligent Agents on a Mobile Device

In delivering the intelligent agent machinery, we commission Agent Factory Lite. This supports the strong agent notion encapsulated in a BDI architeture. Two agents have been designed and implemented for Gulliver's Genie:

Spatial Agent: This agent monitors the incoming data from the GPS receiver. After ensuring that the quality of the data is adequate, it extracts position and orientation readings. Having checked that both are consistent with previous readings, it then deliberates about broadcasting them to the relevant components of Gulliver's Genie. For example, it must consider factors such as the current user activity, the distance traveled since the last reading and how much time has elapsed since the last update.

Cache Agent: Intelligent decisions must be made to anticipate user content requirements. Based on information received from the Spatial Agent, Intelligent precaching is utilised to retrieve content for potential candidate points of interest prior to the user's arrival. As revised user movement data is received, these are incrementally eliminated. In this way, the Cache Agent ensures that content is downloaded in a transparent manner and that the desired presentation is instantly available on the user's arrival.

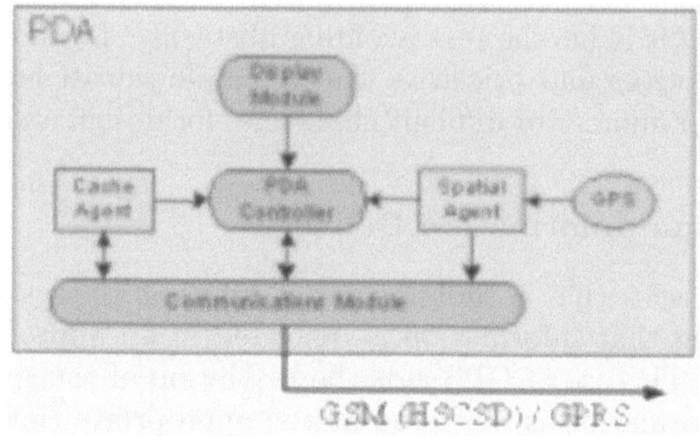

Fig. 1. The PDA-based components of Gulliver's Genie

6 Future Work

An immediate priority is to continue refining the agent's behavior through the addition of extra rules and to conduct some field trials. Secondly, it is planned to incorporate some weak mobility into the agents so that they can determine criteria and situations where it may be appropriate for their migration to another node. Load balancing or depleted battery life are examples that may motivate migration.

References

1. Schilit, B., Adams, N., and Want, R.: Context-Aware Computing Applications. In: Proceedings of the Workshop on Mobile Computing Systems and Applications. Santa Cruz, CA, USA (1994)
2. Schmidt, A.: Implicit Human Computer Interaction through Context. Personal Technologies, Vol. 4 (2&3), June 2000.
3. Adorni, G., Bergenti, F., Poggi, A., Rimassa, G.: Enabling FIPA Agents on Small Devices. In: Proc. of the Fifth International Workshop on Cooperative Information Agents. Modena, Italy (2001)
4. O'Hare, G.M.P., O'Hare, P.T., Lowen, T.D.: Far and A WAY: Context Sensitive Service Delivery through Mobile Lightweight PDA hosted Agents. In: Proc. of 15th International Florida Artificial Intelligence Conference. Pensacola, FL, USA (2002)
5. Lowen, T.D., O'Hare, G.M.P., O'Hare, P.T.: Mobile Agents point the WAY: Context Sensitive Service Delivery through Mobile Lightweight Agents. In: Proc. of First International Joint Conference on Autonomous Agents and Multi-Agent Systems, Bologna, Italy (2002)

Consistent Multi-device Design Using Device Categories

Martijn van Welie and Boyd de Groot

Satama Interactive Amsterdam, Poeldijkstraat 4, 1059 VM, Amsterdam, The Netherlands
{martijn.van.welie,boyd.de.groot}@satama.com

Abstract. Mobile devices differ in many aspects such as screen, keys, browsers, java support and much more. The difficult task designers now face is how to design solutions that take sufficient advantage of specific device characteristics while offering a consistent and similar experience. In this paper we discuss an approach using device categories to tackle this design challenge. Our categorization is based on relevant design considerations rather than device features.

1 Introduction

The mobile industry is traveling fast. New devices are appearing constantly as device manufacturers try to beat their competitors. Operators that offer mobile Internet to their customers continuously update portals to deliver the most attractive mobile experience that is possible. In doing so they face the challenge to design for this large set of devices that is increasingly differentiating. In the early days, most phones had a simple WAP browser, where only the differences in browsers were relevant. Nowadays, the screen sizes have differentiated so much that special designs for large screens are necessary to provide an optimal experience. Some devices also adopted color screens and even touch screens with a stylus. Recently the Japanese I-mode technology introduced a new browser type as some devices now support XHTML and WML 2.0. Although devices are differentiating, operators and other application developers are struggling with their goal to provide a consistent design solution. Especially operators are interested in achieving this kind of consistency over a large set of applications. For end-users consistency provides a recognizable and predictable user experience. In the following sections we propose an approach that uses device categories to manage to the number of different designs that need to be made.

2 Device Characteristics

The first thing to understand is which device characteristics differ between mobile devices. The second thing is to understand to what extent they have an impact on the sort of design solution that should be made, which will be done in the next section.

F. Paternò (Ed.): Mobile HCI 2002, LNCS 2411, pp. 315–318, 2002.
© Springer-Verlag Berlin Heidelberg 2002

- **The browser**. Even if a browser supports a standard such as WAP 1.X that does not mean that the browser interprets the WML files the same way. Some browsers implement browser specific tags that other browsers ignore. Later versions of WML are backwards compatible but lead to unsatisfactory results if a legacy browser views the pages.
- **The markup language**. The markup languages that are used nowadays include WML, cHTML, XHTML of which there exist different versions for each of them. Naturally, any design solution needs to be viewable on the device it is intended for. If not, the solution is not viewable.
- **Output capabilities**. Displays exist in all sorts of sizes and some are in color while others are in black & white or in grayscale. The most direct impact of display differences concerns the number of lines of text that can be displayed. Typically, this ranges from 3 to 9 lines of text. Another factor is the shape of the display. Most of them are vertically oriented but horizontally oriented devices are also common.
- **Interaction style / input capabilities**. The number or keys that are used on the device differs significantly. Some have shortcut keys to directly access the browser while other devices require several actions to start the browser. Most devices also use soft keys in the browser but the number of available soft keys varies between 1 and 3. In I-mode phones the keypad is used for menu shortcuts while this is not supported in most WAP enabled phones. Some phones even have a stylus and touch screen that allows an entirely different way of interaction.
- **Data connection.** Some devices use GSM others HSCSD and nowadays GPRS. The bandwidth and the time to connect differ significantly.

These aspects all influence what the best solution is for a particular device. Theoretically, delivering device-specific solutions would be possible if we could detect the exact device that is used to access the service.

3 Design Issues

The types of design issues that are involved in designing for a wide range of devices are numerous. A few examples include:

- *The use of page titles, brand headers, and section labeling*. On a small screen it is only possible to show a page title in order to leave as much space as possible on the remaining screen area. On larger screens, the content might be grouped and labeled which would be disastrous for small screens. For commercial services space is also required for brand elements such as logos.
- *The number of links that can be presented on one page*. Scrolling is problematic on small devices so it is important to make sure the number of links is matched with the number of lines that fit on one screen. This directly influences way the site should be structured.

- *The layout possibilities.* The larger screens make it possible to use a two-column layout of the information. Doing so may significantly improve the user experience.
- *The text length.* The width of the screen determines how link labels should be for links and other text elements. Using long labels leads to wrapping which may cause problems in distinguishing one link from the next.
- *The use of soft keys.* Good use of soft keys is important feature for speeding up the interaction. Will there always be a soft key for "back" navigation or should a back link be included in the design? If the number keys are used as menu shortcuts the menu items should be numbered.
- *Use of images.* Now that some devices have color displays, new types of applications can be envisioned. Images may also enhance the user experience by way of icons etc. However, using images dramatically increases the amount of data that needs to be downloaded. Adding one image may typically lead to a 4 times increase of data while the connection speed may only be 2 times as fast compared to GSM.

The bottom line is that "design-once run-everywhere" does not apply for the current mobile market. When designing for mobile devices it is simply a necessity to take the targeted device types into account.

4 Designing Using UI Device Categories

At present it is at least partially possible to detect the device from a service point of view. The exact solution that is sent to the device can be delivered using techniques such as XML in combination with device-specific XSLTs. However, it is usually not feasible to design and build a XSLT for every device on the market. The logical conclusion is to design for a specific class of devices rather than for every particular device. The difficult part is define the classes and how to design for them. Ideally, the devices class would not have to change when new devices come out.

In choosing the categories it is important not to define it in any criteria that is very precise or easily outdated (e.g. the screen resolution). A possible more time-independent approach is to define it using the way the phone is typically used and by whom. Based on the current roadmaps of handset manufactures, we estimate that our approach should work for at least the next 1 or 2 years. Phone manufactures also develop their phones for particular market segments [1]. Nokia uses four categories [1] for its devices (series 30, 40, 60 & 90) which are centered around device features and cost for manufacturing. For designers, a categorization should be focused around the issue of *designing separate solutions* rather than classifying device features. In our case, the screen characteristics and the interaction style have the most impact on the design. For example, when designing for the Nokia series 30 & 40, the same design can be used since the differences are not large enough to justify separate designs.

These four basic categories will each require a "compromised" design solution coded in all markup variants. This way, the number of designs that must be made comes down to 4 while the number of *coded* solutions comes down to 8 or less. For the PDA categories it is nowadays not necessary to code solutions in multiple markup-languages, HTML is the defacto standard.

Table 1: A UI Device Categorization

Category	Definition	Usage
Basic	Small screen around 3,5x2,5cm, only the most basic navigation possibilities using 1 or 2 softkeys. Often b/w screen but can also color devices are emerging. Technology is typically WML. Typically 4 to 6 lines of text can be displayed	Trendy youths where the phone has a high social value. Phones must be small.
Advanced	Larger screen around 3,5x5cm, typically color screens and feature rich phones using 2 or 3 softkeys and sometimes even extended keypads. Technology is color WML, cHTML or XHTML. Typically 6-9 lines of text can be displayed. Two-column layouts can be used for icons.	Demanding users that require more than just a phone. Phones should be easy to operate and are hence larger than basic phones.
Horizontal PDA	Phone/PDA with a horizontal screen layout and qwerty keyboard. Technology typically XHTML and also WAP. Also exist in touch screen variant. Typically 6-9 lines of text can be displayed.	Heavy business users that need to write a lot of messages and need PIM features.
Vertical PDA	Phone/PDA with vertical screen layout. Has touch screen and possibly a mini qwerty keyboard. Typically uses XHTML. Typically 15-20 lines of text can be displayed.	Business users that do not need a keyboard bit still require a feature rich phone/PDA combination.

The problem with any categorization is that there will always be devices that do not fit exactly in the category's description. For example there are devices with a tall but narrow display that still belong into the basic category because of all other device characteristics.

5 Conclusions

In order to deliver a high quality experience on a mobile device, it is necessary to take device differences into account. Since designing solutions for every device on the market is not feasible, devices need to be categorized. In this paper we propose a possible categorization based on the type of usage.

References

1. Lee, L.: Nokia Vision on Mobile Java. (2001) Sun Nokia Conference, December 2001, Singapore, http://www.sun.com.sg/events/presentation/files/sun_nokia_conference/pres_luke.ppt

New Generation of IP-Phone Enabled Mobile Devices

Michimune Kohno and Jun Rekimoto

Interaction Laboratory, Sony Computer Science Laboratories, Inc.,
3-14-13 Takanawa-Muse Bldg., Higashi-Gotanda, Shinagawa-ku,
Tokyo, 141-0022 Japan
{mkohno,rekimoto}@csl.sony.co.jp

Abstract. This paper describes how IP-phone communication and real-world user interface can become a new standard for mobile terminals. Current wireless broadband networks such as the 802.11a/b will eventually lead to IP mobile phones. As "smart" appliances emerge on the market, mobile terminals can play a new role in the ubiquitous computing environment. This paper integrates IP-phone communication capability and intuitive user interface into a mobile terminal. It explains the use of the "pick-and-drop" technique as a controlling interface for voice sessions and music streams, enabling both usability and security, important in a practical ubiquitous computing environment. Finally, a prototype implementation will be briefly described.

1 Introduction

The rapid development of computer networks provides us with the ability to communicate with one another, wired or wireless, all over the world through broadband network in the forms of web technology, VPN, and host-spot services.

Information appliances will soon emerge on the market by introducing protocols such as INS, UPnP, and Jini[2,3]. The main purpose of these protocols is to control appliances remotely by using PCs, PDAs, or mobile phones.

Mobile devices receive broadband network communication capability through IEEE 802.11a/b. As wireless infrastructure (such as hot-spot locations) increases, more users are expected to access private data from a remote site with mobile devices in the near future.

A mobile device with network communication can be an IP-phone terminal device using current VoIP technologies[4], which allows us to make phone calls without carrier lines. Despite obstacles with mobility support[5], fast handover, and power consumption, integrating voice communication into a mobile terminal will add to its value by combining the conveniences of a mobile phone and a computer to create a new generation of handheld devices.

This paper describes the effects of this terminal in a ubiquitous computing environment, presents some practical uses, and proposes a solution.

F. Paternò (Ed.): Mobile HCI 2002, LNCS 2411, pp. 319–323, 2002.
© Springer-Verlag Berlin Heidelberg 2002

2 Mobile Device with Voice Communication Capability

The fusion of a mobile phone and a computer goes beyond the idea that the two is mounted on a single component. For instance, provided with the current technology, when sending a file to the person on the receiving end of a phone call, you have to attach it to an e-mail with the correct e-mail address by using the mail service on your PC. However, if the PC has telephone capability, no e-mail address needs to be selected because *the PC is already connected* to the receiver through the phone call. Ideally, IP voice communication does not limit the number of callers, allowing the capability to collectively send data to all callers easily.

Although this feature exists in many instant messaging applications, it has yet to be applied to mobile devices. Furthermore, these applications do not offer any functions to access private data stored at a remote site.

By meshing appliances and mobile terminals, new applications for current technologies such as projectors, tablet-type computers, audio devices, and car gadgets will emerge for the ubiquitous computing environment.

The following scenarios show concrete applications of this technology:

- You are talking with a mobile phone while walking. You can transfer the call to the microphone and speaker in the car or office.
- You have just received a file sent by a peer. You can display it on a nearby projector.
- You are listening to music in your room with a stereo. You can transfer the music session to your mobile phone and go out.

Although security problems and access control rights may arise in these example scenarios, which are important issues for practical use, this paper only focuses on user interface that provides usability for such applications. We will discuss security issues in another paper.

3 User Interface Technique as Session Control

Several approaches can be taken to utilize a mobile terminal as a universal remote control in a ubiquitous computing environment; however, most work in this area focus on how to display the controlling interface on a window[7,8].

This paper proposes a different approach from the existing research by:

- integrating the IP-phone function, and
- using the pick-and-drop technique[9] as the user interface for the session control

The user interface technique that Rekimoto et al originally proposed uses direct manipulation for transferring files between computers. We adopted it to switch the receiving end in a voice communication or a music stream from one device to another, namely, *physically bringing the terminals closer together* by mobilizing the endpoint of a session. In a similar way, if you want to see a

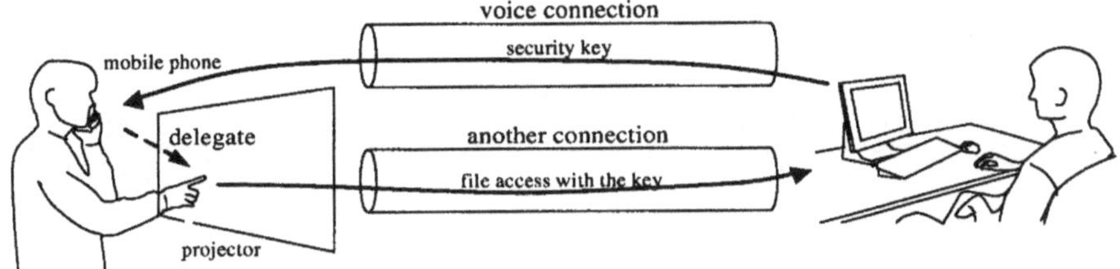

Fig. 1. An example scenario: using a mobile phone as a controlling interface for a wall projector. By bringing the mobile phone closer to the wall, the contents of the received file is projected on the wall. Note that security credentials are implicitly transferred to grant access rights to the file

received file on a certain display, you can bring the mobile terminal closer to the display. The display would then grant temporary access rights to the file allowing the viewer to open the file. The viewing device can either be controlled via the terminal or the opened voice session.

As mobile phones merge with computer networks in the future, the mobile terminal will become increasingly important as a controlling device in the ubiquitous computing environment. This characteristic is further enhanced by adding such intuitive user interfaces as having the capability to *physically* bring objects closer, pointing at a device (like a remote control), and so on. Furthermore, from a technical perspective, it is important that such intuitive user interfaces can *implicitly* resolve security issues that usually degrade usability.

4 Prototype Implementation

We have created a prototype incorporating the characteristics. Fig.2 illustrates the overview. We used a Linux PDA as the handheld device with an 802.11b network interface card. We then implemented voice communication capability using a simple protocol and H.323. This same system was also implemented running on Windows2000.

We used UPnP as the underlying protocol. Therefore if a terminal enters an area in which wireless communication is available, any appliance with this type of communication enabled can be detected by the terminal.

As the target appliance, we used a PC with speakers and a microphone. In order to detect the approach of an existing terminal to the speakers, an RFID tag and a reader/writer device were used. The reader/writer, which is connected to the PC via RS232C, was attached to the speaker, and an RFID tag was attached to the terminal. The detection of the RFID tag activates the session control procedure, which opens communication between the PC and the terminal.

RSA, a popular public key encryption mechanism, is used to verify security information. When communication is opened by detecting the RFID tag, digital certificates are exchanged with each other and verified. If authentication is

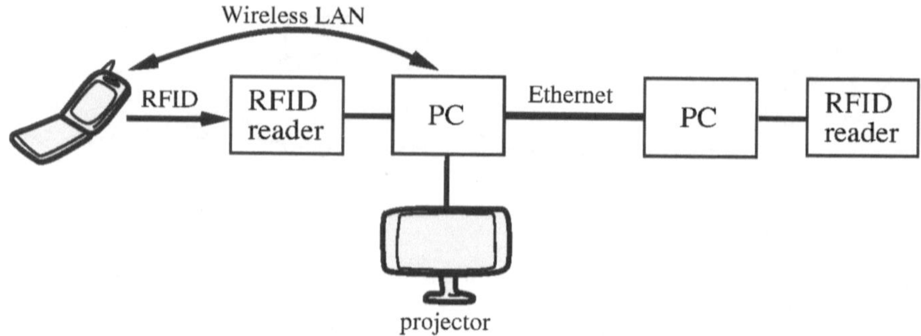

Fig. 2. The prototype system overview

correct, then a secured session channel is opened between the PC and the terminal to prevent other appliances from accessing. Next, the PC gives the terminal temporary access rights via the secured session.

Thus, this system allows us to open a connection without touching the LCD screen of the terminal.

5 Conclusion

This paper explained how voice communication and real-world user interface can create a new generation of mobile terminals, which can also implicitly resolve security issues.

Passing security information in a session with voice communication already established is still a fairly new concept, and critical for mobile devices to become actual controllers of smart appliances.

We believe that numerous applications that combine with the ubiquitous computing environment will emerge in the near future, thus we plan to develop more practical systems. The details of how such systems can become more secure will be discussed in another paper.

References

1. Waldo, J.: The Jini Architecture for Network-Centric Computing, in Communications of the ACM, vol. 42, No. 7, July 1999, pp. 76–82.
2. Adjie-Winoto, W., Schwartz, E., Balakrishnan, H. and Lilley, J.: The design and implementation of an intentional naming system, In Proc. of the 17th ACM SOSP, Dec. 1999.
3. Microsoft Corportaion: Understanding Universal Plug and Play: A White Paper, `http://www.upnp.org/download/UPNP_UnderstandingUPNP.doc`.
4. IETF: Session Initiation Protocol, `http://www.cs.columbia.edu/sip/`.
5. Perkins, C. and Johnson, B.: Mobility Support in IPv6, in Proc. of MOBICOM'96, ACM, 1996.

6. Kunishi, M., Ishiyama, M., Uehara K., Esaki, H. and Teraoka, F.: LIN6: A New Approach to Mobility Support in IPv6, in Proc. of the Third International Symposium on Wireless Personal Multimedia Communications, Nov. 2000.
7. Buyukkokten, O., Garcia-Molina, H., Paepcke, A. and Winograd, T.: Power Browser: Efficient Web Browsing for PDAs, in Proc. CHI2000, Apr. 2000.
8. Zimmermann, G., Vanderheiden, G. and Gilman, A.: Prototype Implementations for a Universal Remote Console Specification, in Proc. CHI2002, Apr. 2002.
9. Rekimoto, J.: Pick-and-Drop: A Direct Manipulation Technique for Multiple Computer Environments, in Proc. of UIST'98, ACM Press, pp. 31–39.

A Content Provider-Specified Web Clipping Approach for Mobile Content Adaptation[1]

Seomin Yang[2], Hyukjoon Lee[2], Kwangsue Chung[3], and Hwansung Kim[3]

[2]School of Computer Engineering, Kwangwoon University
447-1 Wolgye-Dong, Nowon-Gu, Seoul, 139-701, Korea
Fax: +82-2-915-5127
uniload@explore.kwangwoon.ac.kr, hlee@daisy.kwangwoon.ac.kr
[3]School of Electronics Engineering, Kwangwoon University, Seoul, Korea
447-1 Wolgye-Dong, Nowon-Gu, Seoul, 139-701, Korea
{kchung,hskim}@daisy,kwangwoon.ac.kr

Abstract. In this paper a new mobile content adaptation method based on web clipping is introduced. In this method, a clip is automatically extracted from a source page based on a clip specification provided by a content provider and transformed into a target page according to a set of conversion rules. A new XML-based language, WCML (Web Clipping Markup Language) is defined to store the clips in an intermediate meta-language, which are later transformed into multiple presentation pages in different target markup languages. A clip editing tool is designed and implemented to allow the content provider to easily provide clip specifications and preview and work with the layout of target pages through a graphic user interface. Transcoding of image objects in major image file formats is also supported.

1 Introduction

As the number of nomadic users carrying portable mobile communication devices increases rapidly, the demand for Internet or web access through these devices also grows tremendously. However the rendering capabilities of these devices are limited and networks (mostly wireless) offer unstable connectivity and scarce bandwidth.

Transcoding proxy server approach has been widely accepted as a practical means to enhance user experience in wireless mobile web access. Transcoding provides users with accelerated network access speed through a lossy compression approach which reduces the amount of transferred data through the wireless link at the expense of the lower image quality in terms of resolution, size, the number of color depth, etc. [1,2]. Although transcoding provides a framework for mobile content adaptation, they are not quite good enough for practical use — the readability of transformed pages is too low due to the lack of the function of page layout adaptation.

As an alternative approach, several mobile presentation language standards have been developed for display on small mobile terminals. WML (Wireless Markup

[1] This work was supported by Grant No. R01-2001-00349 from the Korea Science & Engineering Foundation and the Research Grant of Kwangwoon University in 2002.

F. Paternò (Ed.): Mobile HCI 2002, LNCS 2411, pp. 324–328, 2002.
© Springer-Verlag Berlin Heidelberg 2002

Language)[3], CHTML [4], XHTML basic [5] are some examples of such languages. These mobile markup languages, however, brings a new overhead to the mobile content providers — they must now create and maintain multiple copies of the same content in different markup languages. The notion of transcoding can be extended to include markup language conversion to remove this overhead. However, the readability problem remains with a simple tag conversion scheme.

Web clipping is a technique used to extract and present parts of a HTML document for a device with a small-size display like PDA's [6]. One of the biggest advantages of web clipping is that a source page becomes highly manageable for layout adaptation without loss of any of its information. Therefore, a better transformation result can be achieved if the source page is clipped first. In this paper we present a new mobile content adaptation method based on web clipping. In this method a source page is automatically clipped and transformed according to the clip specification made by the content provider using a clip editing tool. Each clip is transformed into an intermediate meta-language document, which in turn is transformed into the presentation page in the target markup language. Image transcoding functionality for most major image formats is also supported.

The rest of this paper is organized as follows. In section 2 we discuss the main features and key techniques used in our system. In section 3 we describe the clip editing tool. Finally, in section 4 we conclude our discussion.

2 Web Clipping-Based Mobile Content Adaptation

In our web clipping-based mobile content adaptation method, the basic unit of content transformation is a clip. Each clip is specified by a pair of beginning and ending tags and represented by a pair of nodes in a DOM tree. This information is used in the clipping step to extract the clip from a source page. A clip can be specified by the content provider using the clip editing tool described in the next section.

A clip may contain HTML errors as a HTML document often contains errors. These errors must be corrected before markup conversion proceeds so that the clip becomes a well-formed document. The error correction step consists of inserting missing tags and relocating incorrectly positioned tags.

After the errors are fixed, the clipped document in its source markup language is converted into the document in the intermediate meta-language WCML by a set of conversion rules specified in XSLT. The conversion rules are determined a priori so that the resulting intermediate document remains as close to the source document as possible in terms of structures and attributes, while most of the elements and attributes that require high-level browser capabilities are removed. The intermediate document is then converted again into the target language document. Here, the conversion rules are also specified in XSLT. If the target markup language has smaller page units such as the cards and decks of WML, the source texts are decomposed accordingly. A separate XSLT stylesheet is specified for each source and target markup language. The overall processing steps described above are illustrated in Fig. 1.

WCML is defined as a XML application. It is both minimal and inclusive in the sense that the elements for good appearance are excluded, while those for the representation of basic document structures are kept (Table 1). WCML is not a new markup language for either source pages or converted pages, but is an intermediate language for internal use only. Image transcoding converts an image object into another for better presentation on a mobile display using size reduction, quality degradation and format conversion.

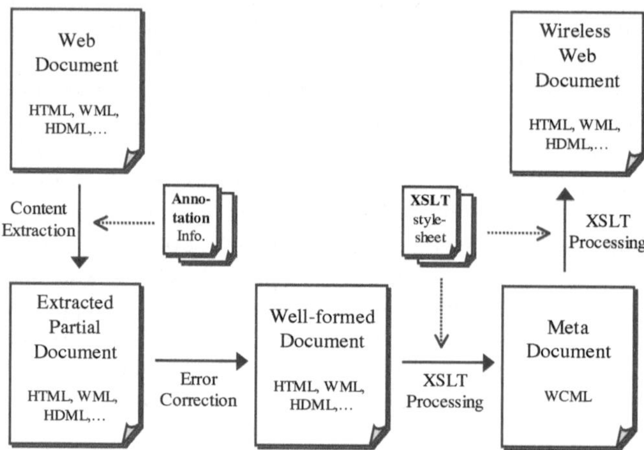

Fig. 1. Processing steps of web clipping and transformation from a HTML page to a target page

Table 1. The summary of WCML elements. Only the 1st-level and 2nd-level elements are shown and their descriptions are shown.

1st-Level Elts.	2nd-Level Elts.	Description
wcml:clip-info		Information about a clip
	wcml:clip-title	To identify a clip
	wcml:clip-desc	Brief description
wcml:content		Actual content of a clip
	wcml:clip-div	To define a block-level text
	wcml:heading	For headings
	wcml:para	For paragraphs
	wcml:anchor	For anchors
	wcml:list	For lists of information; Must contain one or more list elements
	wcml:table	To arrange data into rows and columns of cells
	wcml:object	To include embedded objects, such as images

3 The Clip Editing Tool

An automated process of clipping HTML pages based on a simple analysis may produce a misleading result because the usage of HTML tags does not directly reflect the underlying content of a page. An example of this is a large image divided into several pieces and pasted together using a table tag. Thus, the best possibility might be to have the content provider specify the clips.

The key features of our clip editor can be summarized as follows:

- A GUI with intuitive control by mouse clicks
-
- Layout control and addition of new objects
- The preview function of a converted page
- An interface for the content transformation server for upload, download, modification and management of annotations

Associated with each clip area are five attributes, i.e., *clip ID*, *title*, *brief description*, *level*, and *clip location information*. The *clip ID* is used to identify each clip. The *title* and *brief description* of a clip are displayed on the transformed page. The title is given a hyperlink pointing to the clip so that upon clicking the title, one can move to the transformed clip. Several clips can be grouped together to form a larger clip or a composite clip. A clip can be further clipped (multi-level clipping). This is useful when a clip chosen for a PDA is too large for the display on a mobile handset and hence need be divided into several smaller clips. The *level* specifies at which level of clipping the clip belongs to and its value increases by one as a clip is clipped further. The *clip location information* specifies the path from the root to the clipped nodes in the DOM tree for the entire page. Fig. 2 illustrates an example of specification of clips in a HTML page.

A clip can be selected by clicking a mouse on the part of a page that is delimited by a pair of beginning and ending tags and displayed by a surrounding box. The selected part is then alpha-blended and the values of clip attributes are assigned as described above. The values of these attributes are stored as an annotation file along with the URL mapping information in the transformation server to be used in the clipping step. This information is also used to update an existing specification or to map it to another page of the same structure. The transfer of an annotation file between the clip editor and the transformation server can be executed within the clip editor.

A dynamic page generated by web programming facilities such as CGI, PHP, Java servelet, JSP and ASP usually has a framework that is repeatedly used to display content in the same pattern. Inbox mail lists, bulletin boards, and shopping mall pages are some examples. This framework can be specified as a dynamic clip and each occurrence of this framework in a page can be clipped separately by searching for the same pattern in the DOM tree. Other kinds of dynamic pages cannot be clipped.

The preview feature allows the user to preview the result of content adaptation using terminal display emulators. The content adaptation server performs the actual transformation to generate the preview pages. The preview feature also allows the user to edit the clip page while previewing it. New elements such as images, texts and hyperlinks can not only be added, but the layout can also be adjusted.

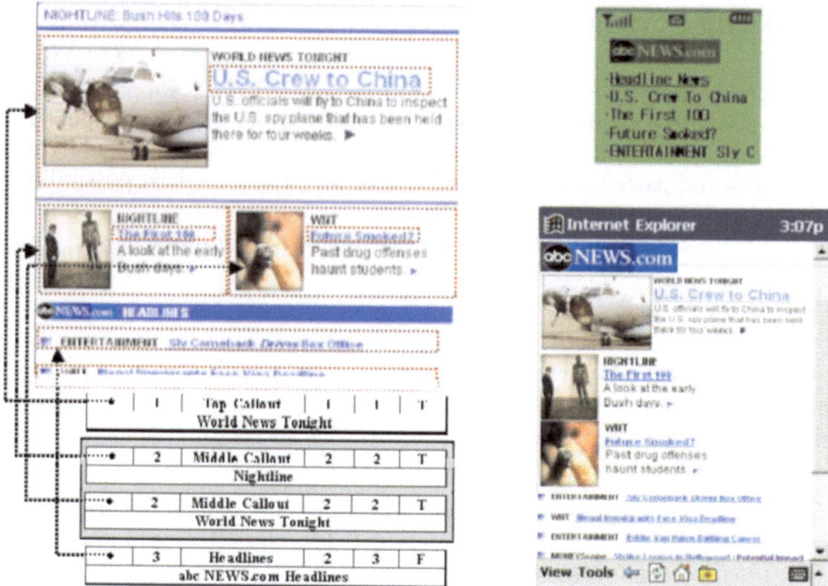

Fig. 2. An example of specification of clips in a HTML page and mobile adaptation. The shaded box at the bottom represents group clips. On the right hand side, the content adaptation results for a mobile handset and a PDA is shown.

4 Conclusion

We found that input from the content provider is crucial in obtaining satisfactory results in mobile content adaptation because it provides the most accurate information on which parts of the source document must be transformed and how it must be transformed. Web clipping is an efficient means for mobile content adaptation as well as the content provider's intervention. We presented in this paper a mobile content adaptation method based on web clipping and a clip editing tool for clip specification through a graphic user interface.

References

1. A. Fox, S. D. Gribble, E. A. Brewer, E. Amir: Adapting to network and client variability via on-demand dynamic distillation. Oper. Sys. Review, vol. 30, no. 5 (1996), pp.160-180
2. B. Housel and D. Lindquist: Webexpress: A system for optimizing web browsing in a wireless environment," in Proc. of the 2nd ACM/IEEE Intl. Conf. on Mobile Computing and Networking (1996), pp.108-116
3. WAP Forum: WML 1.1. http://www.wapforum.com/ (1999)
4. Compact HTML. http://www.w3.org/TR/1998/NOTE-compactHTML-19980209/ (1998)
5. W3C: XHTML basic. http://www.w3.org/TR/xhtml-basic/ (2000)
6. Pedro Gomes, Sergio Tostao, Daniel Goncalves and Joaquim Jorge: Web Clipping: Compression Heuristics for Displaying Text on a PDA. Mobile HCI 01, France (2001)

Case Study: A PDA Example of User Centered Design

Harald Lamberts

Microsoft, Mobile Devices Division
98052 Redmond, WA, USA
haraldl@microsoft.com

Abstract. A common complaint about software design is that it is often based on the underlying technology rather than designed for end-user tasks. Especially for software for more technical tasks, such as setting up a data connection, designing for user tasks might seem hard since the user's goal is rarely setting up the connection itself but usually performing a task that requires a connection. This paper describes how the Pocket PC connectivity User Interface was redesigned by using scenarios and personas based on field research. The success rate for the task of setting up a data connection in internal usability studies with the resulting design was 90%.This paper focuses on the process of creating and using scenarios to design and evaluate a user interface that supports the core user goals and tasks. Creating the personas from field research data has been a critical factor in making and using realistic and meaningful scenarios.

1 Technology vs. User Tasks

The software industry is a fast paced industry where products have to get to market as quickly as possible. It's easy to make the mistake to try to gain time by immediately starting to build the software and only do a usability evaluation just before shipping. In that case, software engineers would be designing the product, but software engineers have more experience and knowledge of the system and are therefore incapable of assessing whether a piece of information would be easy to understand for users [7]. As a result, the design is often based on the underlying technology rather than designed for the users' tasks and goals. To design for usability instead, there should be an early focus on users and tasks [4]. In addition, Cooper [3] says: "No matter who is designing, if the coding is underway, the effect will be negligible. A fundamental premise of bringing design to the software development process is that it must occur before programming begins".

1.1 Connectivity Tasks

When designing for more technical tasks it is likely that the difference in knowledge of the system, between developer and end-user, is even bigger. For example, when designing software for setting up data connections to the Internet or a corporate

F. Paternò (Ed.): Mobile HCI 2002, LNCS 2411, pp. 329–333, 2002.
© Springer-Verlag Berlin Heidelberg 2002

network, an obvious approach is to build just the UI that is needed to enter the necessary information to establish a connection such as the network, the modem, and the dial-up number. However, the user's goal is usually not setting up the connection itself but for example to browse the web, sending an email or an instant message. But, to accomplish these goals the user first has to complete tasks or provide a type of information that are technical and difficult to relate to.

Internal Usability studies with multiple PDA's have also shown that users have problems setting up connections because they don't know which connection type they need or what network they need to connect to for the task at hand. For example, users don't know if they have to connect to the Internet or the corporate network to use Instant Messaging. So, designing usable connectivity UI is difficult, mainly because a large part of users are not familiar with the underlying technology. Is there a way to solve this design problem?

2 From Field Research to Implementation Using Scenarios

This section describes the process that was used to design the Pocket PC connectivity UI to support the core user tasks. When designing for user tasks, it is important to know up front what the real core tasks and goals of the target end-users are and in which context. This is why field research has to be completed before the product team starts defining the design.

2.1 Field Research and Personas

In the Mobile Devices Division, and other divisions in our organization, persona creation starts with quantitative market segmentation. For the identified market segments, field research and market research studies are conducted to learn who our users are, what their goals and core tasks are and what their work context is. By integrating these quantitative and qualitative data points collected on targeted Mobile Professional customers, a detailed view of how, why and what these mobile professionals do is created. Finally, personas are created to represent the defined customer segments.[1]

The personas support user centered design because they personify the research data and thereby enable the product team to focus on the data through these virtual end-users. For this reason it is important to base the personas on real data, not on assumptions, anecdote and appeals to reason [5].

[1] A detailed explanation of how the market segmentation and the field studies were done and how the personas were created is beyond the scope of this paper.

2.2 Define the Core Scenarios

After gathering the field data and creating the personas it is important to make sure both are used to design the product. But, going from data to design is hard because user data never dictates exactly what to design; it takes a creative leap to do this [1]. Therefore, the responsibility of User Researchers is not just to provide a study report but also to enable the product team to use the data to define the product feature set and design the UI for it. To get a sense of the granularity of the personas data, here are some examples:

- Mobile Professionals use mobile devices when outside the work building, somewhere else in town or out of town for no more than a few days. For long and international trips they will bring a laptop.
- Information Managers go on business trips regularly and want access to data.
- Sellers are away from their office more than half of their work time, and want to process, send and receive information after a client visit.
- Community managers connect when outside between meetings and at home after a day at work.

Although these examples may seem quite accessible at first, it is rather factual and easy to forget. According to Grudin and Pruitt [5], using personas in the scenarios provides the engagement necessary to remember them.

Besides the field data that was gathered, also usability problems were found in internal usability lab studies. For example, some users don't know which setting to change when connecting from a different location. Usability problems like these can be used to prioritize or slightly adjust the scenarios to focus on, especially when redesigning an existing UI.

For the Pocket PC connectivity UI, the scenarios were created by using the field data to answer the following questions for each persona: 'What task does the persona want to create a connection for?', 'In which context does the persona want to connect?', and 'How is the persona connecting, using what hardware?'. Based on this, the following scenarios for the three personas were the result:

1. Kate traveled from London to New York and wants to synchronize her device from the Hotel without dialing an expensive international number.
2. Ray just met with a client and now wants to connect from the car to synchronize his schedule and e-mail using a wireless modem.
3. Steve purchases a PDA modem and wants to read his work email from home.

2.3 Designing UI for the Core Scenarios

Once the scenarios are defined, they become the design object [2]. The product feature team meets for several brainstorm sessions to come up with a variety of UI ideas to support these scenarios. These ideas are then clustered and the designer sketches a few design alternatives. It is good to briefly explore different design alternatives before settling on a single approach, also known as parallel design [7]. If available, known usability problems can be used to make design trade-offs. When the sketches are completed, the feature team runs through the scenarios to see which design best supports the core scenarios. To get a better understanding of these steps, the design process for one of the three scenarios is described below.

Example. The design process for scenario 3 mentioned above, was as follows:
After Steve buys a PDA modem there are three likely actions he might start with: 1. he inserts the modem, 2. he navigates to the e-mail application, or 3. he navigates to the connection settings screen. From internal usability studies it was known that not all users know where in the UI to setup a connection. So, if Steve starts with plugging in the modem or by navigating to the application he wants to use, the device should help him find the connection settings screen he needs. The final design does this by bringing up a dialog asking him if he wants to setup a new connection when Steve inserts the modem. When he taps 'Yes', the first page of the connection setup wizard is displayed. If Steve's first action is opening the browser, a similar dialog is shown: "The browser needs a connection, do you want to create a connection now?"

Once Steve is at the connection settings screen, the device should also help him because many users are not familiar enough with the underlying technology to accomplish their goal. Therefore, besides helping the user get to the right place in the UI, it was decided to create a task based UI for the connection settings to help users understand what they need to setup. The UI displays a list of tasks, such as "Add a modem connection to the internet". Selecting any of those tasks will display the appropriate wizard or settings screen.

The key characteristic of the chosen design direction is to walk the user through the necessary steps to complete the task, without having to understand the underlying technology. This makes sense when considering that, as stated before, the user's goal usually isn't setting up the connection itself but more likely a task for which they need a connection, such as browsing or checking e-mail.

For the other two scenarios exactly the same process was used. It is possible that different scenarios need different UI. In that case, using known usability issues or determining which scenario is more common can be helpful to make design trade-offs.

2.4 Conduct Iterative Testing Using the Core Scenarios

Although doing field research and using scenarios in the beginning of the design process is effective, it is still possible that potential usability issues were underestimated or overlooked. Iterative design evaluations help you further refine the UI by identifying these usability issues [6]. Immediately after the design direction for the Pocket PC connectivity UI was chosen, it was prototyped using Macromedia Director and evaluated in the usability lab on a PC. Building such a prototype is very low-cost and low-risk compared to actually implementing the design. The recruited participants for this study were representative for the end-user and the tasks they needed to complete were almost identical to the three core scenarios. This way, it is possible to measure how well the design supports these scenarios.

If all participants can successfully complete the tasks, then the usability goals are met. The results of this first usability study showed a success rate of 90% for the task of setting up a connection. Based on these results it was decided to start implementing this design direction. If the results show a low success rate or reveals many issues, it is probably necessary to do another design and test iteration before deciding to start the implementation.

Later, when the real implementation on the Pocket PC was almost completed, the exact same usability study was repeated but now using the actual PDA hardware. This is important for mobile devices because when using a PC prototype, usability problems related to form factor, performance, hardware interaction and the participants' PC experience might have been missed or misleading.

3 Conclusion

Based on the design process described in this paper, using scenarios has proven to be a successful way of designing for user tasks. Scenarios built on personas enable the product team to focus on the user data and communicate design ideas during all stages of the software design process. We believe that creating the personas from field research data has been a critical factor to the success of this project. It enables the creation of realistic meaningful scenarios. Although personas based on assumptions still provide end-user focus, they do not necessarily reflect the real user, their context, tasks and goals.

References

1. Beyer, H., Holtzblatt, K.: Contextual Design: Defining Customer-Centered Systems. Morgan Kaufmann, San Francisco (1998).
2. Carroll, J.: Scenario based design. Wiley (1995)
3. Cooper, A.: The Inmates are running the asylum. SAMS, Indianapolis (1999).
4. Gould, J. D., Lewis, C.: Designing for Usability: Key Principles and What Designers Think. Communications of the ACM 28, no.3 (1985) 360-411
5. Grudin, J., Pruitt, J.: Personas, Participatory design and Product Development: An infrastructure for Engagement. To appear in Proceedings of PDC 2002 (Conference on Participatory Design).
6. Mayhew, D.J.: The Usability Engineering Lifecycle. Morgan Kaufmann, San Francisco (1999).
7. Nielsen, J.: Usability Engineering. Academic Press, Boston (1993).

Leveraging the Context of Use in Mobile Service Design

Boyd de Groot and Martijn van Welie

Satama Interactive Amsterdam, Poeldijkstraat 4, 1059 VM, Amsterdam, The Netherlands
{boyd.de.groot,martijn.van.welie}@satama.com

Abstract. User-centered design for mobile devices and services is becoming increasingly difficult. Not only do the devices have inherent usability constraints and widely divergent characteristics, but also the mere mobile context of use presents an array of design challenges. Besides challenges, the mobile context of use also offers opportunities for truly helping people accomplishing their goals more effectively. In this paper we discuss the value of understanding the mobile context of use and propose a structured approach to improve the design of, what is ultimately an user's "ecosystem of connected terminals".

1 Introduction

The mobile industry is on the brink of its second wave into the mass consumer market: the mobile Internet. Whether this second wave will be as successful as the first one, which led to the mass adoption of the mobile phone for voice communication, is currently the subject of heavy debate amongst large groups of people inside and outside the industry. One side of the debate points at the demise of WAP to underline their argumentation while the other side points at the success of SMS and I-mode.

Looking in hindsight from a HCI perspective, the success and failure of SMS and WAP respectively, can be explained by looking at the balance between the *value* to the user these services offer versus the 'basic' usability. Apparently SMS offers people so much value in terms of social interaction and communication that the user is willing to invest time in a poor user interface. At this moment WAP is obviously at the wrong side of the balance: poor value on top of poor usability. These examples illustrate that success in the mobile industry is more a matter of addressing "killer values" than offering "killer applications".

Because of the inherent usability constraints of mobile devices, the strategy for designing successful services for the mobile internet seems to be clear and simple.

1. Try to 'design' services which offer a high enough value to the user to overcome the usability constraints;
2. at the same time, try to minimize usability problems within the device constraints.

In the following sections we will focus on the first part of the above mentioned strategy: how to design services that offer true value to the user and how understanding the mobile context of use can broaden the design method, especially in the early conceptual design phases.

F. Paternò (Ed.): Mobile HCI 2002, LNCS 2411, pp. 334–338, 2002.
© Springer-Verlag Berlin Heidelberg 2002

2 Leveraging the Mobile Context of Use

A famous saying by Dutch soccer player, Johan Cruyff, goes like this: "Every disadvantage has its advantage". At first sight the mobile environment only offers disadvantages: small screens, limited input possibilities, limited browsers, poor data-transfer performance, distracting environments of use, etc. However, mobile technology potentially also has some unique and distinct advantages in the fact that a lot of information about the context of use is available without requiring user-input. Context information that could be detected are the user's *identity*, the user's *location*, the local *time* and the *device* being used. By utilizing this information in a creative way, new innovative services can be developed that can support users accomplishing their goals in a more effective way.

- **Identity.** Can be detected by using the user's unique phone number. This makes it possible to offer highly customized services, based on user preferences.
- **Location.** Not widely supported by operators yet, but given the essence of the mobile context of use, location based services can potentially offer tremendous value to users. For example, locating the nearest taxi stand without having to punch in a postal code (which the user does not know anyway).
- **Time.** Already used in currently available services, for example in weather and traffic information services. Creatively linked to the other properties like identity and location, services can become much more relevant for the individual user and thus more valuable. Consider a service that alerts a particular motorist about an upcoming traffic jam on his/her route and immediately suggests alternatives.
- **Device.** At a certain moment a service will need to "push" information to the user's device. As the range of different devices is growing rapidly, the service will need to detect what type of device is used in order to provide an optimized user experience.

Especially in the early conceptual phases of the design process, it can be very inspiring and useful to match a specific context of use to user goals and generate service ideas and concepts on the basis of this. Table 1 shows an example of this approach to discover potential service ideas for "mobile movie information" [1]. For each of the possibilities of identity, location and time, an example is given. A possible next step would be to flesh out the examples further using another design method like personas and scenarios.

As can be seen, this approach enables designers to quickly generate a large set of possible service and design concepts. The context property "device" will have to be taken into account as soon as the high level concepts are more detailed. For small phones, for example, the service might send an SMS to the user containing which movies are playing at the nearest cinema. For mobile PDA's the service might deliver the same information as a Flash object with movie trailers.

Also after the initial idea generation, of course, the current technological constraints and possibilities need to be taken into account. Many ideas might be unrealistic at the moment, but this should not interfere with the initial idea generation.

Table 1. Generating service ideas.

Service example: **Movie Info**	**Identity**	**Location**	**Time**
General Movie Info			
Give me information about movies based on my favourite genre, director, and actors.	X		
Which cinemas are in the neighbourhood?		X	
Which movie plays at 21.00?			X
Is there a movie that I would like playing in my neighbourhood?	X	X	
Can we still go to a movie tonight, playing in a cinema nearby?		X	X
When is my favourite movie coming out?	X		X
Is there a movie that I would probably like, playing tonight in a cinema nearby?	X	X	X

3 The Next Level: The User's Ecosystem of Connected Terminals

In the previous section the focus has mainly been on the context of use of a *single* device. With the rapid growth of the mobile Internet, however, industry players are more and more offering services through different channels: web, WAP, phone resident applications, PC resident applications, etc. Operators, for example, are currently vigorously setting up mobile portals besides their already existing web portals. Usually, the services are only copied and adapted to the smaller screens of mobile devices. Mobile phone manufacturers have since long offered phone resident PIM services, like an address book and a calendar. PC applications are also added to support synchronization with PC applications like MS Outlook. More recently, phone manufacturers are also offering services through own web and WAP portals. Mainly CRM services, but recently also download services for ring tones and phone display graphics.

In this highly competitive market a lot of overlap is occurring which is not beneficial to the clarity and usability of mobile services as a whole. Consider the messaging service. People can use PC-based Outlook for email, phone-based SMS, web-based email and SMS through many web portals from operators, manufacturers, etc. And very soon, all this will also be offered through different mobile devices.

In this emerging setting the competitive edge will eventually also have to come from offering more value by taking the context of use into account. Even though the industry likes to talk about "virtual services", people with their millions of years of history of handling artifacts, will probably view the mobile Internet as a, what we call, "ecosystem of connected terminals", where the interaction with each terminal is dependant on the context of use. Multi-channel services in this ecosystem should also try to take advantage of the specific context of use of each channel in such a way that the different channels reinforce each other. User tasks, and more important mindsets, differ according to the context of use. Fig. 1 illustrates this principle for a mobile calendar.

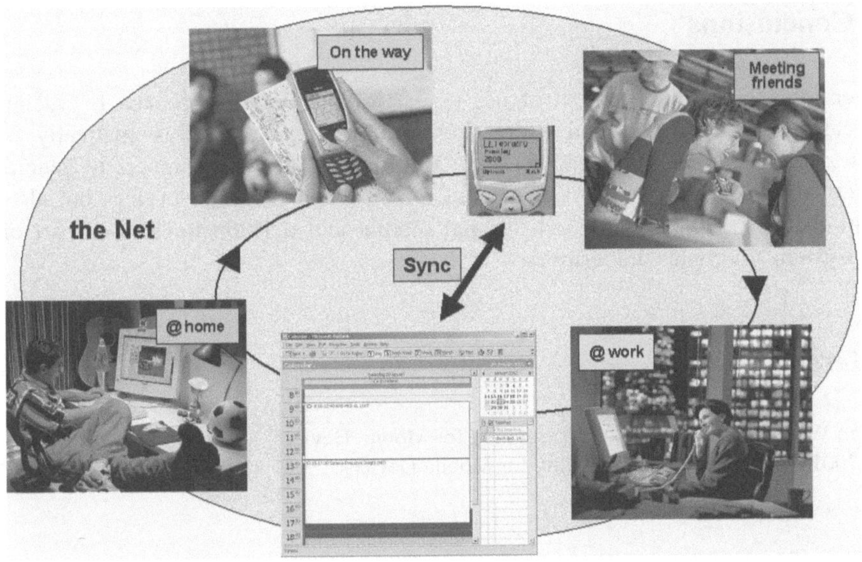

Fig. 1. A mobile calendar service within the user's ecosystem of connected terminals.

Mobile calendar services with synchronization features can offer a much higher value when taking into account the context of use in different situations and how this influences or even determines user tasks and user mindsets. The following overview lists a few context dependent characteristics that can serve as input for further identifying user requirements.

- **@home**. Private; focus on social life; often less time-critical; less concurrent tasks; large UI. Context is especially fit for managing social appointments and setting mobile reminders.
- **@work**. Business; focus on professional life; more time-critical; more concurrent tasks; large UI. Context is especially fit for managing business appointments and setting mobile reminders.
- **On the way**. Private/business; focus on social/professional life; time-critical; more unrelated concurrent tasks, ad-hoc; small UI. Context is especially fit for receiving mobile reminders; there is also limited room for simple calendar management (for example in a train).
- **Meeting friends**. Private; focus on social life; highly time-critical; many unrelated concurrent tasks, mainly ad-hoc; small UI. Context is especially fit for receiving mobile reminders; there is hardly room for calendar management, quick note-taking should be supported, for example a quick SMS-like reminder to oneself, which can be dealt with at home or work.

4 Conclusions

Leveraging the mobile context of use as part of the design approach is crucial for increasing service value and decreasing usability problems. Systematically using context elements such as location, time, and identity, allows designers to generate a variety of concept alternatives. Not only for "isolated" mobile services but also for multi-channel services delivered through mobile and desktop devices as part of an "ecosystem" of connected terminals.

References

1. van Welie, M., de Ridder, G.: Designing for Mobile Devices: a Context-Oriented Approach. (2001) IBC Conference "Usability for Mobile Devices", 9-11 May 2001, London, UK

A Ubiquitous Computing Environment for Language Learning

Maximiliano Paredes[1], Manuel Ortega[2], Pedro P. Sánchez-Villalón[2], and J. Ángel Velázquez-Iturbide[1]

[1] Escuela Superior de Ciencias Experimentales y Tecnología, Universidad Rey Juan Carlos 28933 Móstoles, Madrid
[2] Escuela Superior de Informática, Paseo de la Universidad s/n, Universidad de Castilla La Mancha 13071 Ciudad Real
Manuel.Ortega@uclm.es

Abstract. This work presents a piece of research on the methods and mechanisms necessary to bring the Information and Communication Technologies into the traditional classroom. This will be achieved by putting collaborative and Ubiquitous Computing paradigms together to integrate both fields into the educational environment. As a study case we have developed a system for language learning, in particular English as a Foreign Language (EFL).

1 Introduction

Computers in educational environments has not been successfully integrated into the classroom yet, though great efforts have been made to achieve this. Computers are usually used as a standalone device in certain learning areas (mainly laboratories and practical classes), without taking full advantage of their great potential. Occasionally, they have been used interconnected in a local network area just to share tools, such as databases and information resources of different kinds. Communication among learners has been neglected, and only communication between learners and teachers has been provided via email facilities in remote learning environments, such as distance education. When the student uses the computer, he/she is isolated facing the task assigned over the computer environment. This distinct separation of the "two worlds", on the one hand the computer and on the other hand the classroom, hinders the learning process. Up to now, we have tried to adapt the classroom to the computer, but why not try just the other way round? The paradigm of ubiquitous computation allows us to sweep this borderline away fading the computational environment into the classroom. This way, the student interacts with the system in an unnoticeable way.

So far, ubiquitous applications applied in the classroom scenario are scarce and they provide little collaborative work in group support. The existing applications satisfy some of the scenario requirements we can detect, but not all. The described system in Roth et al. [1] is a robust system that supports collaborative and information

F. Paternò (Ed.): Mobile HCI 2002, LNCS 2411, pp. 339–343, 2002.
© Springer-Verlag Berlin Heidelberg 2002

management tasks in a ubiquitous environment. However, the system does not provide context information, such as being constantly aware of the student's location in the classroom at every point. The system described in Myres [2] provides synchronization among all the devices efficiently, but it has the following inconvenience: the user has to employ wire-bound devices, which seriously limits physical movement and actions around. Danesh [3] shows an educational application for a group of children using an infrared connection system. Here the children have to synchronize their PDAs to exchange information. This process of synchronization is carried out via infrared communication, which has the following disadvantage: the student has to line up the devices so that the communication can be established. In addition, this system is not very powerful because the users' information is stored in the PDAs' small memory as there is not a powerful centralizing server or device. Zurita [4] solves the inconveniences of infrared networks using a wireless network. The paper presents two educational applications for a group of children. These applications do not satisfy our domain's needs because of two reasons: the application is not sensitive to the user context and the collaborative tools do not support the discussion process of the diverse proposals. *Classroom 2000* [5] is a Ubiquitous System intended to be a useful help after the class but our approach should consider the whole learning process, before, during and after the class.

2 AULA: A Ubiquitous Computing Environment for the Collaborative Composition of Written Texts

The AULA system is intended to improve communication abilities in a learning environment in order to achieve the necessary skills to develop a project in group by writing reports in a collaborative task. AULA is composed of: a two-way projection/edition whiteboard, a database server, the Location Manager (providing context awareness to the system), the Session Coordinator and the mobile devices (PDAs and possibly TabletPCs). Every student has his/her own device. The communication technologies used are: a wireless network (RF) and infrared communications. The system architecture is described with more detail in [6].

The learning task to develop is writing a text (an essay, a report, a news article, etc.) in a collaborative way, since this is a usual activity in a great number of jobs in real life. When the class starts, the teacher sets up the main topic and other appropriate characteristics or properties of the document to write. The teacher can make these indications in several ways: verbally, using whether the synchronous or the asynchronous collaborative tools that AULA provides or combining both facilities. After this initial phase, a brainstorming process about the composition topic starts. The system facilitates the structuring of all the information resulting from the brainstorming process into the so-called aspects, which are separated into several ideas. These aspects and ideas are blocks of partial information, which constitute the initial framework of the document. This must be worked out by the students.

The brainstorming phase generates a discussion process among the group members, in which the learners discuss which aspects and ideas are most appropriate to be part of the written composition and their relevance to the topic. At this stage the students make use of the collaborative tools: chatting, discussion boards, exchanging pieces of information. Throughout this process, the teacher plays a monitoring role. This way, the teacher will be present throughout the process. In any case, the teacher can give brief hints or clarifications about some proposals, which would be projected in the projection whiteboard. But it is the students that play the project manager role. The process of discussion finishes when the group of students come to an agreement, accepting some of the aspects and ideas suggested and refusing others, through a voting process that the system provides.

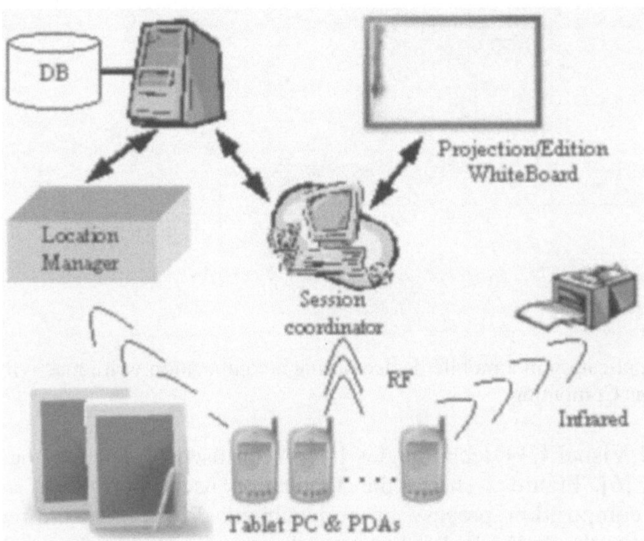

Fig. 1. Infrastructure of the ubiquitous classroom

We have carried out practical classes of text composition in a traditional environment in which one student writes the proposals of the classmates on the blackboard ready for discussion. In an experimental way we have been able to detect problems in this type of scenarios: the students repeated the writing of the proposals to reorder them appropriately. Erasing, modifying or writing a new proposal was tedious and time-consuming. The structuring that the system carries out in aspects and ideas solves these problems, being suitable from the pedagogical point of view (by providing a visual organization clearer for the students) and it facilitates the dynamics of the activity (erasing, modifying, etc).

Once the partial components of the document have been built, the students will focus on rewriting them adding correct structural elements to make coherent sentences or paragraphs, organizing them in a complete text using different structural and lexical connectors of the language. Later, the text is suggested for revision and acceptance to the group. Once this has been carried out, the teacher revises the document and explains possible mistakes. Then, the document is saved in a case database as a portable text file. Finally the teacher makes an evaluation of each student's participation and of the working process and the learners' solution.

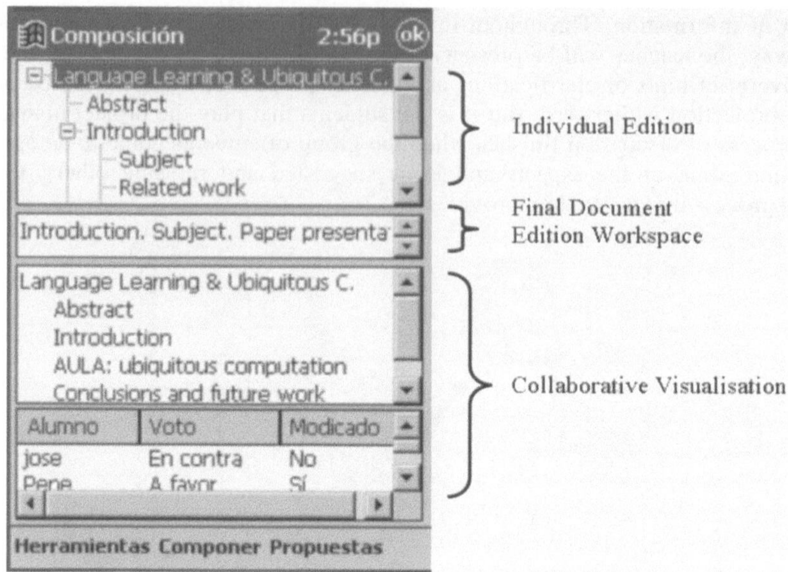

Fig. 2. The application in a mobile device during a composition writing activity about Learning and Ubiquitous Computing

Java and Visual C++ technologies have been used to develop the current system applications [6]. Figure 2 shows the application user interface of a mobile device during the composition process of a document. We can see three well defined working areas: the Individual Edition workspace (in which the students will write their individual contributions), the Collaborative Visualization workspace (in this area all the contributions of the group will be displayed) and the Final Document Edition workspace (it will show the proposals that have been accepted, which will be part of the document). The display in the PDA is big enough so that the student can see all the information in a comfortable way. Nevertheless, at present we are evaluating the implementation of the system in a TabletPC.

The aspects and ideas are stored in a persistent way in the student's individual workspace allowing the student to work in their proposals even when the PDA is not connected to AULA. In this case, there are synchronous collaboration functionalities not accessible in that moment, but many others do, as for example the creation and modification of aspects and ideas in the personal workspace. When the student enters the classroom and he/she starts his/her PDA, a synchronization process updates the information automatically. These mechanisms make AULA system characteristic and different from other learning systems, providing support to the student before, during and after the class.

3 Conclusions and Future Work

In this article we have shown a model for learning through ubiquitous computing as the interaction paradigm. This model is an ideal frame to investigate and to extract the implications of the ubiquitous computing in computational systems of collaborative learning in group. Our work has focused on the teaching of English as a Second Language through a process of composition writing in group. The students will feel they are in a traditional classroom. In our model, the students continuously take notes and write in what seems their small notebooks and frequently direct their attention to the whiteboard, where the teacher presents the information or a wider view of the resulting writing process can be observed. This way the traditional elements do not disappear, but they evolve thanks to electronic devices.

In the future we will try to analyze and compare the results of our system with the results obtained through traditional methodologies, as to their effects on the development of communication abilities in English.

References

1. Roth, J., Unger, C.; Using Handheld Devices in Synchronous Collaborative Scenarios., Proceedings of the Second International Symposium on Handheld and Ubiquitous Computing, pp.187-199, Springer. Bristol, U.K., Sept. 2000.
2. Myres, B.A., Stiel, H., Gargiulo, R.; Collaboration Using Multiple PDAs Connected to a PC., Proceedings of the ACM 1998 conference on Computer Supported Cooperative Work, pp. 285-294, Seattle, Washington, USA, 1998.
3. Danesh, A., Inkpen, K., Lau, F., Shu, K., Booth K.; Geney™: Designing a Collaborative Activity for the Palm™ Handheld Computer., Proceedings of the SIGCHI'01, pp. 388-395, Seattle, Washington, USA, 2001.
4. Zurita, G., Nussbaum, M.; Mobile CSCL Applications Supported by Mobile Computing., Pontificial Catholic University of Chile, Ret. Feb. 14, 2002 from http://julita.usask.ca/mable/Accepted.htm#Submission6
5. Abowd, G.D.; Classroom 2000: An experiment with the instrumentation of a living educational environment., IBM Systems Journal, Special issue on Pervasive Computing, 38 (4), pp. 508-530, 1999.
6. Ortega, M., Redondo, M.A, Paredes, M, Sánchez-Villalón, P.P., Bravo, C., Bravo, J.; Ubiquitous Computing and Collaboration: New paradigms in the classroom of the 21st Century., Computers and Education: Towards an Interconnected Society, M. Ortega and J. Bravo (Eds.), Kluwer Academic Publishers (2001), p 261-273.

XHTML in Mobile Application Development

Anne Kaikkonen and Virpi Roto

Nokia Research Center, P.O.Box 407, 00045 Nokia Group, Finland
anne.kaikkonen@nokia.com, virpi.roto@nokia.com

Abstract. Nokia Research Center conducted a usability test for two XHTML Mobile Profile (MP) applications: a news application and an auction application. The goal of the test was to find out how XHTML MP components should and should not be used in order to build a usable mobile application. To compare different user interface solutions, both applications were designed in three different user interface styles. The findings on user performance, perception, and preference were used to make Mobile Application Development Guidelines.

1 Introduction

The language of WAP 2.0 will be XHTML Mobile Profile [1]. This is a big step towards one Internet for both fixed and wireless worlds, since the same site can be browsed both from a PC and a mobile phone. For an ordinary end user, the change from WML to XHTML MP is not that dramatic, but, depending on the browser, the "mobile Internet" may look more like the Internet that people know from the fixed world. A familiar look and feel is likely to increase user satisfaction when using WAP services for the first time.

At the time of writing this paper, there are few WAP 2.0 devices and services on the market. Also, there are no guidelines for developing usable mobile services with XHTML MP. It would be beneficial, however, if there were usability guidelines for WAP 2.0 already at this early phase. Usability guidelines or recommendations for WAP 1.x have been presented in several publications [2,3,4,5,6].

Nokia Research Center conducted a usability test of 20 test sessions for XHTML MP applications. The goal was to find the user interface solutions that users prefer and find easy to use as well as the solutions that users find difficult to use and thus should be avoided. We also wanted to find out whether there are cultural differences between the preferences of different nationalities on the user interface solutions. Based on the results, as well as the results of hundreds of earlier WAP 1.x usability test sessions, we compiled a list of usability guidelines for WAP 2.0 service development [7, 8].

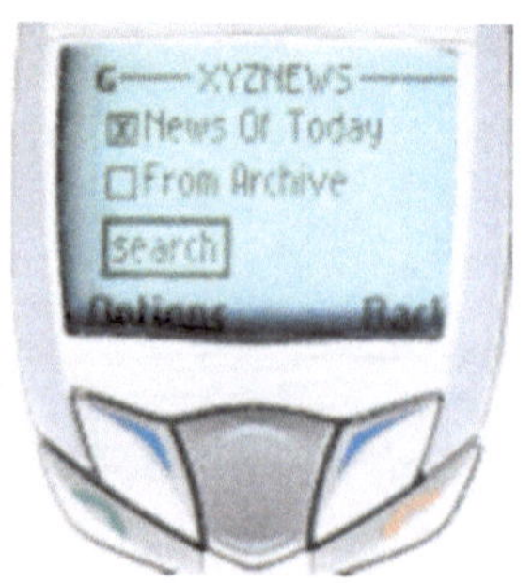

In this paper, we concentrate on the differences found between WML and XTHML MP usage, and between different user types. At the end, we list some differences between designing applications for fixed and mobile devices.

F. Paternò (Ed.): Mobile HCI 2002, LNCS 2411, pp. 344–348, 2002.
© Springer-Verlag Berlin Heidelberg 2002

2 Procedure

The tests were carried out with Nokia 65xx mobile phones, running a prototype version of the Nokia XHTML browser. The applications were used via a GPRS connection as WAP services, and the implementation was pure XHTML Mobile Profile.

The number of subjects was 20: 12 in Helsinki, Finland and 8 in Boston, USA. The subjects in Finland were from various European countries or from Japan. The subjects varied from active users of the current mobile Internet to ones that had never used it. Each user knew, however, at least the principle of either WAP in Europe/USA or Japanese mobile Internet (i-mode, J-sky or EZ-web). All subjects had a mobile phone in daily use and they knew how to type with a mobile phone keypad; they had either been writing SMS messages or inserted names to the phonebook of the phone.

The user interfaces used in the test were specified and developed just for this test. There were two services in the test: a news service for information retrieval and an auction service with interactive forms. Because we wanted to make general guidelines, we needed to compare the usability of different kinds of user interface solutions. We selected basic user interface solutions used in WAP, the web, or Windows today and built the services in three different ways. The three user interfaces contained the same data, but the way the data was presented, navigation, and the usage of the elements varied in each user interface style.

Different users tested the applications and user interface styles in different orders. To avoid associations that would make one style better than other, we named the styles according to fruits:

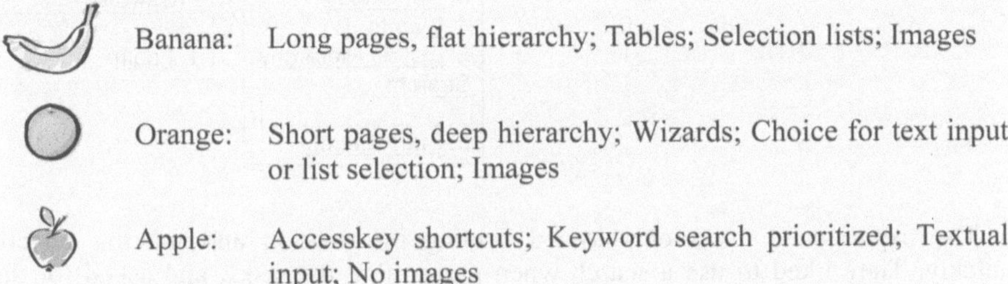

Banana:	Long pages, flat hierarchy; Tables; Selection lists; Images
Orange:	Short pages, deep hierarchy; Wizards; Choice for text input or list selection; Images
Apple:	Accesskey shortcuts; Keyword search prioritized; Textual input; No images

When designing the user interface styles, we targeted the Banana style for novice users, Orange for intermediates, and Apple for experts. In Banana and Apple, we placed one or two logical entities of data on one page, whereas in Orange we even split one logical entity onto several pages to minimize scrolling. Examples of these different user interface styles follow:

3 Main Results

It seems that the differences in the users' preferences on different user interface solutions were not between the subjects from different countries, but between the experienced mobile application users and novice users. There were differences between countries in the experience level, however: all Japanese subjects were very experienced, whereas in Boston, the subjects were not that familiar with the mobile Internet.

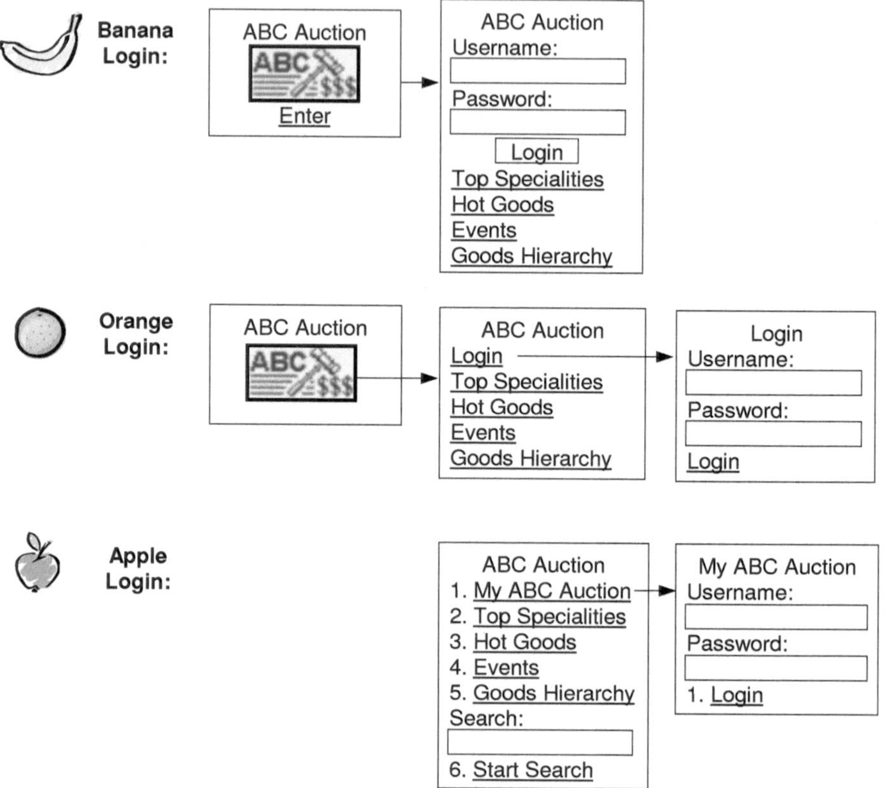

The experienced users emphasized finding information and making selections quickly. They liked to use a search when performing the tasks, and asked for direct links to the main pages of the service. Text input was less painful for experienced users, and they could better keep in mind the selections they had made in forms. Experienced users also more easily understood the structure of the application, which made it easier for them to know where they were. They were also more patient in waiting for the page to download. Because these services were being used for the first time, many expert users liked the simple user interface styles and felt the Apple style, designed for them, was too complicated. A longer study period would be needed to find out whether frequent use would increase the success of the Apple style.

Novice mobile Internet users had to learn to browse and use the pages without a mouse. They did not always scroll down enough to see the whole page. Novice users preferred a tree hierarchy to a search functionality in navigation, because seeing the list of sections helped them to understand the structure of the application and to increase the feel of control. The users that were not familiar with text input with the phone keypad did not like to input text, but preferred selection lists. They were uncertain whether they used the system as they were expected to, and when a page was downloading too slowly, they wanted to cancel the download. These users preferred the Banana and Orange styles, but the ones who used the search function in the Apple style liked it.

3.1 Some Differences between XHTML Mobile Profile and WML

The new user interface elements provided by XHTML MP include push buttons, check boxes, radio buttons, multi-line text fields, animated image format, and tools for sophisticated page layout. XHTML MP does not provide decks of cards, timers or other events, softkey assignments, input filtering, or multi-page forms [1,9]. What does this mean for the user interface design?

Page Length

In WAP 2.0, data is downloaded as single pages, not as decks. This has an effect on the optimal page length: in many cases, it is better to have a few long pages rather than the same information split on several short pages. In our tests, subjects felt that the download time of a short page was longer than that of a long page (time perception related to amount of information expected). The appropriate page length depends on the information on it: target pages containing the data that the user was searching for may be even 20 screenfuls long, whereas the navigation pages normally should not contain more than 10 one-line links.

Forms

It seemed that users started to lose control if the form was longer than two screenfuls or if the form was split onto several pages. A multi-page form could work, if the form was short, familiar, and there was no need in any situation to go back and change the values. Users did benefit from viewing all the selections on one page. If the users did not remember the selections they made in a peaceful lab environment, they are even less likely to do so in a disturbance-prone mobile environment.

Users also seemed to think that the input they give is immediately saved, without a need to send the data to server. If there are several ways to proceed from the form page, the input may not get saved even if the user meant to do so.

Push Buttons

Users seemed to understand well the command buttons to execute the final actions for the tasks. It was better if the labels in buttons were not technical terms; terms like Update or Submit are harder to understand than Save Changes or OK. Push buttons should not be used for cases where more information is needed before the operation can be executed. In those cases, use hyperlinks instead of push buttons.

Text and Number Input

The standard XHTML MP does not provide a way to define whether an input field is numeric or textual. This makes it hard for the end users to enter values, since they have to do the mode change from text to numbers, or vice versa. We recommend avoid requiring a mix of numbers and letters in one field, but putting the device in the correct input mode would need a change in the XHTML MP language. Nokia devices will understand a cascading style sheet (CSS) property "`-wap-input-format`", which defines the input mode and prevents input errors already on the terminal side.

4 Conclusion

Although the move from WML to XHTML Mobile Profile helps the application developers in multi-platform development, there still remains a possibility to make the application usage difficult for the end user. Lack of decks means the site structure must be different than in WML, and the developers must use extensions in order to have smart input fields. The developers must learn to use push buttons and hyperlinks in a consistent way.

WAP 2.0 is not the only change that the mobile Internet will face in the near future. At the same time, quicker networks (like GPRS), more WAP capable devices, and better displays and input devices are coming into use. These are likely to make WAP more usable also for the end users.

References

1. Wireless Application Protocol Forum Ltd.: XHTML Mobile Profile Specification. http://www.wapforum.org/what/technical.htm
2. Kaikkonen, A. & Williams, D.: "Here, there, everywhere": Designing Usable Wireless Services. In Tutorial Notes of Interact 2001 (2001)
3. Kaikkonen, A & Törmänen, P.: User Experience in Mobile Banking. In Proceedings of HCI2000, BCS (2000)
4. Kaikkonen, A. & Williams, D.: Designing Usable Mobile Services. In Tutorial Notes of CHI2000, ACM (2000)
5. Kaasinen, E., Kolari, J., Laakko, T.: Mobile-Transparent Access to Web Services. Interact 2001 Proceedings (2001)
6. Chittaro, L., Dal Cin, P.: Evaluating Interface Design Choices on WAP Phones: Single-choice List Selection and Navigation among Cards. In Proceedings of Mobile HCI 2001 (2001)
7. Nokia Corporation: Mobile Application Development Guidelines. http://www.forum.nokia.com/ (2002)
8. Nokia Corporation: XHTML Guidelines. http://www.forum.nokia.com/gfe1033 (2002)
9. Wireless Application Protocol Forum Ltd.: XHTML Mobile Profile Specification. http://www.wapforum.org/what/technical.htm

Understanding Contextual Interactions to Design Navigational Context-Aware Applications

Nicholas A. Bradley and Mark D. Dunlop

Department of Computer and Information Sciences
University of Strathclyde, Glasgow, Scotland, G1 1XH, UK
Tel: +44 (0) 141 552 4400, Fax: +44 (0) 141 552 5330
{Nick.Bradley,Mark.Dunlop}@cis.strath.ac.uk

Abstract. Context-aware technology has stimulated rigorous research into novel ways to support people in a wide range of tasks and situations. However, the effectiveness of these technologies will ultimately be dependent on the extent to which contextual interactions are understood and accounted for in their design. This study involved an investigation of contextual interactions required for route navigation. The purpose was to illustrate the heterogeneous nature of humans in interaction with their environmental context. Participants were interviewed to determine how each interacts with or use objects/information in the environment in which to navigate/orientate. Results revealed that people vary individually and collectively. Usability implications for the design of navigational context-aware applications are identified and discussed.

1 Introduction

Context-aware technology has stimulated a growth of research and development into novel ways in which to support the user in a wide range of tasks and situations. Context-aware applications are designed to discover and take advantage of contextual information such as a user's location, time of day, nearby people and devices, and user activity [1]. Examples include (i) context-aware mobile tourist guides [2] and (ii) location-aware shopping assistants [3]. However, it is becoming increasingly paramount that this proliferation of context-aware technologies is matched by a suitable and sufficient analysis of context research issues. A sound understanding and appreciation of context can lead to improved usability [4] and is regarded as the key to unlocking the true value of business applications on handheld devices [5]. Dey & Abowd [6] state that context can 'increase the richness of communication in Human-Computer Interaction (HCI) and make it possible to produce more useful computational services'.

The notion of context has been discussed in many disciplines (e.g. psychology, linguistics, computer science) and has recently triggered more multi-disciplinary interest (partially due to the inconsistencies and ambivalent definitions across different research specializations). While a general and unifying theory or formalisation of context is still in its infancy [7], commonalities do exist, e.g. context is broken down into components/variables and their *interactions* are addressed. Zetie

F. Paternò (Ed.): Mobile HCI 2002, LNCS 2411, pp. 349–353, 2002.
© Springer-Verlag Berlin Heidelberg 2002

[5] emphasizes the importance of understanding contextual interactions and explains that task analysis is critical for a suitable assessment.

This study involved an investigation of contextual interactions required for route navigation on foot. A preliminary study revealed that existing navigational location-aware applications (e.g. PocketMap City Guide) use a generic representation of contextual information (i.e. same types of information for all users). Our purpose was to illustrate through a series of interviews the heterogeneous nature of users in interaction with their environmental context. *The study hypothesis is that people vary individually and collectively in their use of surrounding environmental objects/information to navigate.* It is anticipated that this will have an impact on usability design requirements for navigational context-aware applications.

2 Methodology

After a pilot study of four participants, the main interview study was finalised and consisted of 24 participants (12 males and 12 females). Four participants (2 males and 2 females) fell into each of the six age categories: 18 or under, 19-25, 26-35, 36-45, 46-65, and 66 or over. All participants were resident in Greater Glasgow and their professions ranged from school pupils to a retired lecturer. The interview study was comprised of three parts:

1. *Pre-interview questionnaire*: Information on participants' personal details, familiarity with Glasgow centre and knowledge of context-aware computing.
2. *Interview:* The main interview consisted of four destinations that participants had to describe how to reach on foot. Two destinations were described verbally and two in writing.
3. *Post-interview questionnaire*: Information on participants' opinions on the importance of different types of contextual information for route navigation, design issues relating to usability and their mobile needs/requirements.

For part 2, well known destinations and a suitable starting point were chosen. The selection criteria were that each destination must be a similar distance from the starting point (approx. 10 minutes), and in different directions. The order by which participants were presented with each destination was randomised and the verbal/written order was alternated (with an equal balance of those who had to write first with those who had to verbalise first). Each interview was recorded in full.

In line with techniques used for verbal protocol analysis [8], participants' descriptions from part 2 involved a subjective categorization of different types of contextual information into nine categories: *directional* (e.g. left/right, north/south), *structural* (e.g. road, monument, church), *textual-structural based* (e.g. Border's bookshop, Greave Sports), *textual-area/street based* (e.g. Sauchiehall St., George Sq.), *environmental* (e.g. hill, river, tree), *numerical* (e.g. first, second, 100m), *descriptive* (e.g. steep, tall), *temporal/distance based* (e.g. walk until you reach… or just before you get to…), and *sensory* (smell/hearing/ touch) (e.g. sound of go-kart engines while passing ScotKart Centre or smelling hops near a brewery). Accumulated scores were calculated each time a participant mentioned/wrote a word/phrase relating to one of the listed contextual categories.

3 Results

The results of the interview study are represented in figures 1-6.

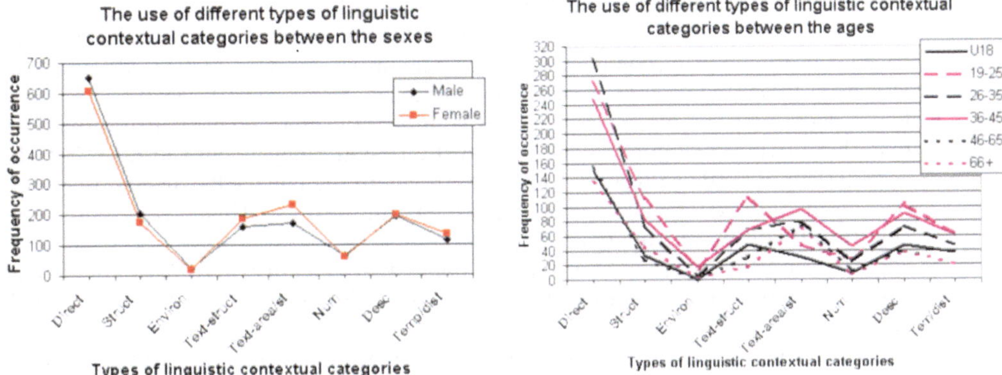

Fig. 1. and Fig. 2. Use of contextual information between the sexes and ages

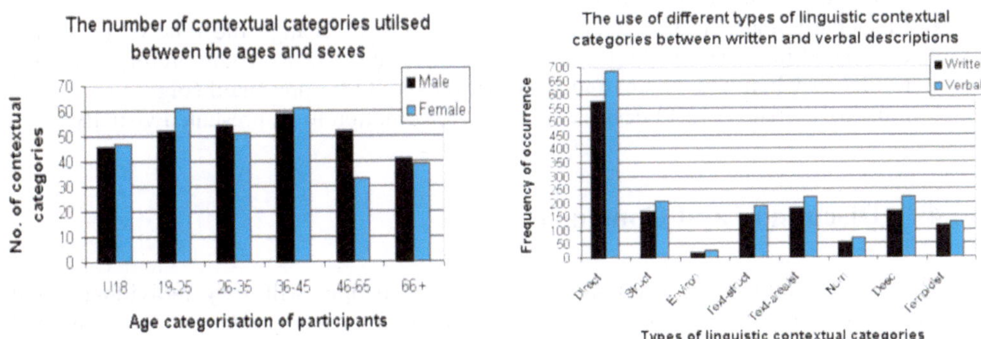

Fig. 3. Number of contextual categories in total for each group by age and sex

Fig. 4. The use of contextual information for written and verbal descriptions

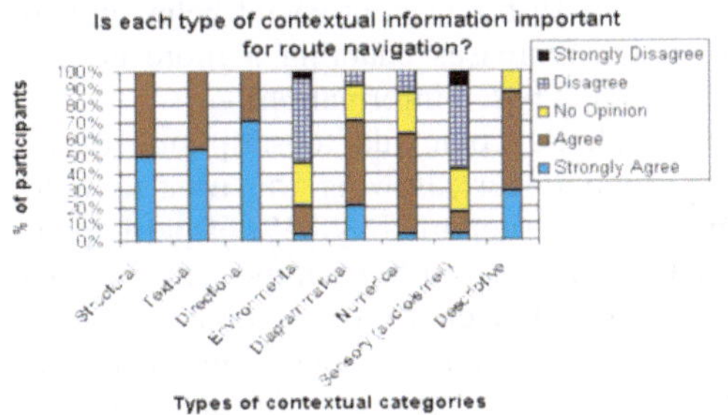

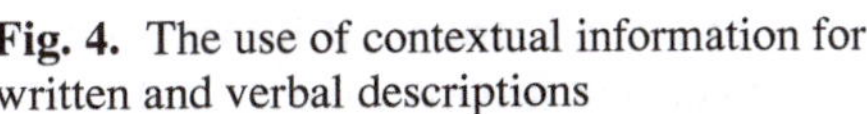

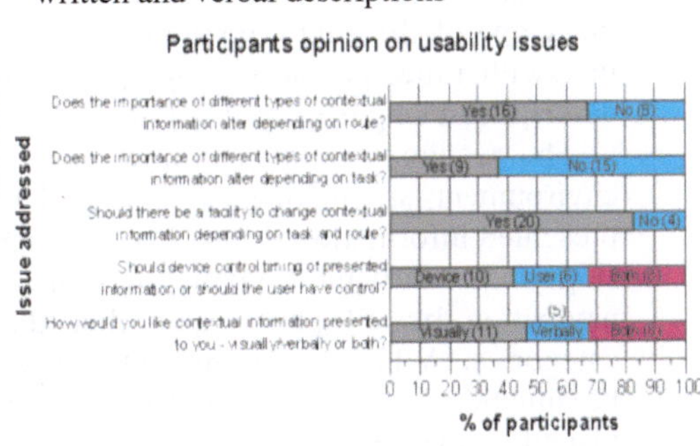

Fig. 5. Participants' opinions on the importance of contextual information

Fig. 6. Participants' opinions on usability issues

The key findings from graphs 1-6 are:

- There is little difference between sexes in use of different types of linguistic contextual categories (Figure 1). Though, female participants used marginally more textual-structural and textual-area/street based information, whereas males used slightly more directional and structural information.
- The age groups 26-35, 36-45, 46-65 and 66+ all used more textual-area/street based information in comparison to textual-structural information (Figure 2). However, this trend was reversed for age groups under 18 and 19-25.
- Participants over 18 but under 45 used significantly more contextual information than those over 45 (Figure 2) (1% sig-level on 2-tail independent t-test, $t = 7.4$, $df = 38$). This is further illustrated in Figure 3 where the under-18s, 46-65s and over 66s used less contextual categories than the others. Under-18s were not significantly different from over 45s.
- More words/phrases from each type of contextual category were used for verbal descriptions in comparison to written descriptions (Figure 4) (1% sig-level, 2-tailed correlated t-test, $t = 4.01$, $df = 23$).
- Most participants either agreed or strongly agreed that structural (100%), textual (100%), directional (100%), diagrammatic (71%), numerical (63%) and descriptive (88%) are important for route navigation (Figure 5). Whereas most participants disagreed or strongly disagreed that environmental (54%) and sensory (58%) contextual information were important.
- Most participants (83%) would desire a facility to change the type of presented contextual information (Figure 6). Most participants (42%) would prefer the device to dictate when new contextual information should be presented (Figure 6). Also, most participants (46%) desire contextual information to be presented visually.

4 Discussion and Conclusions

The results support the original hypothesis that people will vary individually and collectively in their use of contextual information to navigate/orientate. Figure 2 illustrates significant differences between ages, the most noticeable being a greater use of textual-structural information than textual-area/street based information by the younger age groups (under 25s). Although the explanation for this trend is outwith the scope of the study, possible reasons could be differences in social behaviour, or that with time, people's geographic knowledge increases enabling a more concise description using mainly street names (textual-street based information).

The results demonstrate how each participant's contextual descriptions of the environment are unique, and so indicates support for allowing the user to tailor presented information for his/her own needs; a view supported by 83% of participants (Figure 5). Structural, textual and directional were viewed to be important for route navigation by all participants, but there were differences of opinion for other categories. While environmental and sensory information were rated low, there may be situations were this would change (e.g. we are currently repeating the study for visually impaired people).

In line with Dey [4], the results also emphasise the need to understand contextual interactions in order to maximise usability. There were differences in information presentation styles (verbal vs. visual) and clear preferences for control over contextual information. The main usability implication/issue is therefore that the design of the application must allow an element of user control in order to present contextual information that is appropriate to a user's task and situation. For example, some participants described of scenarios where speech output would be better for reaching a destination promptly (minimising visual checks) involving concise information (i.e. directional, textual-area/street based and/or textual-structural based information). Whereas, visual presentation involving additional information (e.g. descriptive, numerical, etc) may be preferred (or used in conjunction with speech output) when touring a city for the first time in order to provide a greater spatial orientation and awareness of surrounding environmental features/landmarks. Lastly, another usability implication, based on the results from Figure 4, may be that more contextual information needs to be provided for speech output than for visual presentation.

It is anticipated that the study results will facilitate the design of future navigational context-aware applications. By understanding the dynamic nature of a user's contextual interactions, application designers can better determine which behaviours to support and how to support them. The next stage of our work involves designing a multi-category mobile navigation tool for controlled user experiments while developing a model of contextual interactions encompassing a multidisciplinary appreciation.

References

1. Chen, G. & Kotz, D. *A Survey of Context-Aware Mobile Computing Research.* Technical Report. TR2000-381. Department of Computer Science, Dartmouth College. (2000).
2. Cheverst, K., Davies, N., Mitchell, K., Friday, A. & Efstratiou. Developing a Context-Aware Electronic Tourist Guide: Some Issues and Experiences. *Proc CHI 2000.* (April 2000), 17 – 24.
3. Asthana, A., Cravatts, M. & Krzyanowski, P. An indoor wireless system for personalized shopping assistance. *Proc IEEE Workshop on Mobile Computing Systems and Applications*, Santa Cruz, California, (December 1994), 69-74.
4. Dey, A.K. Supporting the Construction of Context-Aware Applications. *Dagstuhl seminar on Ubiquitous Computing*, (September 2001).
5. Zetie, C. Unwired Express website: Market Overview - The Emerging Context-Aware Software Market. (2002) http://www.unwiredexpress.com
6. Dey, A.K. & Abowd, G.D. Towards a Better Understanding of Context and Context-Awareness. *Proc CHI 2000 Workshop on The What, Who, Where, When, and How of Context-Awareness.* The Hague, Netherlands, (April 2000).
7. Benerecetti, M., Bouquet, P. & Ghidini, C. On the Dimensions of Context Dependence: Partiality, Approximation, and Perspective. *Modeling and Using Context: Proc. 3rd International Conference, CONTEXT 2001,* Dundee, Scotland (July 2001) 59-72.
8. Bainbridge, L. Verbal Protocol Analysis. In Wilson, J.R. & Corlett, E.N. (Eds.), Evaluation of Human Work: A practical Ergonomics Methodology. London: Taylor and Francis, (1991), 161-179.

Towards Scalable User Interfaces in 3D City Information Systems

Teija Vainio[1], Outi Kotala[1], Ismo Rakkolainen[2], and Hannu Kupila [3]

[1] Hypermedia Laboratory, 33014 University of Tampere, Finland
{teija.vainio,outi.kotala}@uta.fi

[2] Institute of Signal Processing, Tampere University of Technology, 33101 Tampere, Finland
{ismo.rakkolainen@tut.fi}

[3] Laboratory of Geoinformatics, Tampere University of Technology, 33101 Tampere, Finland
{hannu.kupila@tut.fi}

Abstract. A 3D revolution has taken place during the last few years, and it is shifting towards hand-held devices. In this paper, we adapted our 3D City Info for mobile users and built a demonstration of future mobile services. Our main purpose was to study navigation and way finding in a three-dimensional city model that is connected in real-time to a map of the same area and to a database, which contains information from the same area. We have built a fully working mobile laptop version of the 3D City Info with an integrated GPS receiver for our field tests. The three-dimensional model appears to illustrate motion and change of location more clearly than two-dimensional map alone. In the future the possibility to scale, zoom and drag modules and components of the interfaces seems to be useful for different contexts of use.

1 Introduction

Web-based city information systems and city guides are widely known. It is also possible to visit through the World Wide Web many cities by navigating in the streets of three-dimensional city models. Also city guides for mobile devices already exist and studies e.g. about mobile "wearable" computer as a personal tourist guide [5] and the meaning of context-awareness and adaptive hypermedia in an electronic tourist guide [2] have done. Wireless networks, positioning, displays, 3D hardware and software for hand-held devices are evolving rapidly, enabling many new kinds of 3D and VR applications and mobile 3D devices.

In this paper, we describe our experiences of developing and evaluating the 3D City Info for mobile users. We study usability of the city information system, which utilizes three-dimensional model, a map and a database. We created a city information system to provide location-based information to users. The model of the city of Tampere is connected to a map of the same area and to a database, which contains

F. Paternò (Ed.): Mobile HCI 2002, LNCS 2411, pp. 354–358, 2002.
© Springer-Verlag Berlin Heidelberg 2002

information from the same area. When the user is navigating in three-dimensional model, he can verify his location from the map and through other means. A photo-realistic visualization is chosen to describe the real world because the interface design should convey the naturalness of the real world in order to provide appropriate tools for navigation, orientation, and feedback for the user [8]. One of our hypotheses is that a result of a query from a database could be easy to perceive when visualizing a result both in a map and a model.

We have earlier implemented a web-based 3D City Info for wired users [10]. We have also studied earlier the usability of 3D City info for mobile users by mock-up version. In this study we have implemented a fully working mobile laptop version of the 3D City Info with an integrated GPS receiver. The principles we investigate in our system are based on navigation and way finding. Way finding tasks can be classified into naïve search, primed search, and exploration [4]. Purposeful, oriented movement during navigation improves with increased spatial knowledge of the environment. Spatial knowledge can be described as three levels of information: landmark knowledge, procedural knowledge and survey knowledge [3], [4]. The landmarks are defined as the predominant objects in the environment [7], [1]. Our assumption is that landmarks are easier to perceive from 3D model than from a map, but a map is still useful when we are navigating in 3D model and in a real world.

1.1 Model and Database

Modelling is an important part of any three-dimensional environment, it should be as fast, automatic, and cost-effective as possible, to make 3D city models feasible. In this project we used two different methods to create graphical VRML model. Both methods produce a model, which is in real coordinate system. In the first method, we utilized the source material from ground information database of the city of Tampere. The other method we used based on a stereographical aerial measuring. This method is fast and reliable, if new stereo images from model area are available.

High-performance 3D graphics will be integrated into multimedia cellular phones and hand-held PDAs with wireless Internet, etc. Mobile 3D graphics will be feasible for consumers within a few years and 3D models can be accessed with mobile devices in many ways like wired downloads, at WLAN hotspots, over UMTS, etc.

We combined a relational database of services to the VRML city model. The system is platform independent and requires only a standard Java-enabled web browser and a VRML browser. A user can make queries about services in Tampere city centre, e.g., the locations of hotels or restaurants. The system visualizes the query results both in the VRML world and on the interconnected 2D map. Our 3D city info uses a novel mechanism to connect a database to a VRML model. A Java 2 applet communicates with the database through a standard JDBC API. It provides cross-DBMS connectivity to a wide range of SQL databases. The applet communicates with the VRML world through the EAI interface. Standard relational database technologies are used on the server side and this enables easy updates to the data content.

2 User Interface Design

Because the screen of a mobile device usually has a limited space, all the elements of the service are not visible at the screen at the same time. The map, the 3D city model and the search are each put inside a module or palette of its own. Only two modules are visible at the screen at the same time. By clicking on the interleaf on the right the user can view different modules. The metaphor of an address book interleaf is used to help the user to orientate to the multi-module interface and keep in mind all the possible parts of the service. In the mobile user context, the interleaf-like menu seems more understandable than ordinary menus; the hand-held device is easily perceived as a book-like aid for everyday life, so the metaphor is entitled.

Designing the user interface for mobile devices is also challenging because in mobile usage the context may be constantly changed. Thus, two extremes of mobility can be defined: semi mobility and fully mobile. In semi mobile environments there are some aspects that are not constantly changing, such as in a train, otherwise fully mobile environments may be physically unstable [10]. In our study we concentrate to semi mobile environments.

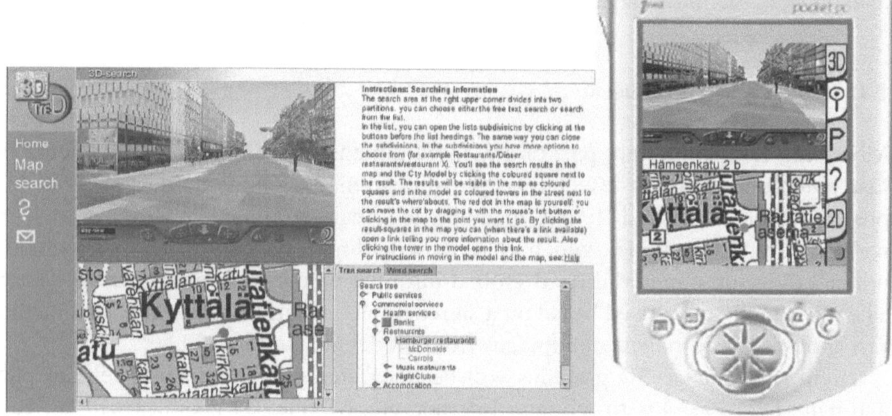

Fig. 1. The main user interfaces of 3D City Info for wired and for mobile users

2.1 Experiment

Eleven volunteers participated in an experiment in which we have built a fully working mobile laptop version of the 3D City Info with an integrated GPS receiver. The city info database and model run locally on a 650 MHz PIII laptop capable of fast 3D graphics. An online database connection could easily be built with, e.g., GSM, GPRS, or WLAN, but we used database locally. Because of the heavy equipment the experiment was done in a car and the user interface we used was the one for wired users (figure 1) due to the equipment and software solutions. By using the GPS receiver a user did not have to find his/her location from a map or a model, the information system was updating the location automatically when moving from place to place during the test.

The participants were assigned three tasks and afterwards they were interviewed. The first task was to find the nearest bookshop with the help of the 3D City Info and to show the right direction to get there. The second task was to describe to the driver the current location during a short drive and to tell what they could see at the screen. The last task was to find all the theatres at the centre of the city with the help of the information system and point out the way there. The participants could choose which they preferred to use, a map or a 3D model or both, to complete the tasks.

3 Results

Our former studies of 3D City Info for mobile users appears to support the conclusion that users prefer to use a combined mobile system rather than to use only the map or the model alone and that they more likely recognized their own position and the landmarks from the photo-realistic model than from the map [9]. In this study we used a fully working city information system, which includes a 3D city model, a map, a database and a GPS receiver. Based on our preliminary user study, the three-dimensional model illustrates motion and changing the location more clearly than two-dimensional map alone. The visual similarities with reality enable the participants to find the places from the real city environments when he or she is moving, e.g. walking or driving a car.

3.1 Discussion and Future Work

In the future the mobile user interface might be based on the idea of the components, which users could scale, zoom and drag in different contexts of use. The same idea is already utilized in many different applications for desktop computers. Former studies has stated that in designing multi-scale interfaces the designers have created radically new environments in which such interaction principles that are without precedents in the real world. Especially during the nineties interface designers have used multi-scale navigation. For example with the zooming interface the user can freely move in the electronic world and is allowed to adjust the scale at which he wishes to interact with the world. It is a new type of user action, which has no counterpart in the real world. [6]

For mobile users there might be occasions when a user might need to see a map larger than the city model (for example in the case of finding a street name) so the interface should be adaptable to meet the needs. In other cases the user doesn't need a large map or a map at all, but a three dimensional model to find a specific place (for example in the case of finding a front door of a building). In 3D City Info for mobile users a draft of user interface was created to illustrate the functionality of such components that could be scaled and dragged in the user interface.

Acknowledgements

The Finnish National Agency Tekes mainly funded this research project. The Laboratory of Geoinformatics and the Digital Media Institute at the Tampere University of Technology, and the Tampere Polytechnic School of Art and Media, made the city model. Arttu Heinonen, Simo Pulkkinen and Lauri Toivonen at the Tampere University of Technology completed the software. We wish to thank Professor Tapio Majahalme, Cai Melakoski and Minna Kulju for their support. We thank also Director Jarmo Levonen and Professor Tere Váden from the Hypermedia Laboratory at the University of Tampere for their helpful comments.

References

1. Charitos, D., Designing Space in Virtual Environments for Aiding Wayfinding Behavior. Department of Architecture, University of Strathclyde. The fourth UK VR-SIG Conference, Brunel University, 1st November 1997.
2. Cheverest, K., Davies, N., Mitchell, K., Friday, and Efstratiou, C., Developing a Context-aware Electronic Tourist Guide: Some Issues and Experiences in Proceedings of CHI´2000, ACM, New York, 17-24, The Hague, Netherlands April , 2000.
3. Darken, R., Wayfinding in Large-Scale Virtual Worlds. Conference Companion ACM SIGCHI '95, 45-46.
4. Darken, R. and Sibert, J. Navigating Large Virtual Spaces. International Journal of Human-Computer Interaction, January-March 1996, 8 (1), 49-72.
5. The Deep Map project. Available at http:// http://www.eml.villa-bosch.de/english/Research/Memory/1
6. Guiard, Y., Bourgeois, F., Mottet, D., and Beaudouin-Lafon, M. Beyond the 10-bit Barrier: Fitts' Law in Blandford, A., Vanderdonckt, J. and Gray, P. (eds.), Multi-Scale Electronic Worldsin People and Computers XV –Interaction without Frontiers. Joint Proceedings of HCI 2001 and IHM 2001. Springer 2001, 573-587.
7. Lynch, K., The Image of the City. 1960, Cambridge: MIT Press. 194.
8. Mills, S. and Noyes, J. Virtual reality: An overview of User-related Design Issues Revised Paper for Special Issue on "Virtual reality: User Issues" in Interacting With Computers, May 1998. Interacting with Computers 11 (1999), 375-386.
9. Rakkolainen, I., Vainio, T., A 3D City Info for Mobile Users. Computers & Graphics, Special Issue on Multimedia Appliances, Vol. 25, No. 4. Elsevier 2001, 619-625.
10. Väänänen-Vainio-Mattila, K. and Ruuska, S., Designing Mobile Phones and Communicators for Consumers' needs at Nokia in Bergman, E., (eds.) Information Appliances and Beyond. Interaction Design for Consumers Products, Academic Press. Morgan Kaufmann Publishers. 2000,169-204.
11. Tre-D city information system. Available at http://www.uta.fi/hyper/projektit/tred/

Mobile Image Messaging – Anticipating the Outlines of the Usage Culture

Anne Soronen and Veijo Tuomisto
Hypermedia Laboratory
University of Tampere
33014 Tampere University, Finland
{anne.soronen,veijo.tuomisto}@uta.fi

Abstract. The mobile phone culture is moving towards the multimedia messaging era. When mobile camera phones and MMS messages are available, users can take photographs and send them to each other. Later on, it will be possible to attach video clips to MMS messages and maybe even transmit real-time mobile video. In this paper, we present some future outlines of the use of mobile camera phones in the Finnish mobile culture. We study how mobile users conceptualise mobile camera phones and mobile image messaging. Our goal is to identify different meanings that the users relate to mobile camera phones when they do not have actual experience in using them.

Introduction

The definitions and perceptions that people associate with new technical devices are mainly founded on the appellations that manufactures, analysts and the media offer when they talk about the novelties [7]. The framework that the advertising and the media create in the beginning and before an item has been released contribute to the expectations that the consumers direct to the use value of the device.

This paper concentrates on how Finnish mobile users conceptualise mobile camera phones and mobile image messaging. By mobile image messaging, we refer to MMS messages and to real-time video broadcasting through a mobile device. The term 'mobile camera phone' is not an established one. As an appellation, it emphasizes the hybrid nature of the device but can also raise the question: Are we talking about a phone with which you can shoot or a camera with which you can call?

In our study, we have analyzed some presumptions that Finnish mobile users have about mobile image messaging, including both the communication and the shooting aspect. Our aim is to identify what kind of social and functional meanings the users attach to mobile camera phones when they do not have actual experience in using them.

The mobile phone is used very often when taking care of personal relationships, and this rests principally on real-time emotion exchange via human voice [3]. In our informants' mental models, mobile image messaging was considered all the more as an emotional sensitive practice than calling and text messaging. As a consequence,

F. Paternò (Ed.): Mobile HCI 2002, LNCS 2411, pp. 359–363, 2002.
© Springer-Verlag Berlin Heidelberg 2002

both strongly undesirable and strongly desirable aspects of mobile image messaging were stressed, while more neutral and everyday ways of use did not come out so much.

Implementation

The results reported here are based on focus group interviews in which visual usage scenarios were used to support discussion. By a usage scenario we mean a story about the use of the device or the service. Usually usage scenarios include a user or many users, a context of the user(s), characteristics of the device and possible ways to use it [4]. By applying usage scenarios it's possible to illustrate different ways of use and to keep the technology in the background. This is to say that usage scenarios should also be comprehensible to people who are unfamiliar with the technology in question.

The interviews were carried out during the spring of 2002 when mobile camera phones had just started to appear. Our target group consisted of young adults aged from 20 to 38, all experienced mobile phone users. The informants in question were students and employees who lived in households with no children.

Shooting as an Undesirable Practice

The informants were confident enough that some people would shoot other people secretly with mobile camera phones. For the most part, the problem arouse from the possibility of using the "hidden" camera of the mobile phone very unnoticeably. When someone is using his/her mobile camera phone one can not be sure if he/she is taking pictures or doing something else with the device. The result of this can lead to increased paranoia among people. Personally, the informants did not see it as a huge problem for themselves but they felt it could be a significant problem for e.g. celebrities.

With a mobile camera phone one is more likely to take quick snapshots of people than wait for them to get "ready" for the picture. The informants were afraid that people might take pictures at incorrect moments when they are not necessarily showing the best side of themselves. They also brought up the discomfort that many people feel in front of a camera. Some of the informants presented the idea that mobile camera phones should have a red light showing the recording like regular video cameras. If the light were on, other people would know that the user of the mobile camera phone is presently shooting. The light would also help to notice if the camera was left on accidentally.

Kopomaa uses the term 'off-limit' to describe places where the use of mobile phones is not desirable or is strictly prohibited. The restrictions on mobile phone use are expressed by appealing to the control of private spaces, general well-being or safety, and side effects that may create a health hazard [5]. The informants thought that in the places where the use of a mobile phone is now prohibited, the use of mobile images or video should also be prohibited. In the examples given by the informants, the main concern related to the protection of privacy.

Shooting as a Desirable Practice

The informants saw that a mobile camera phone would be comfortable in the sense that then they would always have a camera with them and could take pictures whenever they wanted. They assumed that taking pictures might increase with mobile camera phones because an ordinary camera is more arduous to carry.

Although the informants highlighted that shooting secretly might become a serious problem for some people, they saw as a positive side that the use of the mobile camera phone would not draw too much attention in the shooting situation itself. As Koskinen et al. put it; a user can be more spontaneous when other people do not know that he/she is taking a picture [6].

Communication via Mobile Camera Phone as a Desirable Practice

Mobile image messaging was considered suitable for special social events mainly for two reasons: The first is a situation when somebody cannot be present but remote participating becomes possible through mobile video broadcasting. The second is the possibility to share (audio) visual memories afterwards with friends. The latter relates to the common practice of using photos to support remembering and to reinforce a sense of community [6]. Most of the informants saw that they would share the picture files only in a digital format.

The social contexts where our informants saw themselves using mobile image messaging were predominantly confined to interaction with one's intimates. Sending mobile greetings and congratulations with photos or video clips to friends or relatives was considered attractive compared to sending a mere text message. It was difficult for our informants to imagine situations in which a photograph or a video connection in mobile communication could give any added value when communicating with half-acquainted people or strangers. It seems that the mobile camera phone will reinforce the practice in which communication time is spent more and more in the presence of "those who matter" [2].

The mobile camera phone was seen as a useful device in calling situations when you have to ask for help or advice because it is often difficult to describe certain items verbally. For example, some female informants saw snapshots or a real-time video connection handy during shopping or dressing if you want to ask for someone else's opinion concerning a piece of clothing. These advisory situations reveal our informants' aim to find needs that mobile camera phones might probably full. Seeking arguments for usefulness and uselessness is a common attitude when talking about new devices [8].

Communication via Mobile Camera Phone as an Undesirable Practice

When the informants considered mobile image messaging as an undesirable communication practice they assumed, among other things, that the operating costs will be high and the quality of picture so bad that it cannot give any added value to mobile communication. Anyway, it seemed that a text or a voice message in multimedia messaging services can compensate for a poor quality of photos or video clips. Correspondingly the study by Frohlich and Tallyn [1] proves that people are tempted to attach ambient sounds or voice annotations to bad photos for improving and vitalizing them.

The informants considered it important to be able to preview the mobile video content when they are the receivers. If people have not agreed beforehand on sending a mobile video in a certain situation, then the receiver should get some hint of the content. In our informants' opinions, the content could be defined by a title, a short text, a voice message, or by showing the first frame of the clip. When a message includes a video or a photo without an interpretive verbal part (text or voice), there is a noticeable risk of sending some unwanted signals to the receiver.

The informants supposed that (Finnish) people tend to send unclear and suspicious visual material via mobile camera phones especially when they are drunk. They predicted that these devices might notably increase the existing practice of drunken people sending text messages and calling friends just for fun. Precisely, bantering and the playful use of mobile camera phones have been presented as one characteristic feature of mobile image messaging [6].

The informants also covered the problem of sending messages to wrong numbers. The risk of sending a personal MMS message including an image or a video clip to a wrong person was considered embarrassing, both to the sender and to the receiver. The situation where there is a two-way tone and a one-way real-time video connection was also considered a problematic communication situation. Some informants felt that a communication situation is unbalanced when both parties cannot shoot and show something to the other at the same time.

Conclusion

It appears that, among adult Finnish users, the most noticeable worry concerning mobile camera phones is shooting lurkingly. It is not surprising that the informants mostly highlighted the negative aspects of the use of mobile camera phones. The sceptical attitude is common when people talk about forthcoming technology that they only know at a notional level but, on the other hand, they are, at the same time, prone to see something inviting in a new technology [8]. As the most tempting feature in the use of mobile camera phone, the informants saw the possibility to share specific emotion-sensitive situations with their intimates.

Overall, it seems that it is difficult for adult users to conceive multimedia messaging and mobile video broadcasting suitable for spontaneous and everyday mobile communication. In the first place, they perceive a mobile camera phone as a phone, which makes it possible to store (audio) visual memories of personally important situations. When people do not have actual user experience in mobile image messaging, they assume that one central use value of the mobile camera phone relates to shooting and not to communicating through images. For the time being, mobile image messaging seems to be considered as somewhat remote and luxurious among Finnish mobile users.

Acknowledgements

The study was carried out during the Multimedia Services in Wireless Terminals project at the Hypermedia Laboratory of the University of Tampere.

References

1. Frohlich, D.M., Tallyn, E.: Audiophotography: Practice and Prospects. Extended Abstracts of CHI '99 Conference, New York (1999) 296-297 http://www.hpl.hp.com/research/userstudies/publications/docs/chi99.doc
2. Gergen, K. J.: The challenge of absent presence. In Katz J.E., Aakhus M.A. (eds.) Perpetual Contact. Mobile Communication, Private Talk, Public Performance. Cambridge University Press, Cambridge (2002) 227-241
3. Kasesniemi, E.-L., Rautiainen P.: Kännyssä piilevät sanomat. Nuoret, väline ja viesti. Tampere University Press, Tampere (2001)
4. Keinonen, T.: Pieniä tarinoita pienistä puhelimista. In Keinonen T. (ed.) Miten käytettävyys muotoillaan? Taideteollinen korkeakoulu, Helsinki (2000) 207-220
5. Kopomaa, T.: The City in Your Pocket. Birth of the Mobile Information Society. Gaudeamus, Helsinki (2000)
6. Koskinen, I., Kurvinen, E., Lehtonen, T.-K.: Mobiili kuva. Edita/IT Press, Helsinki (2001)
7. Pantzar, M.: Edison, Kodak ja mobiilin kuvan juurtuminen – ennakoivaa historian kirjoitusta. In Koskinen I., Kurvinen E., Lehtonen T.-K.: Mobiili kuva. Edita/IT Press, Helsinki (2001) 99-111
8. Pantzar, M.: Eettisiä kysymyksiä. In Rainio, A. (ed.): Henkilökohtainen navigointi. Markkinat, teknologia ja sovellukset. Valtion teknillinen tutkimuskeskus, Espoo (2000) 106-124

Remote Control of Home Automation Systems with Mobile Devices

Luca Tarrini[1], Rolando Bianchi Bandinelli[2], Vittorio Miori[2], and Graziano Bertini[2]

[1] Research associate,
oknxsw@vizzavi.it
[2] ISTI – National Research Council - Via Moruzzi, 1 -56124 Pisa Italy
Rolando.Bandinelli@cnuce.cnr.it

Abstract. Remote control based on mobile devices as mobile phones or PDA's, is considered more and more useful in many computerised applications. This paper deals with the implementation of functions, based on mobile devices, for the remote control of commercial home automation systems. Different solutions are considered and some problems concerning their implementation are discussed. A preliminary development of the interface used to control X10 modules or to interrogate a home database of the device state is here described. Some guide-lines for the interface design are also reported.

1 Introduction

The aim of the project is to realize a house that can be remotely controlled with web-based facilities, giving the possibility of interacting with several key actuators. Namely the Home Automation (HA) system [1] will accept local commands as well as remote commands to manage different types of situations (e.g.: close windows and doors and switch on the heating if the temperature is low, switch on/off appliances, etc).

Commercial HA systems, equipped with advanced web-oriented tools, can provide a more complete environmental control, which includes aiding and monitoring functions for disabled and elderly people. As many other computerized applications in the field of intelligent environments, they are based on a local manager (PC or embedded control connected to a telephone network) that executes different tasks such as controlling sensors and actuators or collecting home data and eventually bio-medical data from monitored people. Using Internet facilities, the collected data could be made accessible to specialized organizations which could detect health or home emergency states (gas or water leaks, fire, intrusions, etc.). Moreover, Internet facilities could be exploited by interested people to perform a remote check of the health condition of relatives or of the home state. In addition to Internet, alternative communication channels (such as SMS, GSM and GPRS) will be used to increase the communication network reliability. Mobile devices can provide some improvements to the performance of HA systems. Indeed, they allow people to be connected independently of time and space constraints. In this report the problem of implementing a WAP access and related interfaces for HA systems remote control, is addressed. The implementation is part of a project currently carried out by the ISTI-CNR Institute with the cooperation of NES s.r.l. (a private company).

F. Paternò (Ed.): Mobile HCI 2002, LNCS 2411, pp. 364–368, 2002.
© Springer-Verlag Berlin Heidelberg 2002

2 Integration of the Communication Functions in a HA System

The selected HA system can be considered, in its basic version, an evolved alarm system (against intrusion, gas leaks, etc.) enhanced with many other functions such as in/out vocal messages, the possibility of controlling the telephone line, the easy- to-use programmable features and a sophisticated management of actuators and sensors (see Fig. 1).

The link between the central unit and the other peripheral devices exploits different technologies such as special busses, direct cables, wireless power line communication, infra-red, etc. [2]. The protocol used for communication through power lines is X-10 that is widely used in the USA and in the recently also in Europe. A newer advanced European standard (EHS-Konnex), not yet very spread, is now taking over. In the first prototype we have considered a HA system with an open architecture that we can easily integrate by adding a remote control and other extensions expressly designed by our group. In this report we will omit the description of the extensions details. Moreover the system contains a database with data on the parameters needed to control the appliances, the state of the sensors and of the whole system.

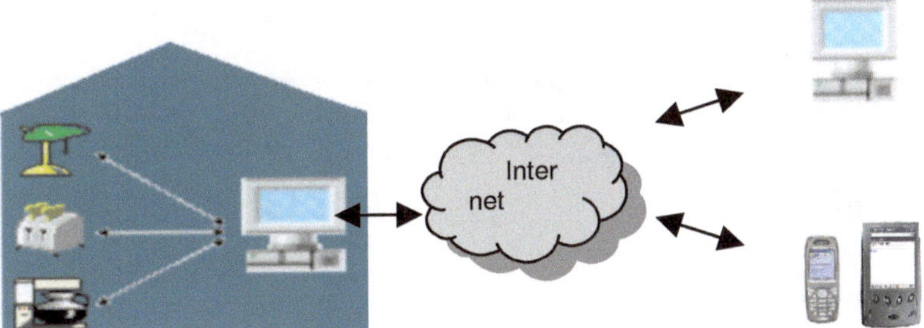

Fig.1. Remote control of the home automation system

3 Interfaces Development for Mobile Devices

Mobile devices (mobile phones, PDA etc.) allow the user to have a continuous control and interaction with HA systems through a small display: it will be possible to write a query and to send a command to a particular appliance (see fig.2. for an example), exploiting the browser capability of a mobile device.

We have designed two different types of interface: one for the computer and the PDA and one for the mobile phone display. All the icons and the graphics needed to make the screen layout appealing for the user were considered when designing the first interface. The second interface does not contain graphics as it makes use of a very small display and in this case it is important to minimize the connections across the HTTP connection.

The computer interface is a very classic one that represents home appliances with icons and the actions that can be chosen with buttons. Many options can be represented in a single screen. In the phone interface instead only one action at a time can be represented in one screen. We have paid special attention to group all the possible coherent actions for one home appliance into a single screen frame in order to enable the simultaneous transmission of related data.

In the example of fig. 2 a data transmission for the first screen and a single transmission for next two screens are depicted. The WAP services can be implemented using WML (Wireless Markup Language) and ASP (Active Server Pages). WML pages containing the results of database queries or a sequence of user commands can be dynamically created.

The recently available "Microsoft.NET" release, that allows the creation of a universal platform, made a quicker implementation of our interfaces possible. Using the CLR resource (Common Language Runtime) of "Microsoft.NET", it is no more necessary to switch among many languages in order to write application code. Another advantage of the ".NET" solution is that different devices (desktop computers, mobile devices) can be used without the necessity of creating a different module for each of them. With ".NET" every device feature can be exploited making use of the suitable interface in a transparent way. In particular by exploiting ASP.NET and NET.Framework we can implement Web "forms" (controls for mobile devices) able to adapt the output rendering to different devices and to generate the most suitable code for each one.

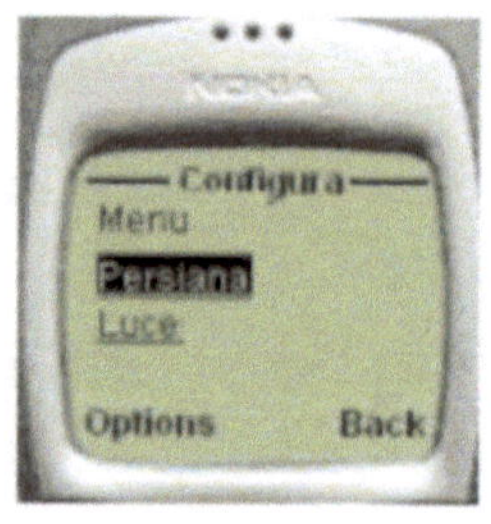

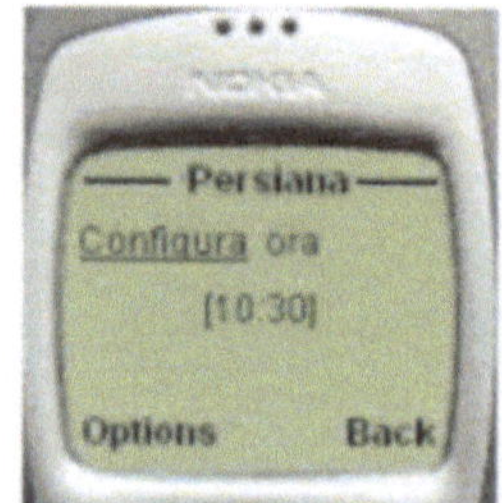

Fig.2. Mobile phone displays with examples of commands

The above controls support WML 1.1, HTML 3.2 and cHTML and can create also useful features like Calendar, etc. Moreover the Web Form, and the Mobile Form allow for the separation of the program code from the rendering code.

ASP.NET pages, as well as the ASP ones, use the request/response model, but the philosophy is event-oriented. In fact through the script at the client side, that ASP.NET automatically generates, the server controls warn the server itself of any event, such as the pushing of a button. Moreover this tool allows the use of VB.NET or C#, avoiding the use of scripting languages such as VB.Script, and it allows to compile pages when accessed for the first time, in a way similar to Java Server Pages.

It is possible not only to execute commands, but also to investigate the state of the controlled system. In fact the information on the status of the devices is kept in the database accessible through ADO.NET (Access Data Objects). ADO.NET is a database access model, based on XML (shown in Fig.4).

During the execution steps the database provides data to a business object at an intermediate level and then to the user interface. The data are transmitted by ADO.NET in XML format.

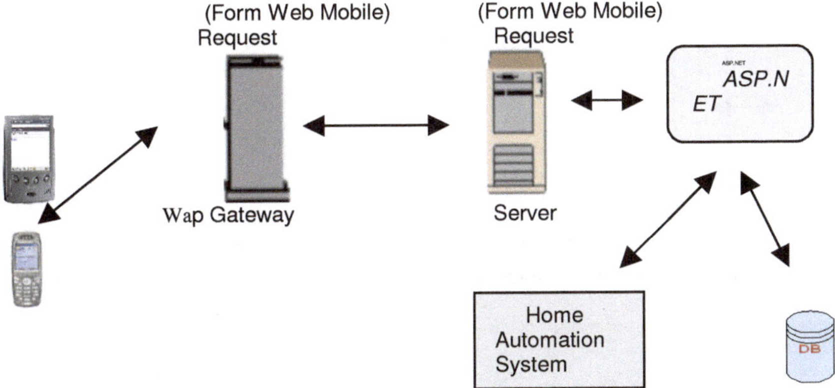

Fig. 3. Tools involved in the remote control services development

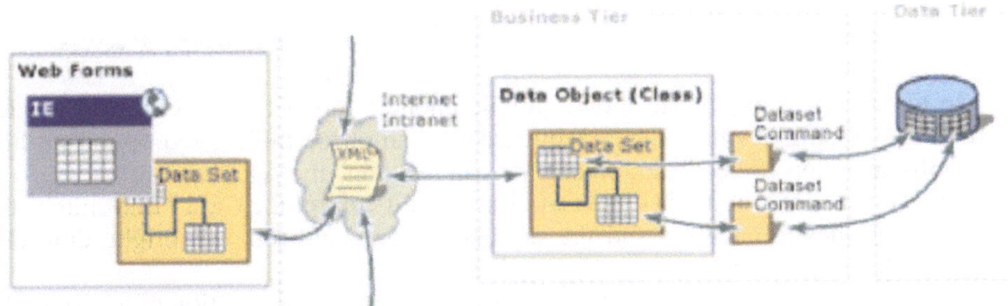

Fig, 4. ADO.NET Architecture

We have also considered the new open source technology offered by the Mono project (http://www.go-mono.com/) and by GNOME (http://mail.gnome.org/archives/gnome-hackers/2002-February/msg00031.html). It is the same approach of Microsoft.NET (or at least similar), but it is not proprietary and the source code can be freely customized. This technology looks promising. It is the result of the work of many people, but it is too recent to be fully reliable and the documentation is not so rich as the one provided for the Microsoft.NET. For this reason we chose to use the technology offered by Microsoft, being aware that in the future a different technology could be used.

4 Preliminary Tests and Demonstrations

In order to test and demonstrate selected system functions, at the beginning we have not used the whole HA system, but we connected directly the computer to the X-10 [3] modules, as shown in Fig. 5.

It is well known that the X-10 protocol uses one-byte addresses: the first half, called "House", is represented by a letter (from A to P) and the second one is represented by a number (from 1 to 16) and is called "Unit".

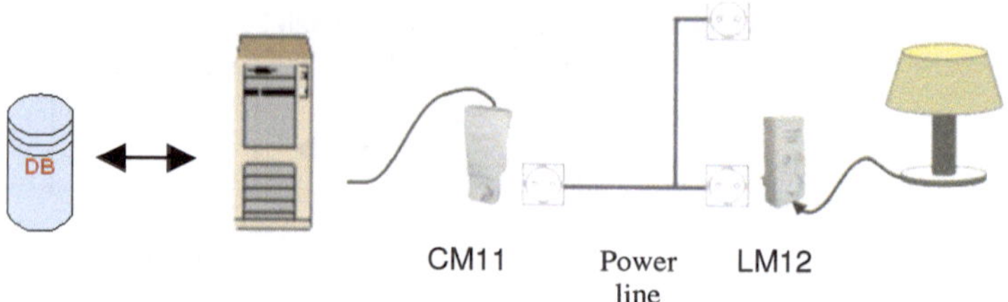

Fig. 5. The CM11 module (X-10 transmitter) directly controls the dimmer LM12 module (X-10 receiver) via the power line

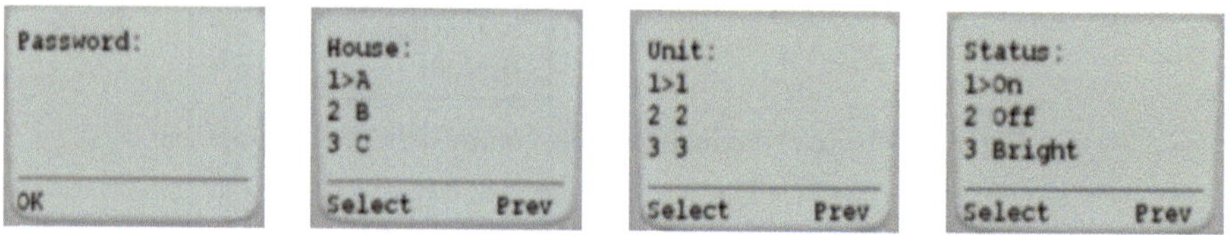

Fig. 6. The display of a mobile phone with an example of a WAP access

In Fig. 6 a possible interface on a mobile phone in order to control a X-10 device, are shown. Using a screen pages sequence we can chose the desired option pushing the softkey button "Select", and finally we can put "on" the A1 device.

The Home server up to date the Data Base and it control the CM11 transmitter; this latter dispatches the address and the associated command (on, off, bright, dim, etc.) through the power line to the selected device.

The main aim of the project, that is still in progress, is the creation of a system that includes a HA control unit that will allow the remote management of a large selection of home appliances. On the base of these preliminary experiences, suitable configurations of the interfaces will be designed in order to facilitate the user interaction with the system services.

Financial Supports

This project was developed in the framework of a collaboration project between the NES srl company (Ospedaletto, Pisa) and CNUCE/IEI-CNR of Pisa and with the Italian government contribution to SME for industrial research (ref. the DDL n° 275 July 22nd 1998, GU n° 186 August 11th 1998).

References

1. Bianchi Bandinelli, R. Fusco, G. Rossi, R Saba, A.: IB (Intelligent Building), Proc. of the I° TIDE Congress, Of Rehabilitation Technology, Strategies for the EU, Brussels, (1993).
2. Dutta-Roy, A.: Networks for Homes, IEEE Spectrum (1999) 27-33
3. Krishnan, C.N. Ramakrishna, P.V. Prasad, T.V. Karthikeyan, S.: Power-Line As Access Medium-A Survey", Int. COMMSPHERE 2000, Indian Institute of Technology Madras, Chennay, India. (2000).

Low-Resolution Supplementary Tactile Cues for Navigational Assistance

Tomas Sokoler[1], Les Nelson[2], and Elin R. Pedersen[3]

[1] Space & Virtuality Studio,The Interactive Institute, Beijerskajen 8,
S-205 06 Malmö, Sweden
Tomas.Sokoler@InteractiveInstitute.se
[2] FX Palo Alto Laboratory, 3400 Hillview Ave
Palo Alto, CA 94304 USA
nelson@fxpal.com
[3] Kraka Inc, 330 Alameda de las Pulgas
Redwood City, CA 94062 USA
elin@kraka.com

Abstract. In this paper we present a mobile navigation device 'displaying' supplementary personalized direction cues by means of a tactile representation. Our prototype, the TactGuide, is operated by subtle tactile inspection and designed to complement the use of our visual, auditory and kinesthetic senses in the process of way finding. Preliminary experiments indicate that users readily map low-resolution tactile cues to spatial directions and that TactGuide successfully can be operated as a supplement to, and without compromising, the use of our existing way finding abilities.

1 Designing for Navigational Assistance

Way finding involves simultaneously reading and piecing together cues from a multitude of information resources distributed in the environment. These resources typically include architectural design patterns, sounds, pictograms, text signs, and the presence of other people willing to guide you on request. When designing for navigational assistance in complex environments it is therefore crucial that interaction with the navigational device leaves the visual, auditory and kinesthetic senses available to the process of 'reading' the environment. Commercially available handheld navigation devices [1] take a traditional PDA approach by displaying navigational information on a graphical display. Operating these devices involve a highly focused mode of interaction that tends to monopolize the user's attention making it difficult to operate the device while at the same time paying attention to other cues in the environment.

We have experimented with the design of a tactile display that complements rather than substitutes the use of our natural abilities and earned skills for way finding. In fact our prototype, the TactGuide, is designed to leverage from the simultaneous use of other inspection mechanisms directed towards the environment. An interesting technical implication from this approach is that the TactGuide only needs to provide the user with low-resolution directional cues: When knowing that the overall direction

F. Paternò (Ed.): Mobile HCI 2002, LNCS 2411, pp. 369–372, 2002.
© Springer-Verlag Berlin Heidelberg 2002

is 'straight ahead' people are fully capable of following that direction while adjusting for physical obstacles and fine tuning the direction taken using cues otherwise perceived from the environment.

Other experimental systems [2][3] have explored the feasibility of communicating personalized navigational cues by means other than visual representation. TactGuide joins this exploration but differs by providing a display mechanism specifically designed to allow for a seamless way to engage/disengage in interaction with the navigation device during way finding. We envision the TactGuide being useful when finding your way to places of personal preference in complex indoor environments. Examples of use situations are: finding your way back to your car in the airport parking garage, finding that bookstore in the shopping mall that your friend told you about and help you locate a particular book in that store.

2 The TactGuide Prototype

The TactGuide design strives for a 'tactful' interaction scheme suited to the task of simultaneously reading and piecing together cues from a multitude of information resources. We hypothesized that a direct and persistent but at the same time 'easy to ignore' quality of the directional cues was important for the successful implementation of such a scheme. We designed the interaction as 'Tactile inspection' where the user on his/her demand uses the thumb to inspect a tangible representation.

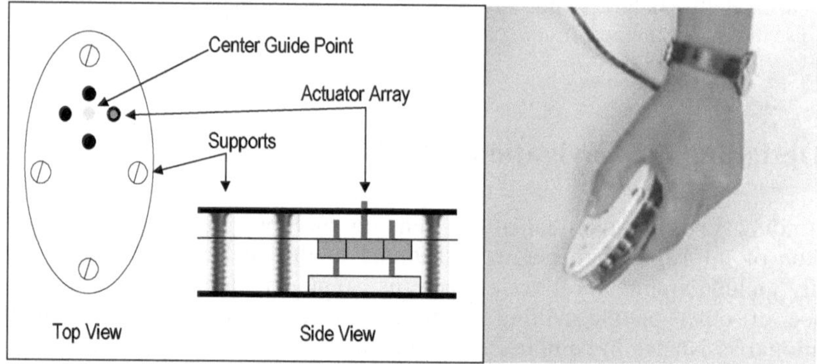

Fig.1. TactGuide display with 1 of 4 pegs raised to indicate relative direction

The TactGuide display (fig.1) has a flat smooth ellipsoidal shaped surface and four holes positioned around a 1mm high raised dot. The shape, smoothness and spatial layout was determined by having users comment on the look and feel of a series of device mockups with different form factors. The total area bounded by the four holes is slightly bigger than that of a thumbprint. Directly underneath each hole is a metal peg attached to a solenoid. A micro controller determines which peg to raise by combining device orientation data from an electronic compass with data on current location and destination. Direction is displayed by raising one of the four pegs through its corresponding hole. When putting your thumb on the TactGuide display there are two sensory inputs to your thumb one from the center dot and one from a peg. The position of the raised peg relative to the center dot provides a vector in one of four directions (Forward, Back, Left, Right). The TactGuide display thereby provides an analogous representation [4] in the sense that a spatial physical representation (center dot to peg vector) is used to communicate a spatial physical relationship (directions in physical space relative to your bodily orientation). The TactGuide display can be used in combination with any infrastructure capable of wireless delivery of position data (Differential GPS or Radio beacons for outdoor and indoor use, respectively).

3 Preliminary User Studies

We used our Tactile Display prototype to see how well users would map the tactile cues to spatial directions while traversing and inspecting a complex indoor environment. We asked 7 subjects to participate in a 'treasure hunt' using directional cues from the TactGuide prototype to track down a number of cardboard boxes (12"x12"x12") placed in our office building. We found that the subjects easily dealt with the tactile inspection and mapping of TactGuide directional cues into real world directions. All subjects indicated that the directness of the representation made the device easy to operate. One subject basically suspended the use of other senses for reading the environment and tried to use the TactGuide as the sole source of directional cues. As expected, using the device in this way caused frustration. The subject never made it to the first doorway en route and stated that the TactGuide cues kept 'making' her bounce the hallway walls. This experience led us to believe that the success of the other subjects indicates that they did constructively combine device cues with cues from the environment.

4 Future TactGuide Prototyping

A fully functional system for navigational aid should incorporate ways of setting up the destination and preferences in terms of the route to be taken. We envision that this functionality could revolve around the placing and retrieving of 'real world bookmarks'. A bookmark is in its simplest form a set of geographic coordinates but one could easily think of more advanced bookmark objects describing in a more rich way the relation between a user and a location. We would like to implement the TactGuide initialization and bookmark manipulations with the use of a general personal device (PDA or cell phone). Provided that these devices are equipped with short-range radio communication capabilities such as for example Bluetooth [5] destination data extracted from the bookmarks could be downloaded to the TactGuide. Bookmark operations that we would like to support are:

- Downloading bookmarks from the immediate environment. Example: Download bookmarks from the directories often found at the entry points in shopping malls or airports.
- Placing bookmarks in the immediate environment. Example: Placing a bookmark at the location where you leave your car in the parking garage.

- Sharing and communicating bookmarks between a group of people. Example: Setting up a face to face meeting between friends or trying to guide a potential customer to your store.

Combining the TactGuide with a standard input and storage device would allow the user to apply his/her preferred personal device for bookmark manipulations and allow integration of bookmarks with already existing personal databases like address books. It would also allow us to more easily embed the TactGuide into everyday physical objects typically associated with way finding and being 'on the move', such as for example briefcase handles and handlebars on shopping/luggage carts.

5 Conclusion

We believe that our preliminary studies strongly support the idea that a tactile representation is well suited as a way to provide supplementary low-resolution directional cues. We also believe that more detailed studies are needed to explore whether TactGuide 'easily' slides between fore- and background and truly allows the users to economize their attentional resources while traversing a complex environment. Finally, we intend to further pursue the more general idea of device interaction schemes that accommodate the users' need to simultaneously deal with large amounts of information presented in different media.

Acknowledgments

The authors would like to thank all the participants who aided us in our user studies. We would also like to thank FXPAL for support of this work.

References

1. Adventure GPS Products, Inc.: http://www.gps4fun.com
2. Tan, H.Z., Pentland, A.: Tactual Displays For Wearable Computing. In proceedings of the first international symposium on wearable computers, (ISWC 97), IEEE, pp.84-88
3. Nemirovsky, P., Davenport, G.: Guideshoes: Navigation Based on Musical Patterns. In extended abstracts of CHI99, ACM press, pp.266-267
4. Norman, D.A.: The Invisible Computer. MIT Press 1998, pp.138-139
5. The Official Bluetooth web site: http://www.bluetooth.com

Mobile Interface for a Smart Wheelchair

Julio Abascal[1], Daniel Cagigas[2], Nestor Garay[1], and Luis Gardeazabal[1]

[1] Laboratory of Human-Computer Interaction for Special Needs
Informatika Fakultatea. Euskal Herriko Unibertsitatea.
Manuel Lardizabal 1. E-20018 Donostia, Spain
{julio, nestor, luisg}@si.ehu.es
[2] Group of Robotics and Rehabilitation Technology
E. T. S. Ingeniería Informática. Universidad de Sevilla.
Avenida Reina Mercedes s/n. Sevilla, Spain
daniel@atc.us.es

Abstract. Smart wheelchairs are designed for severely motor impaired people that have difficulties to drive standard -manual or electric powered-wheelchairs. Their goal is to automate driving tasks as much as possible in order to minimize user intervention. Nevertheless, human involvement is still necessary to maintain high level task control. Therefore in the interface design it is necessary to take into account the restrictions imposed by the system (mobile and small), by the type of users (people with severe motor restrictions) and by the task (to select a destination among a number of choices in a structured environment). This paper describes the structure of an adaptive mobile interface for smart wheelchairs that is driven by the context.

1 Introduction

Smart wheelchairs are designed to improve the mobility of users with severe motor impairments that experiment difficulties to drive traditional electric-powered wheelchairs. The techniques used to automatically drive wheelchairs come from the Mobile Robotics and the Automated Guided Vehicles fields [1]. Smart wheelchairs are usually provided with a number of sensors and the necessary software for control, to be able to automatically follow a path from a starting position to a destination without human intervention [2, 3, 4]. Even if the user of a smart wheelchair does not need to carefully drive it, he or she must be provided with an adequate interface to be able to give the necessary orders for the wheelchair control.

The design of an interface for a smart wheelchair faces diverse problems due to the special characteristics of the user, the system and the task. Smart wheelchair typical users are severely motor –and sometimes voice– impaired people that can not handle a standard interface. On the other hand, the computing capacity of the system is conditioned by the fact that usually it is run by an embedded computer mainly devoted to real time control of the vehicle. In addition, people with disabilities need frequently diverse devices that assist them for everyday tasks. Nevertheless, they can not easily switch from a device to other. Therefore, all these devices tend to be integrated together sharing an interface that gives access to all the functions required by the user: wheelchair movement control, environmental control (usually through a

F. Paternò (Ed.): Mobile HCI 2002, LNCS 2411, pp. 373–377, 2002.
© Springer-Verlag Berlin Heidelberg 2002

domotic system) and, frequently, personal direct or remote communication via computer. Among the diverse tasks that the user can perform form that interface, in this paper we will only analyse the control of the wheelchair movement, because communication and environmental control are well documented [5]. Therefore, the main interface design problem is to give the user the possibility to select a destination among a relatively large number of choices requiring the minimum effort. Next section presents the project were this interface was developed.

2 The *TetraNauta* System

TetraNauta[1] project developed a controller for standard electric-powered wheelchairs that allows users with very severe mobility restrictions (such as people with quadriplegia) to easily navigate in closed structured environments (home, hospital, school, etc.). The main goal of this project was to design a non-expensive automatic driving system to help this kind of users to employ the wheelchair with the minimum effort, but maintaining the user as active as possible –to benefit his or her rehabilitation. For this reason the design of an adequate adaptive mobile user interface was a key factor in *TetraNauta* architecture.

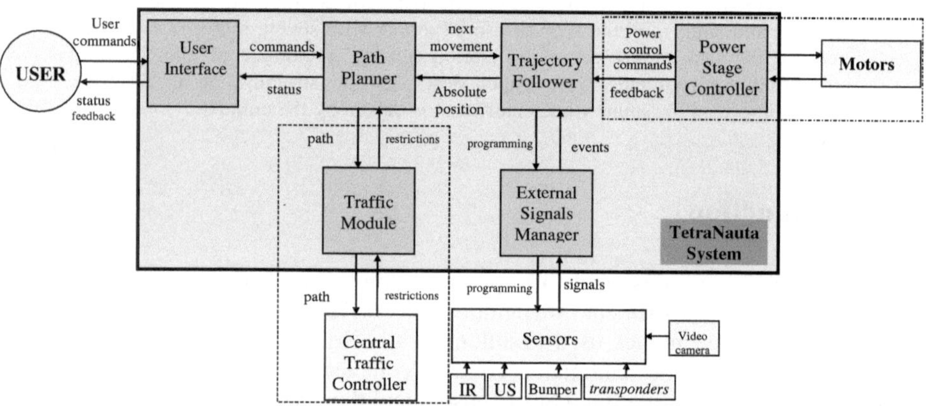

Fig. 1. Architecture of the *TetraNauta* System

From the point of view of the navigation, two main sections can be distinguished in *TetraNauta* architecture (fig.1), the Control Section and the User Interface. The first deals with automatic operations -such as signals handling, control of the motors, etc.- while the second manages user dialogue. Due to already mentioned user motor restrictions, many operations that usually are done by the user, must be transferred to the automatic controller, thereby decreasing the effort made by him or her. Hence, modules for path planning and guidance must be added. The path planning sub-

[1] *TetraNauta* is a research project developed by the National Hospital of Paraplegics; Bioingeniería Aragonesa S. A.; the Group of Robotics and Rehabilitation Technology of the University of Seville and the Laboratory of HCI for Special Needs of the University of the Basque Country.

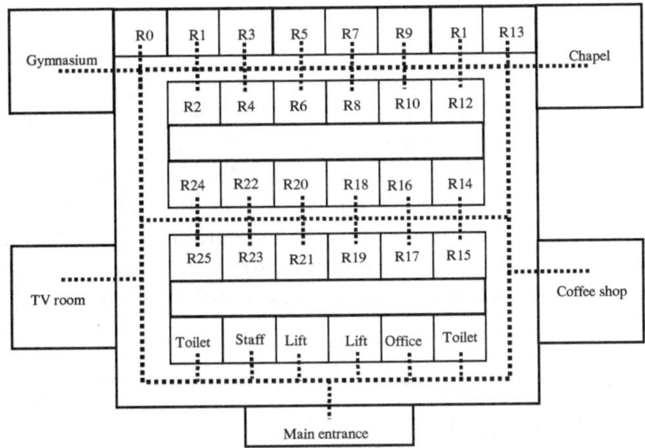

Fig. 2. A graph (dotted lines) representing a plant.

system finds a path free of obstacles between two points on a topological map (a graph, see fig 2), whilst the guidance sub-system drives the wheelchair along the trajectory calculated by the path planning module while estimating its absolute position. The path planner is implemented by means of a search algorithm that finds a path that links two nodes of a graph. A detailed description of that algorithm can be found in [6, 7].

A number of beacons are disseminated in the environment to allow absolute positioning. When the wheelchair approaches one of them it recalibrates its position. There is a command to automatically find a track or beacon to acquire the current position when it is unknown by the system. It is also possible to manually drive the wheelchair to a track or beacon.

3 Interface Structure

The user controls the system through a very intuitive graphical interface that translates his or her orders (e.g. the desired destination) into commands for the Trajectory Planner (that calculates a quasi-optimal trajectory). It also gives feed-back to the user about the current operation.

The input device for *TetraNauta* interface is typically the same used to drive the wheelchair: a joystick, mouth-stick, or any other alternative input system used in Assistive Technology [5]. The output is a small colour display (that currently has been ported to a PDA). Due to the user motor restrictions the selection of the desired command is usually done either by *scanning* a matrix of icons and selecting them with a pushbutton; or by means of *directed scanning* where the user goes through the matrix using a joystick[2].

2 Depending on the user characteristics other input/output devices are possible: for input purposes touch-screens, voice recognition, etc., while, besides a display, synthetic voice can be used for output in some cases. Here we only present the most common choice.

Simplified Task Model

For navigation purposes the user can perform the following tasks (see fig. 3): when the vehicle is stopped, if the current position is not known by the system, the user can automatically find a track (or manually drive the wheelchair to a track). If the current position is known by the system, the user can select a destination or go to a given destination (from the current localization). While the wheelchair is going to a destination, the user can change the current destination (which stops the wheelchair and pass to "select a destination"). In addition, the user can always switch to manual control that stops the wheelchair and leaves the control to a human).

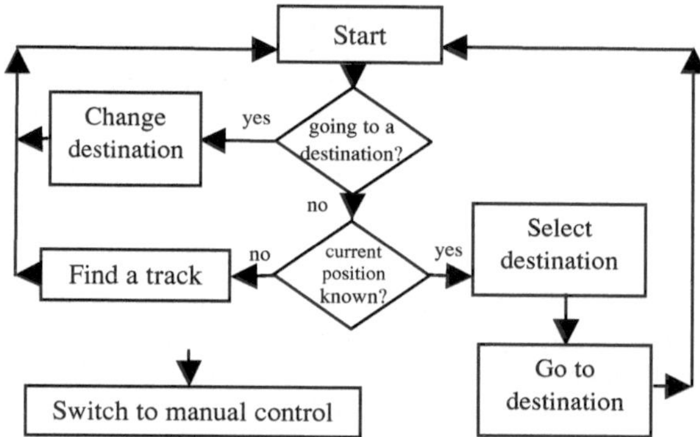

Fig. 3. Simplified task diagram

As it is sown previously, most of the tasks can be represented by icons in a simple interface. Nevertheless, when the user has to select a destination, the number of nodes in a graph representing a structured closed environment (see fig. 2) is usually too large for the display mounted in the wheelchair, making the selection difficult. To avoid this trouble, the interface makes use of the information the system has about the task that it is performing [8]: the current position and the map of possible destinations. Therefore, the same data structure (a multi-layered map) used for trajectory planning is used for display purpose, due to the information management facilities that it provides. In fact, the hierarchical map model is very suitable for compact menu-based displays (do not forget that our implementation of menus is based on *scanning and selection*). Therefore, each abstraction level can be included in a menu and a selection may be carried using a small number of menus.

In addition, user adaptation allows the optimization of the choices. When the user is choosing a destination, the selection set is composed only by the reachable destinations from that point, ordered by frequency of use. In this way, only possible destinations -in order of probability- are offered, minimizing the selection effort.

Other important characteristic of *TetraNauta* interface is that it is not intrusive. The system does not take decisions when the user is able to take it, which is very important to facilitate the user rehabilitation process. Since the abilities of the user can change with the training it is necessary to build an incremental user model, to be able to determine what decisions are in the hands of the user.

More details about the *TetraNauta* interface can be found in [9].

4 Conclusion

The design of an adaptable mobile interface for a smart wheelchair heavily depends on the tasks that this device performs. The knowledge that the system has about the current position and the points that are reachable from this point allows the design of a simple and effective interface that takes into account the context. The interface of *TetraNauta* system is also adaptable and accessible to severely motor impaired people. In addition, it helps to the rehabilitation of the user giving to the user as much decisions as he or she can take.

Acknowledgments

This work has been partially supported by the Spanish *Comisión Interministerial de Ciencia y Tecnología* (CICYT) -contract TER96-2056-C02-02- and the *Ministerio de Trabajo y Asuntos Sociales.*

References

1. Latombe J.-C.: Robot Motion Planning. Kluwer Academic Publishers (1990)
2. Cooper, R.A.: Intelligent Control of Power Wheelchairs. IEEE Eng. in Med. & Biol. (1995)
3. Yoder J.D. et al.: Initial Results in the Development of a Guidance System for a Powered Wheelchair. IEEE Trans. on Rehabilitation Engineering. Vol. 4, No.3, (1996)
4. Bourhis, G. et al.: Mobile Robotics and Mobility Assistance for People with Motor Impairments: Rational Justification for the VAHM Project. In: IEEE Trans. on Rehab. Engin. Vol. 4, No. 1, (1996)
5. Cook A. & S. Hussey.: Assistive Technologies: Principles and practice. Mosby, (1995)
6. Cagigas D.: Un sistema eficiente de planificación de trayectorias en entornos cerrados grandes para robots móviles y sistemas AGV. Ph. D. dissertation (in Spanish). The University of the Basque Country. Donostia, Spain. (2001)
7. Abascal J. et al.: Efficient Path Planning for Autonomous Wheelchairs in Structured Environments. Austrian Journal of Artificial Intelligence (ÖGAI) (2001)
8. Dix A., et al.: Exploiting space and location as a design framework for interactive mobile systems. ACM Trans. on Computer-Human Interaction (TOCHI), 7(3), pp. 285-321, (2000)
9. Abascal J. et al.: Interfacing users with Very Severe Mobility Restrictions with a Semi-Automatically Guided Wheelchair. SIGCAPH. ACM Press. Vol. 63. (1999)

Utilizing Gaze Detection to Simulate the Affordances of Paper in the Rapid Serial Visual Presentation Format

Gustav Öquist [1], Staffan Björk[2], and Mikael Goldstein[3]

[1] Uppsala University, Department of Linguistics, PO Box 527
751 20 Uppsala, Sweden
gustav@stp.ling.uu.se
[2] Interactive Institute, PLAY Studio, PO Box 620
405 30 Göteborg, Sweden
staffan.bjork@interactiveinstitute.se
[3] Ericsson Research, Interaction & Usability Lab, Torshamnsgatan 23,
164 80 Kista, Sweden
mikael.goldstein@era.ericsson.se

Abstract. We present how gaze detection can be used to enhance the Rapid Serial Visual Presentation (RSVP) format, a dynamic text presentation technique suitable for mobile devices. A camera mounted on the device is used to monitor the reader's gaze and control the onset of the text presentation accordingly. The underlying assumptions for the technique are presented together with a description of a prototype, Smart Bailando, as well as our directions for further work.

1 Introduction

On paper the average reading speed for English text is between 220-340 words per minute (wpm) [9]. Reading speed on large screens is today likely to be more or less the same due to improved resolution [12]. Increased resolution will surely improve legibility on small screens as well but readability will remain low due to the limited screen space available [3]. In many cases, a paper copy can simply solve the problem of having to read from a screen at all, but users of handheld devices do not always have access to printing facilities. Developers therefore try to make reading as easy as possible by improving display quality and user interfaces but handheld devices still have an inherent problem in their limited screen space.

This dilemma does however presuppose that the text is presented in the traditional page format. One approach to overcome the size constraint may be to make use of the possibilities actually offered by mobile devices and trade space for time [1]. Dynamic text presentation via Leading or Rapid Serial Visual Presentation (RSVP) requires much smaller screen space compared to traditional text presentation with maintained reading efficiency [1, 6, 8, 11, 12]. Leading, or the Times Square Format, scrolls the text on one line horizontally across the screen whereas RSVP presents the text as

F. Paternò (Ed.): Mobile HCI 2002, LNCS 2411, pp. 378–382, 2002.
© Springer-Verlag Berlin Heidelberg 2002

chunks of words or characters in rapid succession at a single visual location [10]. From a physiological perspective RSVP appears to suit the natural reading process better since the text then moves successively rather than continuously [17].

In a repeated-measurement experiment, Goldstein et al. [4] found that neither reading speed nor comprehension differed from paper text reading for longer texts. However, the NASA-TLX (Task Load Index) revealed significantly higher task load for RSVP conditions compared to paper reading for most factors. One explanation to the high cognitive load may have been that each text chunk was exposed for the same fixed duration of time. Just and Carpenter [7, p. 330] have found that "there is a large variation in the duration of individual fixations as well as the total gaze duration on individual words" when reading text from paper. Adaptive RSVP [5, 17] attempts to match the reader's cognitive text processing pace more adequately by adjusting the exposure time of each chunk with respect to the characteristics of the text being shown. In a usability evaluation Öquist and Goldstein [17] found that adaptation could indeed decrease task load for most factors. In an experiment with a similar approach Castelhano and Muter [2] found that the introduction of punctuation pauses, interruption pauses and pauses at clause boundaries made RSVP significantly more liked. Although these evaluations are not fully comparable they all seem to indicate that the RSVP format has some potential but also some flaws that yet remains to be resolved.

We believe that dynamic text presentation can improve reading efficiency on small screens, moreover we recognize that also relatively new formats like RSVP must adhere to the fundamental principles of reading that has evolved over time in order to be usable. This notion has led us to explore ways of simulating ordinary reading via RSVP using sensors.

2 Paper and Screen Affordances of Traditionally Presented Text

The trade-off between time and space that RSVP offers comes with the additional mental cost that the act of reading for the user is changed. The natural eye movements when reading traditionally presented text involve performing fixation-saccade-fixation patterns including regressions and return sweeps [10]. One inherent difference of RSVP is that it demands the reader to continuously fixate his gaze at one single location in the text presentation window.

It is very common that thought and gaze is frequently diverted from the text during traditional paper reading due to external distractions or periods of reflection. Paper-presented text supports this activity as the text stays in the same place and it is easy to resumes reading. By using Donald Norman's [13] term *affordances,* one could argue that this is an affordance that traditionally presented text on paper or on screen offers. This kind of affordances does not apply to the RSVP format due to its dynamic nature. Thus, readers are forced to continuously monitor themselves when using the RSVP format. If the gaze strays away, the RSVP presentation has to be stopped manually. This may be one reason for the high cognitive demand score obtained in earlier experiments [1, 6, 17] as readers can have felt an urge to fixate on the text continuously since looking away can lead to missing information.

If one enhanced the RSVP application with sensors that register the reader's gaze, *gaze detection*, the application could become context-aware [15] and automatically stop/start the text presentation when the reader looked away from the text. A precondition for this would be that the terminal using the RSVP format would have a built-in camera focused on the reader's eyes continuously during RSVP reading. Mobile phones are currently being released on the market with such a camera integrated into their design (e.g. the Sony Ericsson P800 and the Nokia 7650) and cameras can be bought as add-on modules for PDAs (e.g. the HP Pocket Camera) soon making this requirement very easy to fulfill. Based on the observations presented above, we believe that adding gaze detection functionality to RSVP reading on handheld PDAs and cellular phones is one feasible route to making reading on small devices as convenient as ordinary screen or paper reading.

3 Smart Bailando

In one of our threads to explore the possibilities of RSVP with gaze detection we are currently supervising a Master's thesis [16] where a RSVP application, *Smart Bailando*, is being developed in which the stop/start of the text presentation is controlled by eye movement (Fig 1).

Gaze detection is provided by a software platform for real-time measurements of eye movement developed by Smart Eye AB (www.smarteye.se). The platform allows gaze tracking using a standard PC equipped with one or several digital video cameras

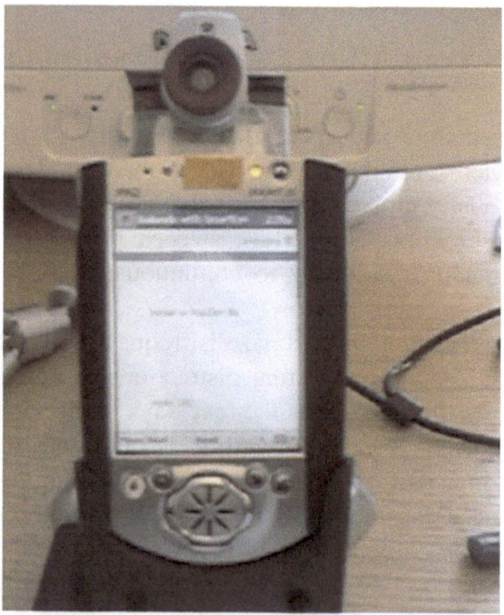

Fig. 1. The Smart Bailando prototype with the gaze detection sensor attached

including web cameras, making it a quick and easy prototyping platform. The platform is written in C++, and as it runs on Windows OS, a Pocket PC version of the system is feasible as soon as PDAs have the required computational powers (which with the current technological development will be within 2 years). As the current gaze detection can only run on a PC, Smart Bailando is built as the client of a client-server application where the Smart Eye platform functions as the server running on the PC. The camera used must be connected to the PC, but by dismantling the camera and mounting it on the PDA a good approximation can be achieved even though it enforced a restriction of how the PDA can be moved.

The prototype application currently runs both on a Pocket PC PDA (a Compaq iPAQ 3630 with a wireless LAN network) and in emulation mode on a PC. The system requires a calibration session and is currently in a refinement phase where the overall performance of the system is being improved by modifying individual parts of the whole system including but not limited to the making the calibration process easier, providing user feedback of the tracking process, improving network performance (all by the students), and modifications of the eye tracking algorithms (by Smart Eye).

Smart Bailando builds upon the Bailando application [5, 17], which was developed at Ericsson Research's Usability & Interaction Lab in Kista, Sweden, to explore Sonified and Adaptive RSVP. Bailando was developed for PDAs running the Pocket PC operating system, which offers the application sufficient processing power for experimenting with resource demanding tasks such as sound playback on a mobile device [5, 6]. The actual text presentation in Smart Bailando is exactly that of Bailando, i.e. it uses the Adaptive RSVP techniques first presented in that prototype.

4 Conclusions and Future Work

We have presented Smart Bailando, a proof-of-concept prototype that uses gaze detection to enhance the RSVP format. We have motivated the use of gaze detection by the desire to mimic the affordances provided by traditional text presentation. Although the current prototype shows that it is technologically possible to use eye movement to control RSVP, we mostly see it as a first step in exploring the feasibility of using real-time eye tracking techniques in combination with RSVP. Based on the prototype, we have identified the following research issues that are currently being pursued: formally evaluating the prototype, building a fully mobile prototype, and evaluating the prototype in a mobile setting.

Acknowledgements

The authors would like to thank Per Åkervall and Roger Granath for all their work with the Smart RSVP prototype. We would also like to thank the people at Smart Eye AB for providing and customizing the gaze detection software to our needs.

References

1. Bruijn, O. and Spence, R. (2000). Rapid Serial Visual Presentation: A space-time trade-off in information presentation. *Proceedings of Advanced Visual Interfaces, AVI'2000.*
2. Castelhano, M.S. and Muter, P. (2001). Optimizing the reading of electronic text using rapid serial visual presentation. *Behavior & Information Technology,* 2001, 20(4), 237-247.
3. Duchnicky, R.L. and Kolers, P.A. (1983). Readability of text scrolled on visual display terminals as a function of window size. *Human Factors*, 25, 683-692.
4. Goldstein, M., Sicheritz, K. and Anneroth, M. (2001). Reading from a small display using the RSVP technique, *Nordic Radio Symposium, NRS01,* Poster session, Nynäshamn, 4-6 April 2001 (full paper available on CD-Rom only).
5. Goldstein, M., Öquist, G., Bayat-M., M., Björk, S. and Ljungberg, P. (2001). Enhancing the reading experience: Using Adaptive and Sonified RSVP for reading on small displays, *Proceedings of 3rd International Mobile HCI Workshop.*
6. Goldstein, M., Öquist, G. and Björk, S. (2002). Immersion does not guarantee Excitement: Evaluating Sonified RSVP *(Full paper submitted to NordiCHI'2002).*
7. Just, M.A. och Carpenter, P.A. (1980). A theory of reading: From eye fixations to comprehension, *Psychological Review,* 87(4), 329-352.
8. Joula, J.F., Ward, N.J. and MacNamara, T. (1982). Visual search and reading of rapid serial presentations of letter strings, words and text. *J. Exper. Psychol.: General*, 111, 208-227.
9. Kump, P. (1999). *Break-trough rapid reading.* New Jersey: Prentice-Hall Press.
10. Mills, C.B. och Weldon, L.J. (1987). Reading text from computer screens. *ACM Computing Surveys,* Vol. 19, No. 4, ACM Press.
11. Muter, P. (1996). Interface design and optimization of reading of continuous text. In van Oostendorp, H. & de Mul, S. (Eds.), *Cognitive aspects of electronic text processing,* 161-180. Norwood, N.J.: Ablex.
12. Muter, P. and Maurutto, P. (1991). Reading and skimming from computer screens and books: The paperless office revisited? *Behavior & Information Technology*, 10, 257–266.
13. Norman, D.A. (1988). *The Psychology of Everyday Things.* Doubleday, NY, USA.
14. Rahman, T. and Muter, P. (1999). Designing an interface to optimize reading with small display windows. *Human Factors,* 1(1), 106-117, Human Factors and Ergonomics Society.
15. Schilit, B.N., Adams, N. and Want, R (1994). Context-aware computing applications. *Proceedings of the IEEE Workshop on Mobile Computing Systems and Applications.*
16. Åkervall, P. and Granath, R. (2002). *Eye Controlled RSVP on Handhelds.* Forthcoming Master's Thesis.
17. Öquist, G. and Goldstein, M. (2002). Towards an improved readability on mobile devices: Evaluating Adaptive Rapid Serial Visual Presentation. *Proceedings of Mobile HCI'2002.*

On Interfaces for Mobile Information Retrieval

Edward Schofield[1] and Gernot Kubin[2]

[1] Vienna Telecommunications Research Center,
Vienna, Austria,
schofield@ftw.at

[2] Signal Processing and Speech Communication Labs,
Graz University of Technology,
Graz, Austria,
g.kubin@ieee.org

Abstract. We consider the task of retrieving online information in mobile environments. We propose question-answering as a more appropriate interface than page-browsing for small displays. We assess different modalities for communicating using a mobile device with question-answering systems, focusing on speech. We then survey existing research in spoken information retrieval, present some new findings, and assess the difficulty of the endeavor.

1 Introduction

The telecommunications industry worldwide is scrambling to bring what is available to networked computers to mobile devices. The degree of proliferation of Web content has not, so far, been matched by content for mobile devices; nor is this necessarily a bad thing. This short paper argues that there are potentially better interfaces for finding information with small devices than through the page-browsing metaphor of the wired web. We review automatic question-answering systems – which respond to questions, rather than keywords, and extract answers, rather than documents – as an alternative, and discuss how mobile user-interfaces could accommodate such systems.

2 Information Needs and Searching Behavior

A vaguely defined need for information is often expressed by "Tell me about Topic X." A more precise need for information is more often expressed as a question. But because current search engines return only documents, not answers, online information-seekers tend to express all their information needs, vague or precise, in the first form. Some studies on Internet searching behavior [5] have revealed that the vast majority of searches are for one or two keywords, after which users must sift through several documents, perhaps many, to locate the information they desire. The modest goals for current search engines impose some obvious inconveniences: sifting through documents, particularly for specialized information, takes time and distracts attention from the task at hand.

F. Paternò (Ed.): Mobile HCI 2002, LNCS 2411, pp. 383–387, 2002.
© Springer-Verlag Berlin Heidelberg 2002

The difficulty of finding online information with a mobile device is much greater. The screens of mobile phones and PDAs range from small to tiny – too small to comfortably read and browse hyperlinked documents. A promising alternative is to specify information-needs more precisely in the form of natural-language questions, and have an automated question-answerer scour the online data for you. Question-answering (QA) systems, originally designed for easy access to structured databases, received boosts in attention from the recent proliferation of Web content and the recent inclusion of a QA-track in the annual TREC competitions [14]. They are now a subject of considerable research.

QA systems offer considerable opportunity for mobile information retrieval. Their are designed to relieve the user of the burden of searching; they therefore circumvent the difficulties of presenting interim search results on small screens. Systems competing in TREC competitions return answers of either 50 or 250 bytes – morsels that can be elegantly shown on a scrollable 5-line mobile display. In contrast, the data sources from which QA systems extract answers can be databases or unwieldy web pages, and need not be formatted for mobile displays.

3 The Qestion of Input

Users can pose questions to mobile devices using a keypad, a stylus, or speech. Text-entry rates for the multi-tap method on older mobile phones are commonly 7-15 wpm; with predictive-text facilities this rate roughly doubles [12]. Key-tapping would therefore allow the entry of a typical 10-word question in 20-40 seconds, with continuous visual attention. Handwriting with a stylus can be doodled at comparable speeds [13]. For some information needs this may suffice. For others, a faster and easier interface would be preferable.

Is speech is a viable modality for posing questions to mobile devices? We now survey some research in spoken information retrieval, describe some new findings, and outline the problems to be solved.

3.1 Posing Questions with Speech

We first outline some fundamental limits on the usefulness of speech as a modality. First, speech is public, potentially disruptive to people nearby and potentially compromising of confidentiality. Second, the cognitive load imposed by speaking can interfere with simultaneous problem-solving tasks like preparing documents [7]. Third, speech recognizers make errors – and will do so for the foreseeable future [4]. Successful spoken interfaces must accommodate these limitations. In particular, they must be robust to errors in recognition by requesting clarification from users in a seamless way.

The advantages of speech as a modality are more obvious. It is rapid: commonly 150–250 wpm [1]. It requires no visual attention. It requires no use of hands. All mobile phones, and some PDAs, are equipped with microphones – and standardized protocols for distributed speech recognition could, in a few years, allow mobile devices to parameterize spoken input and outsource the memory-hungry recognition task to remote servers [3].

The first research into spoken information-retrieval was conducted at Xerox labs in 1993. Kupiec and others [8] observed that intended word-combinations in spoken queries co-occur in documents in closer proximity than erroneous combinations. Based on this idea, they designed a system that simultaneously disambiguated word hypotheses and found relevant text for retrieval. Their prototype recognized isolated words from a vocabulary of 100 000, with encouraging results. More recently, Google Labs deployed a prototype of a voice-operated telephone interface to its popular search engine.

Our research has focused on language modeling for questions, since accurate language models are important prerequisites for accurate speech recognition when vocabularies are large [6]. We have analysed logs of around 500 000 questions compiled from various sources including Usenet Frequently-Asked-Questions and the Excite and Ask Jeeves search engines.

Our first observation is that the lexical structure of fact-seeking questions tends to be highly constrained. To formalize and test this observation we trained bigram and trigram language models on our question-corpus and evaluated the fits of the models on two unseen test sets: the test questions from TREC 10 and a sample from the Project Gutenberg archive. For comparison we trained bigram and trigram language models on a larger corpus of sentences drawn from Gutenberg. The results support our observation. See Table 1.

Table 1. Cross entropy (bits/word) for bigram models

	TREC 10 Test	Gutenberg Test
Question corpus	6.9	9.7
General text corpus	8.1	8.2

A second, partly overlapping, observation is that the syntax of fact-seeking questions tends to be highly constrained. This will be described in detail in a forthcoming paper [11]. Table 2 presents some initial statistics of inter-word transition frequencies, suitable for modeling syntax as a Markov chain between parts of speech.

3.2 The Magnitude of the Problem

The word-error rate of recognized speech is positively correlated with vocabulary size. Open-domain information retrieval would require the largest vocabulary of proper nouns imaginable–perhaps 200 000. Our initial impression is that, in contrast to the proper nouns, the sets of adjectives and verbs in fact-seeking questions is small, even for a large domain, and that verbs commonly appear in few orthographic forms.

A factor that reduces the difficulty of the speech-recognition task is that the words that are most acoustically ambiguous, and most often misrecognized, are 'function words' like articles and propositions that contribute relatively little to

Table 2. Inter-word transition frequencies for a small sample of questions

	article	noun	verb	*wh–*	prep.	adj.	other	END
START	0	.03	.23	.70	0	0	.04	0
article	0	.73	0	0	0	.17	.10	0
noun	.01	.25	.25	0	.16	.03	.06	.25
verb	.31	.39	.06	.01	.06	.01	.09	.07
wh–	0	.06	.83	0	.02	.04	.06	0
preposition	.27	.43	.09	.02	0	.05	.14	0
adjective	0	.52	.04	.04	.09	.04	.04	.22
other	.08	.41	.13	.03	.05	.10	.15	.05

the semantic content of a sentence. Conversely, longer 'content words' present less difficulty for speech recognizers [10]. It is possible that misrecognitions in function words could be ironed out automatically during the deep semantic and syntactic parsing performed by QA systems. A similar principle was verified in [2] for search queries (in formal language), but for questions (in natural language) this has not yet been investigated.

Spoken question-answering on an open-domain is relatively difficult; it can benefit from a restriction on the domain to a specific need, for at least three reasons. First, smaller domains imply smaller vocabularies, with accordingly more accurate spoken interfaces. Second, smaller domains allow tighter linguistic analysis, and accordingly better accuracy in interpreting and finding answers to questions. Third, smaller domains allow more control over the document collections to be searched, permitting semi-automatic indexing and tagging of their content.

4 Conclusions and Future Directions

This paper has described question-answering as an alternative to the metaphor of page-browsing for finding information with mobile devices. It outlined the pros and cons of spoken interfaces for retrieving information, surveyed existing research in the field, and presented some initial findings on the structure of fact-seeking questions. In the light of these findings it then assessed the difficulty of automatically recognizing spoken questions.

The largest hurdle for spoken interfaces for information-retrieval is the inevitability of misrecognitions, and effective systems require some facility to correct these interactively. A few research fields could possibly rise to this challenge. Multimodal interfaces offer some promise in disambiguating input by combining partial hypotheses from different modalities [9]–for example, spoken input could guide the hypotheses made about word-completion during key-tapping, or vice versa. Another possibility is that paraphrases could clarify ambiguities for speech recognizers, as they can for humans. We will address some of these topics in future work.

References

1. D. R. Aaronson and E. Colet. Reading paradigms: From lab to cyberspace? *Behavior Research Methods, Instruments and Computers*, 29(2):250–255, 1997.
2. F. Crestani. An experimental study of the effects of word recognition errors in spoken queries on the effectiveness of an information retrieval system. Technical Report TR-99-016, Berkeley, CA, 1999.
3. H. Hirsch and D. Pearce. The AURORA experimental framework for the performance evaluation of speech recognition under noisy conditions.
4. M. Huckvale. 10 things engineers have discovered about speech recognition. In *NATO ASI Speech Pattern Processing*, 1997.
5. B. J. Jansen and U. W. Pooch. A review of web searching studies and a framework for future research. *Journal of the American Society of Information Science*, 52(3):235–246, 2001.
6. F. Jelinek. *Statistical Methods for Speech Recognition.* The MIT Press, Cambridge, Massachusetts, 1998.
7. L. Karl, M. Pettey, and B. Shneiderman. Speech-activated versus mouse-activated commands for word processing applications: An empirical evaluation, 1993.
8. J. Kupiec, D. Kimber, and V. Balasubramanian. Speech-based retrieval using semantic co-occurrence filtering, 1994.
9. S. L. Oviatt. Mutual disambiguation of recognition errors in a multimodel architecture. In *CHI*, pages 576–583, 1999.
10. L. Rabiner and B.-H. Juang. *Fundamentals of Speech Recognition.* Prentice Hall, Englewood Cliffs, NJ, USA, 1993.
11. E. Schofield and G. Kubin. Language models for questions. *Forthcoming*, 2002.
12. M. S. . K. P. Silfverberg, M. Predicting text entry speed on mobile phones. In *Proceedings of the ACM CHI 2000 Conference on Human Factors in Computing Systems*, The Hague, 2000.
13. W. Soukoreff and I. MacKenzie. Theoretical upper and lower bounds on typing speeds using a stylus and keyboard. *Behaviour and Information Technology*, 14:379–379, 1995.
14. E. M. Voorhees. Overview of the TREC 2001 question answering track. In *Proceedings of the Tenth Text REtrieval Conference (TREC-10)*, 2000.

User Requirement Analysis and Interface Conception for a Mobile, Location-Based Fair Guide

Fabian Hermann and Frank Heidmann

Fraunhofer Institute for Industrial Engineering IAO, Business Unit Usability Engineering, Nobelstr. 12, 70569 Stuttgart, Germany Fax: ++49 711 970 2300,
{Fabian.Hermann,Frank.Heidmann}@iao.fhg.de

Abstract. This paper describes the first phase of user-centred design for a mobile, location-based fair guide which is developed in the research project SAiMotion. The user requirements analysis based on a usage scenario and a formalized use case model developed in the beginning. A focus group and interviews were conducted to gain empirical insights on the potentials and probable acceptance of a location-based fair guide. Among the results of these user participation activities, are the integration and visualization of spatial data, temporal data and user-specific information as well as the support of activity planning. On this basis, an interface conception was worked out. One of it's central parts are interactive maps. Derived principles for visual layout and interaction design of maps on small screens are presented.

The goal of the SAiMotion[1] is to develop a fair guide as location-based service for mobile devices. The technology to enable indoor-localization, and online data-access, as well as it's integration in a mobile client is one of the research focuses. On this basis, an application supporting users in the highly dynamic and information-rich environment of a business fair is planned.

This paper describes the first phase of the usability engineering process covering the analysis of user requirements and the conception of the graphical user interface. As we followed a scenario-based approach [2], the first steps were to work out a usage scenario, to formalize it in a use case model, and to investigate user requirements empirically. In the derived interface conception, interactive maps visualizing spatial data as well as dynamic attribute data take an important role. The process of assessing requirements and developing the user interface for small screens, in particular interactive maps, is described in the following sections.

---

[1] SAiMotion is funded by the "Bundesministerium für Bildung und Forschung" (BMBF, Funding-Key 01AK900A). The authors are responsible for the contents of this publication. The following Fraunhofer Institutes cooperate for SAiMotion: FIT (coordination), IAO, IGD, IIS, IPSI, IZM.

F. Paternò (Ed.): Mobile HCI 2002, LNCS 2411, pp. 388–392, 2002.
© Springer-Verlag Berlin Heidelberg 2002

1 Scenario and Use Case Model

Starting from the basic technology described above a detailed scenario was worked out to explore possible functionalities and application settings. It described several concrete activities of fictitious users like planning a number of stand visits, joining a product presentation, or navigating to a known person.

The work on this scenario occupied a crucial role in the first phase of the research project. It focussed the communication between project partners on a concrete application setting and lead to a creative but efficient discussion about what would be nice to have and what is achievable. So it fostered the early fixing of a concrete, joint project goal. Furthermore, the description of activities of a fictitious user with the system helped to structure the user participation and served for the use case model and functional specification.

The informal description of activities in the scenario were formalized in a use case model [3] describing in detail user roles and actions with the planned application covering the whole functionality.

2 User Participation to Collect Requirements

In order to get first impressions on the probable acceptance of the planned system and requirements of the target group a focus group and interviews were conducted.

Method. Five potential users working as consultants and researchers in the IT field were invited to a focus group. First, the two moderators gave a brief overview over the project and the planned basic functionality. To illustrate the potentials of a mobile location-aware system, two short parts of the scenario were presented. Afterwards, the participants were asked to write down individually "ideas about useful functions" and "conditions for using and possibly buying such a system". This phase lasted about half an hour. In the next step, the participants presented their ideas and a discussion was initiated by the moderators. This discussion took 70 minutes and was recorded. The proposals written down individually or stated in the discussion were categorized and were used to specify functional and non-functional user requirements.

In addition, a semi-structured interview was made with three visitors of a German book fair. Similar to the focus-group, subjects were given a brief overview over the planned system and were asked about expectations, useful functions, and conditions for usage. The interviews lasted between 35 and 50 minutes.

Results – User Requirements for a context-aware mobile fair guide. In general, the participants were very positive about the planned system and generated a lot of ideas how to use the basic functionality for the scenario of visiting a business fair. The most significant features described were:

– *Information browsing and up-to-date information:* Information typically included in a fair catalogue like lists of exhibitors and their products, events, facilities, services etc. should be available in the system. In order to browse these information and use them e.g. for planning, a decisive feature would be a topic ontology. To get or order brochures or product materials electronically from exhibitors was said to be helpful as well. The participants also stressed the advantage to have also up-to-date information that can be retrieved online.

- *Navigation and use of spatial information:* One of the most valuable features for the users are spatial information. One should be able to display locations in order to get an overview and to be navigated by the system. The integration of spatial and attribute data like the catalogue information, e.g. to offer route planning and navigation to locations selected from the catalogue, was seen as a major strength of a mobile system, especially for visiting business fairs because of difficult locations, and great time pressure.
- *Activity planning and scheduling:* The participants wanted to have a schedule for events, meetings, and other appointments that uses semantic data of the fair catalogue and spatial data. The system also should remind of important appointments and offer navigation to the location in question.

  Furthermore, the system should be able to make proposals how to sequence activities, considering fixed dates as well as activities without temporal constraints like visiting stands. Ideally, an "intelligent scheduler" takes events and activities selected by the user and deduces an optimal sequence of activities using spatial and temporal information like distance, estimated walking time etc. It is noteworthy that some participants where quite enthusiastic about such a planning tool, whereas others (nearly one third) where sceptical about its usefulness.
- *Localization of persons and communication:* An important insight from the focus group and the interviews concerned the importance of groups coming to a business fair together but walking jointly only in part. Nearly all participants claimed that to localize peers and to get navigated to them could be an advantage. However, the danger to be monitored e.g. by superiors was discussed extensively, and people described settings were social constraints would force to use such a service if it was available.

As a non-functional user requirement, the usability of the mobile application was discussed. Some participants were critical about the quality of displayed information, maps or routing information because of previous experience with mobile guides. Also to browse abundant content like lists of exhibitors, events, products etc. in order to plan activities was expected to be rather inconvenient. In this context, the issue of preparation at home was raised: About two third of the participants usually prepare a visit on a business fair in advance. Most of them stated that they would use catalogue data and location-data to plan at home on a PC and expected the mobile client mainly to support the execution of previously planned activities.

3 Planned Functionality

According to these empirical results, the strengths of a mobile, location-aware system on a business fair are browsing of spatial information connected with catalogue data, activity planning, and wayfinding. To support these needs, a heuristic tour planner that proposes routes and temporal sequences for selected activities, an information browser to access catalogue information, and interactive maps are planned.

A number of different data sets have to be integrated in the map visualizations: spatial data organized in a geographic information system, catalogue content like ontologies of exhibitors, products, events etc., and user-specific data like the individual tour. Dynamic data, first of all the location of the mobile client must be used as well. The following functionality is planned for the map views: showing different types of facilities or exhibitors falling into particular topic categories; showing the tour a user has planned, including locations and the route; switching to a navigation mode showing an aligned map of the area surrounding the user's location for wayfinding.

4 Interface Design: Principles for Map Visualization

The use case model formed the major input for the conceptual model of the user interface. In a first step, main objects were identified. Among them are the information browser containing catalogue data, the tour planner, and the interactive maps. For the interactive maps, we specified the visual design, usage of cartographic variables, and the interaction style on the map views. Guidelines for displaying maps on small screens were used for specifying how to use visual variables for symbolization (see e.g. [1]). As the planned visualizations are highly interactive, a major concern was to derive principles about how to organize the user interaction considering the restriction of the small screens of mobile devices:

- *Information focussing with tooltips:* Many situations of map use require the user to read and integrate a lot of spatial and non-spatial information. The most complex example in the business fair scenario is the understanding of a tour, either planned in advance by the user himself or proposed by the system. All the discussed information types—spatial information, attribute data, and temporal information like appointments or walking times—have to be elaborated. A map giving a spatial overview presumably supports this process, but it is impossible to display all required information on a small display at once. We propose tooltips as a simple technique to "zoom" into additional data to map objects without leaving the spatial overview. In the case of a tour view, the name of an tour element, and e.g. starting times of fixed appointments can be presented in tooltips (see fig. 1, a).
- *Control of displayed information:* The user must switch to the list browser to change the displayed information. After selecting items or categories in the list views, a new map can be requested. As an exception, we chose few categories that are offered in an interactive legend (see fig. 1, b) to support actions that are specified in the use case model as most important or frequent.
- *Adaptation of scale and scroll area:* an important usability issue is to display the requested information in an expected way. When a user requests a set of locations displayed on a map the scale and scroll area should be adapted so that all relevant objects and, if known, the own location can be seen. If the objects matching the

request are scattered to the fairground, a small scale results and the user has the responsibility to zoom in.

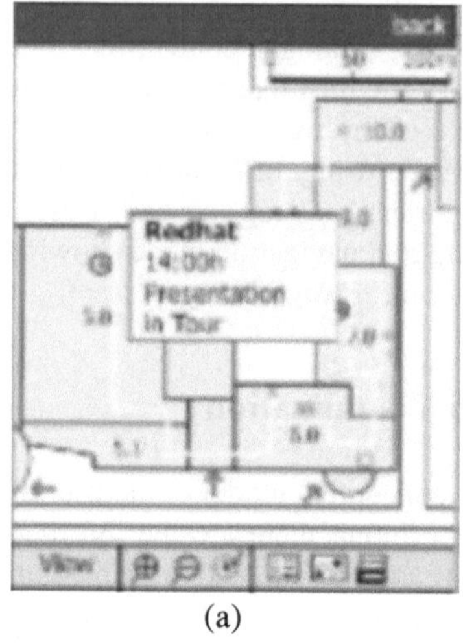

(a)

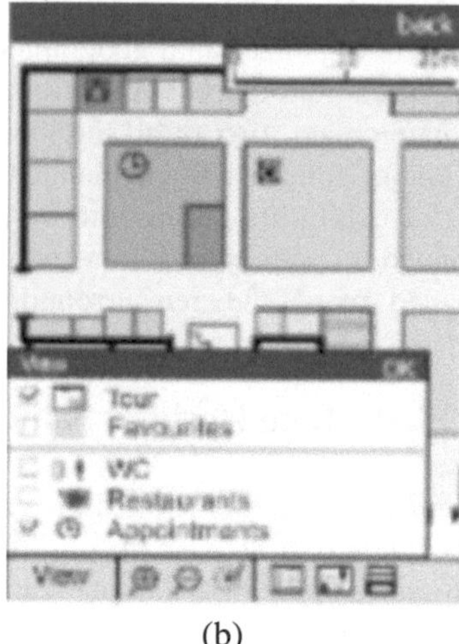

(b)

Fig. 1. Prototype views of SAiMotion interactive maps, showing tour elements and route together with additional information in a object-specific tooltip (a) and an interactive legend (b).

5 Further Steps

The reported principles for interactive map visualizations on small screens are currently investigated by empirical user testing. Based on these results, the prototypes will be integrated in a mock-up prototype that covers the whole graphical user interface and can be used for usability evaluation in a mobile setting.

References

1. Brunner, K.: Kartengraphik am Bildschirm – Einschränkungen und Probleme. Kartographische Nachrichten (2001) 233—239.
2. Carroll, J. M.: Making Use: Scenario-Based Design of Human-Computer Interactions. MIT Press, Cambridge, MA (2000)
3. Constantine, L. L. & Lockwood, L. A. D.: Software for use: a practical guide to the models and methods for usage centered design (1999) Addison Wesley, Reading, MA

A Cross-Modal Electronic Travel Aid Device

F. Fontana, A. Fusiello, M. Gobbi, V. Murino,
D. Rocchesso, L. Sartor, and A. Panuccio

Dipartimento di Informatica, University of Verona,
Ca' Vignal 2, Strada Le Grazie 15, 37134 Verona, Italy,
{fontana,fusiello,murino,rocchesso,panuccio}@sci.univr.it

Abstract. This paper describes the design of an Electronic Travel Aid device, that will enable blind individuals to "see the world with their ears." A wearable prototype will be assembled using low-cost hardware: earphones, sunglasses fitted with two CMOS micro cameras, and a palmtop computer. Currently, the system is able to detect the light spot produced by a laser pointer, compute its angular position and depth, and generate a correspondent sound providing the auditory cues for the perception of the position and distance of the pointed surface. In this way the blind person can use a common pointer as a replacement of the cane.

1 Introduction

Electronic Travel Aids (ETA) for visually impaired individuals aim at conveying information in a way that the user can reconstruct a scenario alternative to the visual one, but having similar informative characteristics, such that the impaired person can make use of at least part of the information that sighted people normally use to experience the world. This information is typically conveyed via the haptic and auditory channels.

Since the mid-seventies, a number of vision substitution devices have been proposed which claim to convert visual into auditory or haptic information with the aim of being mobility aids for blind people [9,6]. Early examples of this approach are the Laser-Cane and the Ultra Sonic Torch [12]. These devices are all based on beams of ultrasonic sound which produces either an audio or tactile stimulus in response to nearby pointed objects, proportional to the proximity of the detected object. In the VIDET project a device converting the information of object distance into haptic cues has been developed [13].

As compared to visual information, sound does not rely on focused attention of the human subject, it is pervasive in a 3-D environment. However, it is difficult to design sounds in such a way that they become effective information carriers. When one aims at using sounds to display spatial features of objects, it should be considered that, while vision is best suited for perceiving attributes of light modifiers (surfaces, fluids), audition is best suited for perceiving attributes of sound sources, regardless of where they are located [10].

F. Paternò (Ed.): Mobile HCI 2002, LNCS 2411, pp. 393–397, 2002.
© Springer-Verlag Berlin Heidelberg 2002

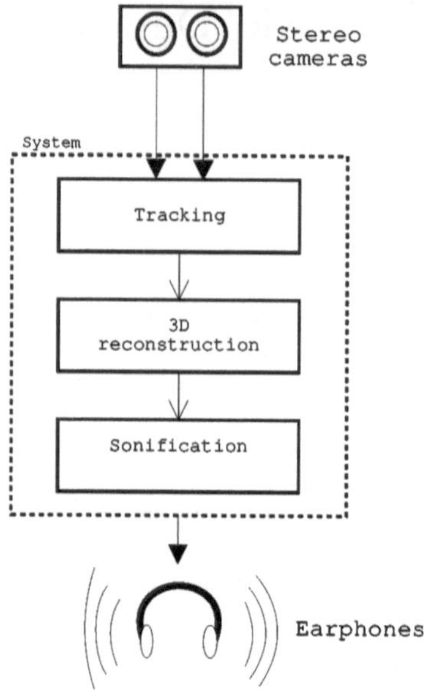

Fig. 1. Overall system architecture. The laser spot is tracked, then its 3D position is computed and sonified.

In this research, we export to ETAs the idea of augmenting the humans' visual perception using acoustic stimuli.
This should lead to novel interface designs and, in particular, to new strategies for the designs of ETAs and other aids for blind people. On the visual side, we have to deal with methods that allow recognition and positioning of the objects that are present in the scene, and, on the auditory side, with techniques that allow blind people to experience the presence of such objects through synthetic sounds.
In spite of increasing power of wearable computers, there is relatively little research on the integration of audio and visual cues for environment perception, although the potential benefits can be very large in the field of the human-computer interaction (HCI). This paper describes a system prototype where such issues are applied.
The system is composed by earphones, two CMOS micro cameras (a stereo head), and a computer (see Fig. 1). The system is able to detect the light spot produced by a laser pointer, compute its depth, and provide suitable spatialized sounds to a blind person.

2 Visual Analysis

The human visual system gains information about the depth basing on differences (disparity) between the images formed on the left and right retina. This process, called stereopsys [7,4], can be replicated by a computer, at a rate of some frames per second, using a camera pair. A well-known problem in computational stereopsys is finding *correspondences*, i.e., finding which points in the left and right images are projections of the same scene point (a *conjugate pair*). Using a laser pointer makes this task trivial, provided that we are able to locate the laser spot in the two images.

In the following the processing applied to each image (left and right) in order to locate the target (i.e. the centroid of the blob produced by the laser pointer) in each image will be briefly described.

The visual analysis starts by applying a red band-pass optical filter on both cameras, in order to enhance the image of the red laser spot. Then images are smoothed in time, by averaging pixels at the same position in the preceding m frames (we used $m = 2$). Only pixels whose difference is less than a given threshold are averaged, thereby obtaining a remarkable increment of the stability of the image, without introducing motion blur. Images are binarized with an automatic threshold and a size filter is applied, which discards the connected components (or blobs) whose size is outside a given range. In order to make the spot detection more robust, we impose the epipolar constraint [8] on the surviving blobs; only those satisfying the constraint both left to right and right to left are kept as candidate targets.

The number of candidate targets is used as a feedback to set the binarization threshold, which is varied using a dichotomic search until the candidate targets are more than zero and less than few units (typically 5). If a minimum threshold value is reached no targets are selected, presumably because the laser spot is not visible. To increase the precision of tracking, a forecast of the position of the pointer in the image is used, assuming a constant speed model [1]. The predicted position will be used to match the current target to one of the target candidates found by the preceding elaborations. In the one-target case the most common choice is to take the closest candidate, as stated by the *Nearest Neighbor Data Association* algorithm [1].

Finally, triangulation recovers the full 3-D coordinates of the laser spot, using the disparity of the targets in the two stereo images, the internal parameters of the two cameras, and the geometry of the stereo setting (obtained from calibration [8]). Each conjugate pair ideally defines two rays that intersect in a point of the space. In real world, for errors in measurement of point positions and in calibration of the stereo camera, the rays don't always intersect, so the triangulation computes the coordinates of a 3-D point nearest to both rays [15].

3 Sonification

Sounds that give directional cues are said to be "spatialized" [2]. In other words, they provide distance, azimuth and elevation cues to the listener with respect to the position of an object, assumed to be a sound source. Despite all research done (see [3]), we have only partial knowledge about how we determine the position of a sound source, and even less about how to recreate this effect. In everyday listening, hearing a monophonic source gives not only information about the source position, but also characteristics about the environment, such as source dimension and shape of the room.

The characteristics that a sound acquires during its path from source to the listener's ear canal determine the spatial cues that are conveyed to the listener by the sound. Our model adopts a versatile *structural* model for the rendering of the azimuth [5]. A model providing distance cues through loudness control and reverberation effects is in an advanced stage of development. Further studies are needed to convey reliable elevation cues to the listener.

Once the model for the azimuth [5] has been adapted and fine-tuned to our application, we need a new, independent model capable of rendering distances. In this way we can reasonably think to implement these two models independently in the system, typically in the form of two filter networks in series, the former accounting for distance, the latter for azimuth.

The listening environment we will consider is the interior of a square cavity having the aspect of a long tube, sized 8×0.4×0.4 meters. The internal surfaces of the tube are modeled to exhibit natural absorption properties against the incident sound pressure waves. The surfaces located at the two edges are modeled to behave as *total* absorbers, to avoid the creation of echos inside the tube [11]. It seems reasonable to think that, although quite artificial, this listening context conveys sounds that acquire noticeable spatial cues during their path from the source to the listener.

Given the peculiar geometry of the resonator, these cues mainly account for distance. The square tube has been modeled by means of finite-difference schemes [14]. In particular, a *wave*-based formulation of the finite-difference scheme has been used, known as the Waveguide Mesh, that makes use of the wave decomposition of a pressure signal into its wave components [16]. Waveguide Digital Filters provide wall absorption over the internal surfaces of the tube [17].

4 Preliminary Results and Conclusions

Tests show that the effectiveness of laser tracking depends mainly on how "visible" the laser is inside the captured image. Therefore, if the scene is (locally) brighter than the laser spot or if the laser points too far away from the camera, then problems arise about the stability of laser tracking, because the signal to noise ratio becomes too low. Better results are obtained indoor, with a 4 mW laser source, and within a depth range of 1 to 6 meters. The power of the laser, the camera gains and the resolution are critical parameters.

We evaluated the subjective effectiveness of the sonification informally, by asking some volunteers to use the system and report their impressions. The overall result have been satisfactory, with some problems related to the lack of perception of elevation. Further usability tests of this device are planned, with a group of both sighted and visually-impaired subjects. Precision and latency of the system will be measured.

The proposed system represents a first attempt to integrate vision and sound in an HCI application and, for this reason, it is open to further improvements. The next version of the system will determine the three-dimensional structure of a scene by dense stereopsys, then segments the objects (or surfaces patches) contained therein basing on depth. From that information, it will synthesizes spatialized sounds that convey information about the scene.

We are also migrating to a new architecture based on a iPAQ 3760 PocketPC under Linux with a digital color stereo head by Videre Design. This new system will provide a better resolution which can be exploited for the detection of colors and textures. Moreover, future developments of this prototype will be devoted to the design of alternative sound models in order to provide better distance cues, together with elevation. Another line of improvement lies in the coupling of sonic and visual information coming from objects surfaces, e.g., to acoustically render material or texture, so to enrich the listener's experience about the surrounding environment.

Acknowledgments

This work is part of the "Sounding Landscape" (SoL) project, supported by Hewlett-Packard under the Philantropic Programme.

References

1. Y. Bar-Shalom and T. Fortmann. *Tracking and Data Association.* AP, 1988.
2. Durand R. Begault. *3D Sound for virtual reality and multimedia.* AP Professional, 955 Massachusetts Avenue, Cambridge, 1994.
3. J. Blauert. *Spatial Hearing: the Psychophysics of Human Sound Localization.* MIT Press, Cambridge, MA, 1983.
4. R. C. Bolles, H. H. Baker, and M. J. Hannah. The JISCT stereo evaluation. In *Proceedings of the Image Understanding Workshop*, pages 263–274, Washington, DC, April 1993. ARPA, Morgan Kaufmann.
5. C. P. Brown and R. O. Duda. A structural model for binaural sound synthesis. 6(5):476–488, September 1998.
6. Jr S. A. Dallas. Sound pattern generator. WIPO Patent No. WO82/00395., 1980.
7. U. R. Dhond and J. K. Aggarwal. Structure from stereo – a review. *IEEE Transactions on Systems, Man and Cybernetics*, 19(6):1489–1510, Nov/Dec 1989.
8. O. Faugeras. *Three-Dimensional Computer Vision: A Geometric Viewpoint.* The MIT Press, Cambridge, MA, 1993.
9. R. Fish. An audio display for the blind. *IEEE Trans. Biomed. Eng.*, 23(2), 1976.
10. M. Kubovy and D.Van Valkenburg. Auditory and visual objects. *Cognition*, 80:97–126, 2001.
11. Heinrich Kuttruff. *Room Acoustics.* Elsevier Science, Essex, England, 1991.
12. Kay L. Air sonar with acoustical display of spatial information. In *Animal Sonar System*, pages 769–816, New York, 1980.
13. The LAR-DEIS Videt Project. University of Bologna - Italy. Available at URL http://www.lar.deis.unibo.it/activities/videt.
14. J. Strikwerda. *Finite Difference Schemes and Partial Differential Equations.* Wadsworth & Brooks, Pacific Grove, CA, 1989.
15. E. Trucco and A. Verri. *Introductory Techniques for 3-D Computer Vision.* Prentice-Hall, 1998.
16. Scott A. Van Duyne and Julius O. Smith. Physical modeling with the 2-D digital waveguide mesh. pages 40–47, Tokyo, Japan, 1993. ICMA.
17. Scott A. Van Duyne and Julius O. Smith. A Simplified Approach to Modeling Dispersion Caused by Stiffness in Strings and Plates. pages 407–410, Aarhus, Denmark, September 1994. ICMA.

Globalization of Voice-Enabled Internet Access Solutions

Cristina Chesta

Motorola Technology Center Italy
Cristina.Chesta@motorola.com

Abstract. As voice-enabled Internet access solutions are obtaining an increasing success among the mobile users and at the same time software globalization is becoming an important issue for companies doing business worldwide, a natural emerging area of interest is represented by the application of internationalization and localization techniques to the Voice User Interfaces. The paper addresses the topic from both the software architecture and the human factor perspective and presents our solutions and findings.

1 Introduction

Over the last decade, characterized by the growing popularity of Internet, tailoring software products for different marketplaces has become an important issue for companies doing business worldwide while implementing globalization techniques for writing the code once, and running it everywhere has proven to be the key factor for product acceptance and success [1], [2]. With respect to the traditional web interfaces, voice solutions present some characteristics, as the sequential interaction, the time-constraints and the strong dependency on linguistic and cultural context, which requires specific usability considerations and affect also the globalization process by introducing specific issues.

This paper describes some findings and ideas that have arisen from our work aiming to globalize the Mya Voice Platforms [3] realized by Motorola[1]. These are a group of fully integrated end-to-end voice-enabled Internet access solutions, which includes hardware, software, development tools and a portfolio of voice applications and services. As illustrated in Fig.1, the heart of Mya Voice Platforms is Motorola Voice Browser, which uses Automatic Speech Recognition (ASR) to understand and process human speech, captures requested information from voice-enabled web sites, and then delivers the information via pre-recorded speech or Text-To-Speech (TTS) software that "reads" the relevant data to the user. The document is organized as follows: section 2 describes the software architecture proposed for supporting efficient global voice interface development; section 3 outlines grammars localization issues; finally section 4 presents our concluding remarks.

---

[1] Motorola and the Stylized M Logo are registered in the US Patent & Trademark Office. Java and all other Java-based marks are trademarks or registered trademarks of Sun Microsystems, Inc. in the U.S. and other countries. All other product or service names are the property of their respective owners.

F. Paternò (Ed.): Mobile HCI 2002, LNCS 2411, pp. 398–403, 2002.
© Springer-Verlag Berlin Heidelberg 2002

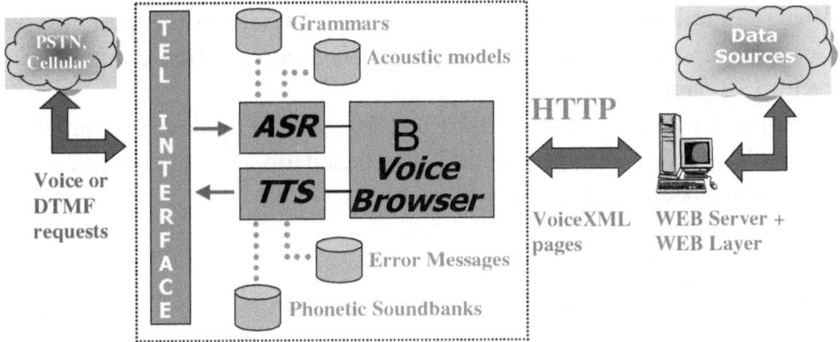

Fig. 1. Mya Voice Platforms Structure

2 Architectural Approach

With reference to the architecture described in Fig. 1, the main component affected by globalization issues is the Web Layer, which constitutes the interface between the voice browser and the data sources. Referring to the definitions of the Localization Industry Standards Association (LISA), software globalization involves two main activities: *Internationalization*, which is the process of designing an application so that it can be adapted to various languages and regions without engineering changes, and *Localization*, which is the process of adapting software for a specific region or language by adding language-specific components and translating text.

The driving principles we followed in the design are directed to allow a clear separation between business and presentation logic, ensure characteristics of flexibility, reusability, portability, and extensibility as well as supporting globalization. The solution adopted is a hierarchical structure, where three sub-layers are responsible to manage specific aspects. The *Core Web Layer* provides the logic related to functionalities common to all applications. The *Internationalization* layer is responsible to guarantee rapid customization for new languages and countries. The *Application* layer must be developed specifically for each application taking advantage of the components and features offered by below layers.

2.1 Core Web Layer

The Core Web Layer architecture design is based on the popular Model View Controller (MVC) design pattern and the implementation has been realized adopting the Java Technology.

The *Model*, which constitutes the server-side business logic, represents the data involved in the application, manages the application flow, and performs all the operations required for the state transition, as managing user input, interacting with data sources, resource bundles, properties files or instantiating Java Beans objects.

The *View* dynamically generates and presents contents to the user, by formatting the output in the appropriate markup language. The presentation component has been realized with JSP pages.

The *Controller* translates user actions and inputs into application function calls to the model and selects the appropriate View based on user preferences and on the Model state. It has been realized through a servlet.

2.2 Internationalization Layer

The goal of the Internationalization Layer is to allow developers to easily individuate and isolate language-dependent resources from the source code. Voice-specific localizable items are mainly TTS/audio prompts and grammars. However, also some general VoiceXML [4] properties, such as timeout settings, caching features, enabling barge-in should be considered and configured according with the particular locale.

We adopted an approach based on the Java ResourceBundle methodology [5] and we developed a set of APIs that extend Java classes to cover specific voice application issues. The ResourceBundle, as its name implies, indicates the place where localization information is maintained and retrieved at run-time. Usually, most of the localization information can be maintained in a text file called property file. The specific language context and country supported by the application are defined by the Locale component. According to the current Locale, the application automatically loads the corresponding string from the property file. As a result, the same code "behaves" consistently but "produces results" differently for different languages. No reworks (rewrite, recompile and rebuild) of source codes are needed for adapting the same application to new languages.

Different approaches have to be adopted in case the textual data is *simple*, *compound* or *complex*. *Simple* textual data consist of static text as for example:

`Hello! Have a good day.`

These data can be retrieved directly from the resource bundles by referencing them through the corresponding identifier. This approach is called *data-driven* because only the localizable artifacts change while the code remains invariant. *Compound* textual data consist of static and variable text, as (changing text is in italic):

`Hello` *`John`*`! Today you have a meeting at` *`2:00 PM`*`.`

In this case, the compound message can be stored in the resource bundles by replacing each variable text item with a placeholder. At run-time, the compound message is constructed by retrieving the message pattern through the corresponding identifier and by inserting the run-time arguments in the placeholders. A special case is constituted by *complex* messages, which might correspond to different message patterns, depending on particular issues such as for example number, gender, conjugation or declension. Here is an example of a complex message:

`Hello John, you have` *`one new mail`* `and` *`three old mails`*`.`

When the application runs, the value for the dynamic arguments may be either singular or plural and the localization developers must provide a way to distinguish the two cases. Since the rules may be different from one language to another, *code-driven* approach should be adopted. The idea is to define a uniform interface for client users. However, behind it, for each case, a specific class will be called to handle the situations like the singular-plural messages. This approach complements the insufficiency of data-driven approach, does not require the rebuild of the core application and is very useful for generating the audio outputs for numbers, dates, times and implementing sorting strategies.

2.3 Application Layer

In order to complete the procedural approach description, we present in this section some considerations related to application development. We refers to a demo application, providing an information service about the locally scheduled movies and the possibility to book tickets which has been realized in 5 languages (English, Italian, Spanish, French and German). The application provides a rich set of interaction modalities (speech, DTMF, recording) and uses different grammars (e.g. cities, dates, times).

The guidelines we have adopted in structuring the dialogue flow and writing the prompts are mainly intended to ensure the system correct reply to the user requests while not forcing the user to spend too much time repeating commands and waiting [6]. In order to improve the effectiveness of the interaction and increase the robustness of the application we found convenient to assume default values/actions in case of consecutive misunderstandings or timeouts, to give the user the possibility to utter global commands, available from any point of the dialogue (e.g. "Help", "Repeat", "Main", "Exit", …) and to write help messages that vary if it is the first, second or third time the user asks for assistance. Fundamental for the usability of application is to guarantee good grammar localization: our recommendations on this subject are detailed in the following section.

3 Grammar Localization

The process of grammar localization presents specific issues as it doesn't involve only textual and audio translation as for most of the others artifacts, but must consider in addition various cultural aspects. We propose in the following some points based on our experience that can be used as a checklist while writing and evaluating grammars, to ensure they are designed according to good human factors and linguistics practice.

- The grammars should conform to the syntactic and lexical rules of the local language. Words should have the correct meaning for the context and there is not necessarily a direct one-to-one mapping between the meaning of a word in two different languages. Sentence structure and words order, as well as world conjugation and declension should be compliant with language rules.

- The grammars should use local cultural names and formats, including: currencies, numbers, phone numbers, times, dates, names and addresses.
- The grammars should have good 'coverage'; i.e. include all the different ways in which local users are likely to say a word or phrase. The types of variation that should be considered include synonyms, phrases with identical meanings, formalities, hesitations, different sentence forms such as imperatives and questions; colloquialisms, over and under wordiness. However, despite good coverage the grammars should be efficient and avoid over-complication as nonsensical phrases or unlikely spoken word combinations.
- The grammars should build in alternative pronunciations for words, where appropriate and ensure coverage of all the dialects of the target users.
- The grammars should be 'universally accessible' by all different types of user, bearing in mind potential differences across the user population according to: age, gender, socio-economic class, geographic region, ethnic background, technical experience and should take cultural norms, biases and sensitivities into account.
- Where possible, the grammars should avoid words that are easily confused by the speech recognizer; i.e. words that have similar phonetic structures to another one.

The test phase should demonstrate that the grammar doesn't produces software errors, but also that it encompass all likely variations in the user's language.

While the first of the above stated points is easily verifiable by the developer itself, the second point require a user trial with "naïve" users (i.e. people who have not been involved in the development of speech grammars and are not experienced in speech recognition), selected in a way to be representative of the target consumer.

4 Conclusions

We discussed some voice specific internationalization and localization issues and we presented our architectural solution as well as an application example.

The field of voice-enabled global solutions is extremely promising and there are still several opportunities for further work. Two interesting research areas that represent a good opportunity to build more effective and appealing human-computer interfaces are the mixed-initiative dialogue, which is the refinement of the dialogue process towards a more flexible and cooperative interaction, and the adaptive interaction, which offers the opportunity to personalize the provided information to the user's preferences, by improving the usability of the service.

Acknowledgements

I would like to thank my colleagues that greatly contributed to this work, especially Cristina Barbero, Alvise Clementi, Simona Ricaldone, Stefano Ricciardi, Jian Yao, and Gaea Vilage. I also would like to thank Kang Lee, Cheung Lo, Nick Allott, Rob Graham and Jim Ferrans for their support and advice.

References

[1] B. Esselink – "A Practical Guide to Localization (Language International World Directory)" – John Benjamins Pub Co, 2000.

[2] S. Martin O'Donnell– "Programming for the world: A Guide to Internationalization" – Prentice Hall PTR/Sun Microsystems, 1998.

[3] http://developers.motorola.com/developers/wireless/products/servers.html

[4] VoiceXML Forum – "Voice eXtensible Markup Language (VoiceXML™) version 1.0", May 2000.

[5] Dale Green – "The Java™ Tutorial. Trail: internationalization".

[6] Bruce Balentine, David P. Morgan – "How to Build a Speech Recognition Application" – Enterprise Integration Group, 1999.

Map-Based Access to Multiple Educational On-Line Resources from Mobile Wireless Devices

P. Brusilovsky[1] and R. Rizzo[2]

[1]School of Information Sciences, University of Pittsburgh,
Pittsburgh PA 15260, USA
peterb@sis.pitt.edu
[2]Institute for Educational and Training Technologies, Italian National Research Council,
Palermo, Italy
rizzo@itdf.pa.cnr.it

Abstract. While large volumes of relevant educational resources are available currently online, almost all existing resources have been designed for relatively large screens and relatively high bandwidth. Searching for the proper interface to access multiple resources from a mobile computer we have selected an approach based on self-organized hypertext maps. This paper presents our approach and its implementation in the KnowledgeSea system. It also discusses the ongoing work on using our approach with very narrow screens of Palm-like devices.

1 Introduction

Large volumes of relevant educational resources are available currently on the Web for the students. Altogether, they well complement course textbooks and enable the students with different level of knowledge or different learning styles to get a better comprehension of the subject. The current model of accessing these resources from computers at home or at the university labs is a restriction - like a requirement to read a textbook always at home or in class, but not outside, in a café, or while riding a bus. The use of mobile wireless handheld devices potentially allows the students to access educational resources really "anywhere". Moreover, the success of mobile e-books shows that the users are quite willing to read predominantly textual sources on mobile devices. The bottleneck here is finding a relevant educational resource that currently demands large screens and high bandwidth (Figure 1).

Searching for the proper interface to access multiple resources on a mobile computer we have considered several options and finally selected an approach based on self-organized hypertext maps. This paper presents our approach and its current implementation – the KnowledgeSea system that can be used successfully with a number of mobile devices with landscape-style screen (such as HP Jornada organizers). It also discusses the ongoing work on using our approach with more narrow screens of Palm-like devices.

F. Paternò (Ed.): Mobile HCI 2002, LNCS 2411, pp. 404–408, 2002.
© Springer-Verlag Berlin Heidelberg 2002

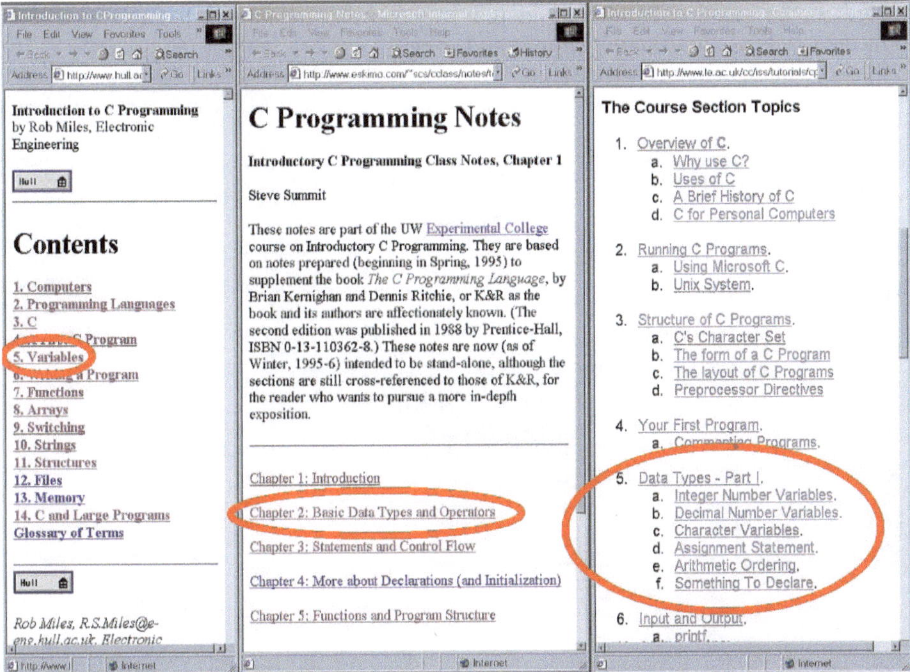

Fig. 1. Selecting relevant reading sections in multiple educational resources is a challenging task that demands good interface skills, large screen and fast Internet connection.

2 The KnowledgeSea Approach

The core of our approach to navigating educational resources is a self-organized hyperspace map automatically built using a Self-Organizing neural Map (SOM) [2]. SOM is a very attractive technology for developing compact maps of large hyperspaces [3] since it builds a map representing the neighborhood relationship between the objects. A two-dimensional 8-by-8 map of educational resources developed with SOM technology is a core of our KnowledgeSea system for map-based access to multiple educational resources (Figure 2). KnowledgeSea was designed to support a typical university class on C programming. In this context, the goal of the students is to find most helpful Web-based material as a part of readings assigned for every lecture in the course. The most easily available Web educational resources are multiple hypertextual C tutorials (see for example http://www.le.ac.uk/cc/iss/tutorials/cprog/cccc.html). In this context, the goal of KnowledgeSea system to help the user to navigate from lectures to relevant tutorial pages in multiple tutorials and between them. The main component of the interface is a KnowledgeSea map – a table in which each cell is used to group together a set of educational resources (Figure 2). The map is organized in a way that resources (web pages) that are semantically related are close to each other on the map. Resources located in the same cell are considered very similar; resources located in directly connected cells are reasonably similar and so on.

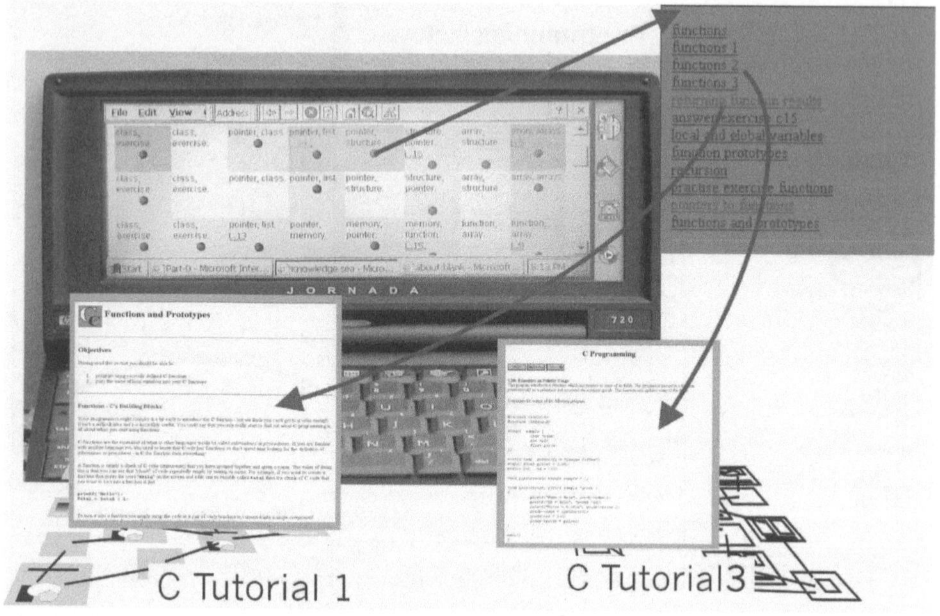

Fig. 2. A session of work with the KnowledgeSea system.

Each cell displays a set of keywords that helps the user to locate the relevant section on the map. Some cells also display links to "critical" resources that serve as origin points for horizontal navigation. The map serves as a mediator to help the user navigate from critical resources to related resources. For lecture-to-tutorial navigation the critical resources are lectures and lecture slides. The presence of related educational resources in the cell is indicated by a red dot. The cell color indicates the "depth of the information sea" – the number of resource pages lying "under" the cell. A click on the red dot opens a cell content window (right on Figure 2) that provides a list of links to all tutorial pages relevant to this cell. A click on any of these links will open a resource-browsing window with the selected relevant page from one of the tutorial. This page is loaded "as is" from its original URL.

The functionality and the usefulness of our map-based information access approach was evaluated it in the context of a real Programming and Data Structures course at the University of Pittsburgh. The characteristics of the SOM, the implementation details of the KnowledgeSea system, and some results of this study were reported in [1]. We were very encouraged to find that about 2/3 of the 21 students participated in the study thought that the system has achieved the goal of providing an access to the online C tutorials completely or "quite well". The overall idea to attach the resources to cells on a map and show where the lecture belongs was considered "very easy" by the 19% of the students and "quite easy" by the 42.8%. Interesting is that when asked in which context they would expect to use the KnowledgeSea system from a mobile device, the majority of the students have selected only home or library. Only few of them have indicated an interest to use the system in a bus, park or from "anywhere when I have some spare time". It is not clear yet whether this answer was caused by the nature of information access task, the nature of the system or simply by the students' prejudice to using computers in a familiar context.

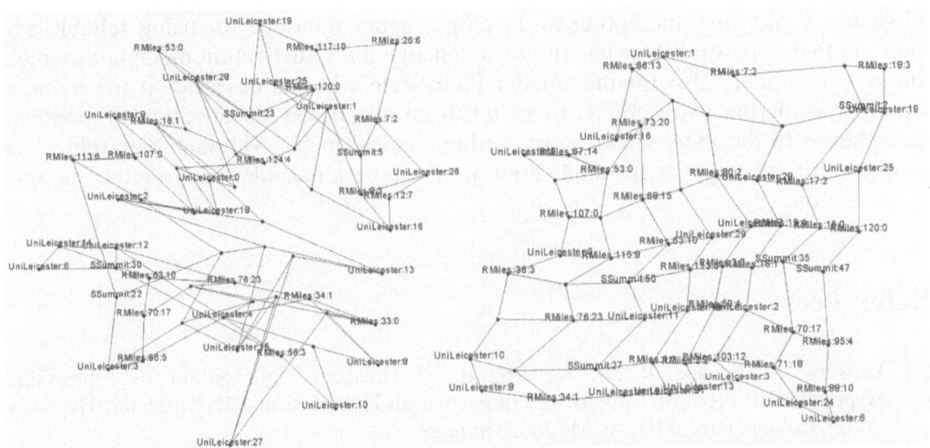

Fig. 3. A representation of the two different map geometries after the learning phase. On the left the 4x15 map and on the right the 8x8 map. The cells are labeled using a reference to the Web pages of the tutorial.

3 Current and Future Work: The Narrow Screen Challenge

The original platform for our experiments was a desktop PC and the HP Jornada organizer with a relatively wide screen. In this context we have been working with an 8x8 SOM map (Figure 2). The "learning" (organizing) stage of the SOM map in this case was not complicated and the standard value of parameter sufficient. Overall, we can conclude that SOM-based access to multiple information resources in the investigated context is a very useful technology. The 8x8 map that we have explored has worked well for the students. This map is large enough to provide a reasonable split of diverse content, yet is small enough to fit a Jornada-like handheld. We plan to continue investigating the same map and the same interface for the context of a larger hyperspace of educational material (6 or more external tutorials instead of 3).

Our new challenge is to make our technology work on a narrow Palm-like screen that is more typical for handheld organizers. As soon as wireless cards become available for the Palm platform, we have started to experiment with Palm devices equipped with Blaser™ Web browser. We have found that with this browser the narrow Palm screen can fit only 3-4 cells of our original text-based map in a row. Since we want to avoid horizontal scrolling, we have started with exploring an unusual 4x15 map geometry. Unfortunately, for this geometry the learning stage of the SOM map was more complicated and the results are not satisfying even with extra efforts to split the learning phase in three sessions and to use non-standard values of the parameters. The four-cell width was apparently not enough to organize an information space without "twisting" as it can be seen on Fig. 3. In this figure it is possible to see the lattice of the network obtained after the learning stage. The lattice is irregular and this is the reason why the resulting map did not look intuitive and contained too many cells with no information items. We concluded that this map could be more confusing than helpful for the students.

Instead of continuing our experiments with narrow maps we are going to stay with wider maps and now attempting to develop a better interface for using reliable 8x8 maps on Palm platform. Our intention is to change the visualization metaphor to make the system suitable also for the smaller Palm-style wireless devices. At the moment we are considering two options: to switch from text-based visualization to graphical metaphor to fit the PDA screen or to use hierarchical maps. Although hierarchies can be difficult to manage in this application they can be a feasible way to guide the user in navigation.

References

[1] Brusilovsky, P. and Rizzo, R.: Map-Based Horizontal Navigation in Educational Hypertext. In: Proc. of 13th ACM Conference on Hypertext and Hypermedia (Hypertext 2002), College Park, MD, ACM (2002) in press

[2] Kohonen, T.: Self-Organizing Maps. Springer Verlag, Berlin

[3] Kohonen, T., Kaski, S., Lagus, K., and Honkela, T.: Very Large Two-Level SOM for the Browsing of the Newsgroups. In: Proc. of ICANN'96, Berlin, Springer Verlag (1996) 269-274

Supporting Adaptivity to Heterogeneous Platforms through User Models

Luisa Marucci and Fabio Paternò

CNUCE-CNR
Via G.Moruzzi 1
Pisa, Italy
luisa.marucci@guest.cnuce.cnr.it, fabio.paterno@cnuce.cnr.it

Abstract. In this paper we describe an approach to providing adaptive support to applications that can be accessed through multiple interactive devices from various locations. It is based on a user model, which can update information about user preferences and knowledge at run-time. Such information is used to adapt the navigation, presentation and content of each user interface also taking into account users' accesses through different interaction platforms.

1 Introduction

The increasing availability of many types of devices and the advent of next generation mobile technologies (such as UMTS), wireless LAN-based solutions, and new types of terminal equipment (such as wearable computers) raise a number of challenges to user interface designers. To this end, there is a need for interactive applications able to adapt to the different contexts of use while preserving usability. We consider the context of use to include the types of devices that support users while performing their tasks and the surrounding environment.

User modelling [1] is an approach that aims to represent aspects regarding users, such as their knowledge level, preferences, goals, position, etc. Such information is useful to furnish user interfaces with adaptivity, that is, the ability to dynamically change their presentation, content and navigation in order to better support users' navigation and learning, also considering the current context of use. Various aspects of the user interfaces can be adapted according to user models. For example, they can adapt their text presentation through techniques such as conditional text or stretch-text and also the kind of presentation from text to speech or vice versa. They can also adapt the user navigation using techniques such as direct guidance, adaptive order of links, hiding of links.

To date, only a few works have considered user modelling to support the design of multi-platform applications. An example is Hippie [2] a prototype that applies user-modelling techniques to aid users in accessing museum information through either a web site or a PDA while in the museum. In our case the use of mobile out-door technologies and user models integrated with task models developed at design time is also considered.

F. Paternò (Ed.): Mobile HCI 2002, LNCS 2411, pp. 409–413, 2002.
© Springer-Verlag Berlin Heidelberg 2002

This paper presents a solution that shows how user modelling can be leveraged to support users accessing an application through multiple interaction devices. The main element is a single user model associated with each user that is dynamically updated when the user interacts with the application through any type of device. We discuss our approach using a case study in the museum application domain.

2 The Method

This approach assumes that a model-based method has been followed to the design of the multi-platform application. Recent developments of the ConcurTaskTrees notation [3] allow designers to develop task models of multi-platform applications. This means that in the same model, designers can describe tasks able to be performed on different platforms, their mutual relationships and what platforms are suitable for each task.

From this high level description it is possible to obtain first the system task model associated with each platform and then the corresponding user interface. The task model can be represented in two ways: a graphical representation that can be edited and analysed with a tool (publicly available at http://giove.cnuce.cnr.it/ctte.html) or in XML format that can be automatically generated (see Figure 1).

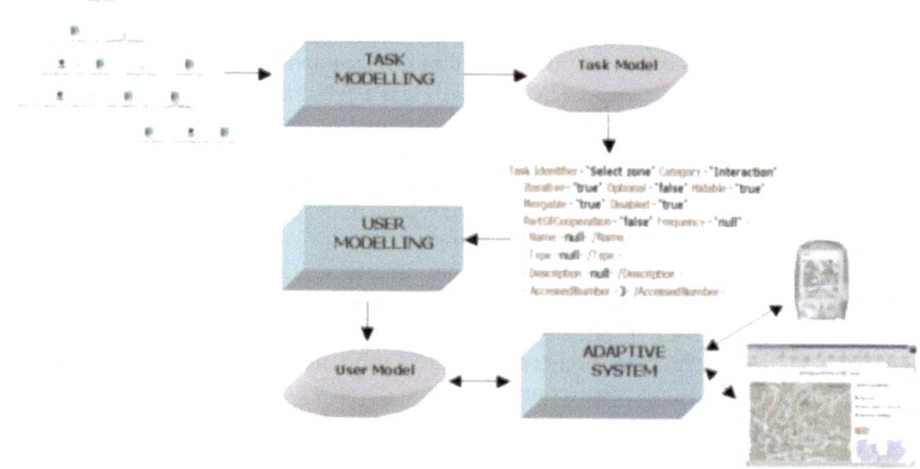

Fig. 1. The method proposed

In our case we use the task model XML specification as input for the creation of the structure of the user model. In addition, in the user model various elements are associated with values that are updated dynamically according to the interactions performed by the corresponding user with any of the available platform. These values are used by the run-time support that modifies the user interface presentation, navigation and content according to some previously defined adaptivity rules.

One advantage of this approach is that the task model developed at design time already provides some useful information for the adaptive support: the temporal dependencies among tasks performed on different platforms, the tasks executable from many platforms, the association of tasks with domain objects and the related attributes (as well as the definition of objects and attributes accessible through a specific platform).

The navigation preferences are detected by analysing the sequence of tasks chosen, the tasks never performed, the task usually performed and so on, as will the presentation preferences (by analysing the objects classes and objects subclasses accessed).

The user model also contains fields that allow dynamic modification of the task availability, indicating whether it is possible to merge more tasks at the same abstraction level in one, whether it is possible to disable a task performance including it in another, more general task, and whether it is possible to completely disable a task for the current user.

Likewise, the supported tasks depend on the interaction platform: there are tasks associated with a desktop virtual visit, others associated with the phone-supported visit, but performance of some kinds of tasks on one platform may depend on the accomplishment of other tasks through other devices (for example the desktop task associated with reviewing the itinerary annotated by phone).

Table 1. Examples of Adaptive rules

Adaptive Navigation Modality	
When: The user always performs the same sequence of tasks in order to reach a goal	**What**: Change of the navigation support in such a way to shorten the achievement of the goal
When: The user performs a specific task in one platform and then accesses the application through other platform	**What**: Modification of the user model state to enable or disable some tasks
When: The user never selects a task (for example, a link selection) during one or more sessions (in any platform)	**What**: Removal of the task support from each platform (for example, remove link)
Adaptive Presentation Modality	
When: The user often (never) selects a domain object subset	**What**: Provide access to this object or attribute in a priory-position (in a non-priority position)
When: The user shows a good knowledge of a certain topic	**What**: The description of the elements of interest become more detailed taking into account the possibilities of the current device

Also the domain objects that can be accessed and manipulated vary according to the device that is available. In general, the domain objects that can be manipulated via phone are more limited than those accessible via desktop computers. In addition, there are some spatial attributes related to the user position that are meaningful only for mobile devices.

Table 1 describes examples of the adaptive rules that are used to drive the adaptivity of the user interface. We explain how they are handled, highlighting the resulting adaptive navigation and presentation modality consequent to the users' interactions with the system through different platforms. In succession, we show *when* a rule comes into force and *what* is the corresponding change in the interactive system behaviour or presentation.

2.1 An Example: Changing Navigation Modality according to Task Dependencies

We will now consider an example of tasks performed in a specific platform which generate a change in the task support provided by another platform. The user first selects a tour through the desktop system, shows preferences for a city zone and then accesses the application through the cellular phone. Figure 2 shows the user interface at three different circumstances: the first user access (a), after the user has selected a tour through the desktop system (b), and lastly after the user has accessed through the desktop system but without selecting any tour.

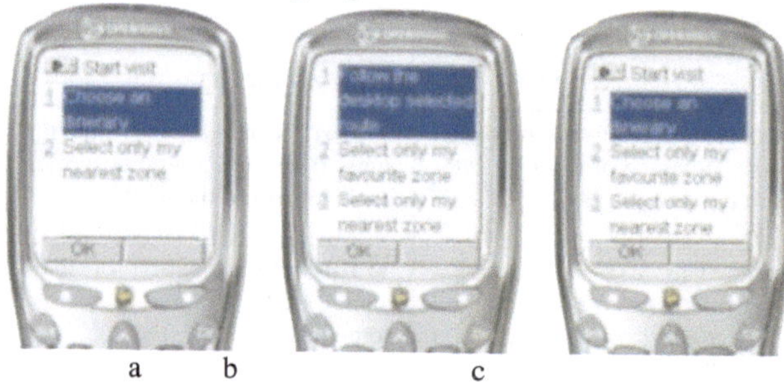

a b c

Fig. 2. Access to the application for the first time (a), after desktop visit with tour selected (b), and after desktop visit and without tour selected (c).

Vice versa, if users access the system first from the mobile phone platform, they can choose the option of selecting the artworks encountered during the actual visit in order to see them better later on, when they access through the desktop system. This will enable the task "More information about artworks visited" in the desktop platform, which has a corresponding element in the user interface. When this element is selected then the desktop application shows the list of artworks that were encountered during the real visit in order to allow the user to receive more detailed information regarding them (Figure 3 shows an example).

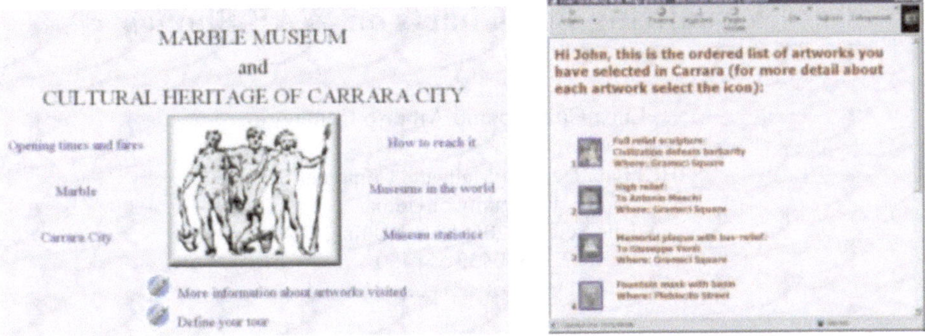

Fig. 3. The user interface in the desktop version after an access through the phone version

3 Conclusions

In the paper we have discussed how adaptive support based on user modelling techniques can be provided when interactions through multiple platforms are considered. Through a case study in the museum application domain we have shown examples of the type of design that can be obtained.

In particular, we have discussed a set of rules that make it possible to change the presentations and dialogues supported by the user interface by taking into account users' interactions through different platforms. This results in greater application flexibility.

References

1. Brusilovsky, P.: Methods and techniques of adaptive hypermedia, User Modelling and User Adapted Interaction, v 6, n 2-3, pp.87-129, 1996 URL: http://www.cntrib.andrew.cmu.edu/plb/UMUAI.ps
2. Oppermann, R., Specht, M.: A Context-sensitive Nomadic Information System as an Exhibition Guide. Proceedings of the Handheld and Ubiquitous Computing Second International Symposium, HUC 2000, pp.127-142.
3. Paternò, F.: Model-based Design and Evaluation of Interactive Applications, Springer Verlag, ISBN 1-85233-155-0, 1999

Visualizing Bar Charts on WAP Phones

Luca Chittaro and Alberto Camaggio

HCI Lab, Dept. of Math and Computer Science
University of Udine
via delle Scienze 206, 33100 Udine, ITALY
+39 0432 558450
chittaro@dimi.uniud.it

Abstract. This paper begins to explore the problem of graphically visualizing numerical data on the very small displays of WAP phones. In particular, we consider visualizations in bar chart format, proposing two possible solutions and testing them with time-series data. A controlled experiment has been carried out to point out possible differences in user performance between the two considered visualizations, and is described in the paper.

1 Introduction

At the midpoint of 2000, approximately 50% of the world's mobile phones shipments were WAP capable, and more than 40 million WAP browser-equipped handsets had been shipped [4]. Since then, the percentage of shipped WAP phones has steadily increased, and today most new models of cellular phones are WAP capable. This is spawning the need for and the development of services specifically tailored to mobile users employing WAP phones. Unfortunately, most WAP phones have a very small, black-and-white screen that makes it hard to design usable and effective services. Moreover, very little has been published on the usability of WAP sites. This situation has prompted a few researchers [1,2,3,5] to focus on the user evaluation of WAP interface design choices.

While the above mentioned evaluations have concentrated on interfaces that mainly displayed textual information (such as menu options, news headlines, movie reservations,...), this paper starts to explore the graphical display of quantitative information on WAP phones. In particular, we consider the problem of graphically displaying a time-series on a WAP phone. Time-series are ubiquitous in several domains ranging from engineering to medicine. The stock market is a particularly suited motivating example, since online stock trading is considered to be one of the most relevant applications for mobile commerce [7]. Figure 1 shows the WAP card returned by a commercial stock market site to a user who is looking for a specific share: the top line shows the name of the share ("Autogrill Spa"), the first displayed number is the time ("10:21"), the second number ("10,11") is the closing value ("Chiusura:"), and the number for change ("Var.%:") falls outside the WAP phone screen. Figure 1 is representative of how many WAP sites (for this and other services)

F. Paternò (Ed.): Mobile HCI 2002, LNCS 2411, pp. 414–418, 2002.
© Springer-Verlag Berlin Heidelberg 2002

provide information to users. As one can easily notice, screen estate is used suboptimally: (i) the user is able to see only one change value at a time for the share (making it difficult to assess how that share is actually going over time), and (ii) she has even to scroll the screen to be able to see that single value. Using tables could be a first step to show some more numbers on the same screen, but graphical solutions are typically used in time-series data analysis to make the task easier for humans. Historically, bar charts are a familiar solution to the problem and are known to allow for an easy comparison among the data values of a time-series (this property may not hold in bar charts that include unnecessary graphic elements to look more impressive, see [6] for a discussion).

In the following, we propose and discuss two possible solutions for displaying bar charts on WAP phones, also evaluating them with a controlled experiment.

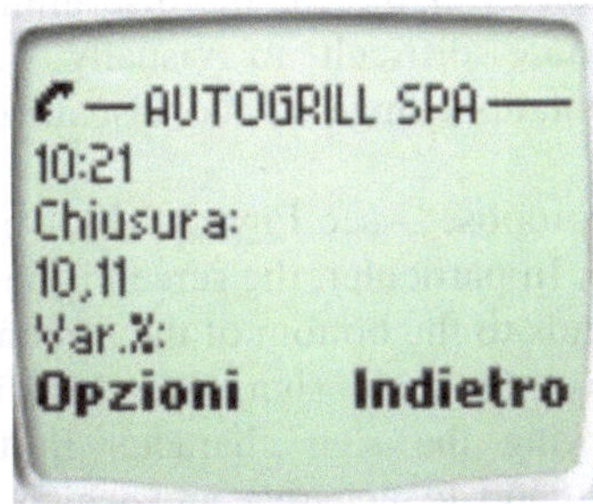

Fig. 1. Share information visualized as text on a stock market WAP site.

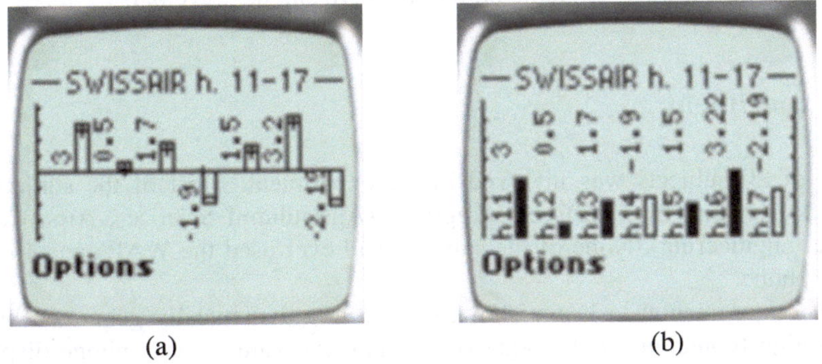

(a) (b)

Fig. 2. Visualizing information as: (a) bar chart, (b) bar chart with color coded sign.

2 The Considered Solutions

The first problem we had to face in our research was of a technical nature: to display an image, most WAP browsers currently require to use the WBMP file format. We thus had to find a solution for dynamically generating WBMP files encoding the visualizations of the desired data on a WAP site. We faced this problem by resorting to GD 2.0 (a publicly available graphics library that is able to produce files in various formats, WBMP included) and embedding calls to GD in PHP scripts to allow for a dynamic image generation on the WAP site.

The first visualization we propose attempts to keep the traditional bar chart organization as much as possible, with positive bars growing upwards, and negative bars downwards. Figure 2(a) shows an example of it on a WAP phone (we are using the Nokia 7110 in our work, but many other phones on the market have similar displays). Although this representation is likely to be very familiar to users, it has one serious drawback on small screens when the displayed bar chart includes *both negative and positive values*. In this case, the considered visualization splits the scarce screen space into two smaller areas (one reserved to positive and the other to negative values) and forces one to choose a mix of the two following undesirable limitations: (i) decrease in the resolution available for the discrimination of differences in size among bars, making it more difficult to visually detect such differences, or (ii) decrease in the representation range of the visualization to keep an acceptable resolution.

The second solution we propose – see Figure 2(b) – trades some familiarity for a better usage of screen estate. In particular, the screen is never split into a negative and positive area: zero corresponds to the bottom of the screen and we display all the bars growing upwards regardless of their sign. We highlight with a color coding mechanism (besides displaying the sign character in the textual version of the numbers) which bars correspond to negative and which to positive values. To this purpose, most current WAP phones offer only two (achromatic) colors (black and white): in the proposed visualization, black bars correspond to positive values and white bars to negative values. Unlike the previous visualization, the full screen resolution is now always available both to positive and negative bars.

3 Experiment

A total of 20 subjects was involved in the experiment. Most of the subjects were university students from different programs (Agricultural Sciences, Arts, Computer Science, Engineering). None of the subjects had ever used the WAP capabilities of a cellular phone.

In the experimental task, we showed subjects a dynamically generated bar chart visualization (containing both negative and positive values) on a phone display and asked them to determine: (i) what was the highest value, (ii) what was the lowest value, and (iii) how many times the values displayed in the chart were negative. Before carrying out the task, subjects were briefly instructed about the meaning of the various graphic elements in the two chart types.

One independent variable and two dependent variables were considered in the study. The independent variable is chart type (in the following, we will refer to the more traditional solution as TRADNL and to the solution employing color coding for the sign as CCSIGN). The dependent variables are: (i) time needed to complete the task, (ii) correctness of the provided answer (we employed the strictest requirement for correctness: all three values reported by the subjects had to be correct).

A within-subject design was followed. Every subject performed the task in the two possible conditions. Therefore, two different time-series were needed and we chose them to be of the same complexity. The assignment of subjects to time-series and condition was carried out following a counterbalancing scheme where the order for the assigned time-series and the order for the assigned condition were varied independently.

Table 1 summarizes the obtained results. First, carrying out the task with CCSIGN required about 20% more time on average than TRADNL, and the t-test indicates that the effect is significant ($p<0.05$). Second, most of the 20 subjects provided the correct answer in both conditions. Only a few subjects made errors: in particular, one subject failed in the TRADNL condition, and three in the CCSIGN condition. This result was analyzed using the Wilcoxon test for dependent samples; this difference between the two conditions is not statistically significant (p=0.32).

Table 1. Experimental Results.

	TRADNL	**CCSIGN**	Significance
Average time to carry out the task	21.1 sec	25.4 sec	p<0.05
Number of subjects who succeeded	19 of 20	17 of 20	not significant

4 Conclusions and Future Work

The experiment has shown that, although the CCSIGN visualization has the advantage of using screen space in a wiser way than TRADNL, it (moderately) increases the time required to examine it. Some considerations must be added. First, since this result might be explained by the fact that CCSIGN was less familiar (in informal discussions with the subjects, it came out that most of them had seen or used bar charts of the TRADNL type, while a number of subjects explicitly commented that they had never seen a bar chart of the CCSIGN type and/or asked for clarifications about it, such as "Why does black stands for positive and white for negative?"), it will be interesting to repeat the experiment with the same subjects over a period of time to contrast the possible differences between novice and expert users of the visualizations. Second, since the drawing algorithm did not make size adaptations of the bars to the available screen space, it will be interesting to test if adding this possibility could favour CCSIGN thanks to the larger space it can offer.

We are also planning a more thorough experimental activity that will include other alternative visualizations of bar charts and other types of tasks. Another dimension to explore is the addition of more interactive features to the charts: although current WAP browsers severely limit this possibility (a new WBMP file must be generated for any graphic update), there are some useful extensions of our bar chart generator that we are considering (e.g., the possibility of scrolling longer time-series by choosing "left-right" options without having to explicitly formulate another query to the WAP site). Finally, if a number of different visualizations is developed for WAP phones, an interesting research direction will be the inclusion of knowledge-based mechanisms to automatically select the most appropriate visualization for a given set of data.

References

1. Buchanan, G., Jones, M., Thimbleby, H., Farrant, S., Pazzani, M.: Improving mobile internet usability. Proc. 10th Internat. WWW Conf., ACM Press, New York (2001) 673-680
2. Chittaro, L., Dal Cin, P.: Evaluating Interface Design Choices on WAP Phones: Single-choice List Selection and Navigation among Cards. Proc. Mobile HCI 2001: 3rd International Workshop on Human Computer Interaction with Mobile Devices, IHM-HCI (2001) 7-13
3. Ericsson, T., Chincholle, D., Goldstein, M.: Both the Cellular Phone and the Service Impact WAP Usability. Joint Proc. of IHM 2001 and HCI 2001. Springer Verlag, Berlin (2001)
4. Gillott, I.: Exploding the Myths of WAP, IDC Research Report, http://www.idc.com/ITAdvisor/press/itp001106a.htm (2000)
5. Schmidt, A., Schroder H., Frick O.: WAP – Designing for small user interfaces. Proc. CHI2000 Conf. Human Factors in Computing Systems, Abstracts Volume. ACM Press, New York (2000) 187-188
6. Tufte, E.R.: The Visual Display of Quantitative Information, Graphics Press, Cheshire, CT (1982)
7. Turban, E., Lee, J., King, D., Warkentin, M., Chung, H.M.: Electronic Commerce: A Managerial Perspective, 2nd Edition. Prentice-Hall, Upper Saddle River, NJ (2002)

Structured Menu Presentation Using Spatial Sound Separation

Gaëtan Lorho[1], Jarmo Hiipakka[1], and Juha Marila[2]

Nokia Research Center
[1] Speech and Audio Systems Laboratory
[2] Visual Communications Laboratory
P.O. Box 407, FIN-00045 Nokia Group, Finland
{gaetan.lorho,jarmo.hiipakka,juha.marila}@nokia.com

Abstract. This paper describes a technique to support user interaction in a hierarchical menu, based on spatial sound separation. A complex menu structure is represented in space using a limited number of sound positions obtained by stereo panning or 3-D audio processing techniques. Spatial organisation of menu items can be designed in a logical way to provide navigation cues to the user, independent of the menu item nature. Two different strategies for menu presentation and interaction are described and compared in this paper. Finally, an application of this technique to the navigation in a large music collection is considered. This case study is an interesting example of usage situation for which eyes-free interaction would be useful, for instance on a portable audio player using headphones and a small remote control.

1 Introduction

User interface design for the visual modality is essentially based on spatial organisation of information due to the naturalness of multiple visual object display and management from a screen. Space has also a large potential for auditory display, since humans can localise and segregate sound events or audio streams in three dimensions. However, the reproduction of 3-D sound sources around a listener in an accurate, controlled and efficient way is challenging and computationally expensive. Despite these limitations, 3-D audio has evolved recently with interesting applications to human-computer interfaces. For instance, Ludwig *et al.* [1] developed an audio window system in which spatially separated sound items can be monitored and manipulated by the user. Spatialisation of audio streams has also been investigated widely, with the principle of the "cocktail party effect".

In this paper, the idea of spatial sound separation is applied to a hierarchical menu structure. Due to the independence between the audio presentation of the menu structure and the sound items, this technique can be applied to different types of menu presentation, including speech, non-speech sounds (i.e. earcons and auditory icons) and music. The number of sound items used to represent the menu is intentionally

F. Paternò (Ed.): Mobile HCI 2002, LNCS 2411, pp. 419–424, 2002.
© Springer-Verlag Berlin Heidelberg 2002

limited to few positions for simplicity, but an efficient spatial organisation of the menu items can be designed to provide navigation cues to the user.

A generic description of hierarchical structures is given first. A solution for spatial presentation and interaction in a menu is then proposed. Both stereo panning and 3-D audio processing techniques, using Head-Related Transfer Functions (HRTF), are considered for spatial separation of sound reproduced over headphones. Finally, an application example is described to illustrate the potential of this technique for eyes-free selection of music on a portable audio player using headphones.

2 Navigation in a Complex Menu Structure

Menu selection is an important mode of human-computer interaction that offers a simple form of user control. Menu-based interfaces are well suited for devices with a restricted visual display and are commonly used in non-visual interfaces, such as telephone-based systems. The principle of a menu selection system is to present a list of options from which the user can choose. Hierarchical menus are often represented as an inverted tree structure leading to an increasing degree of choice refinement, e.g. a system of options, sub-options and so on, in which the user navigates to complete a task [2]. This type of structure allows a variety of tasks to be performed, such as data retrieval, file system management or device control. Very different functions can also be considered within the same menu, as for instance in mobile phones, which often include phonebook, messages, and settings. The hierarchical menu in fig. 1 is presented as an inverted tree structure with a depth extending four levels and a breadth (i.e. the number of alternatives at each level) varying throughout the structure.

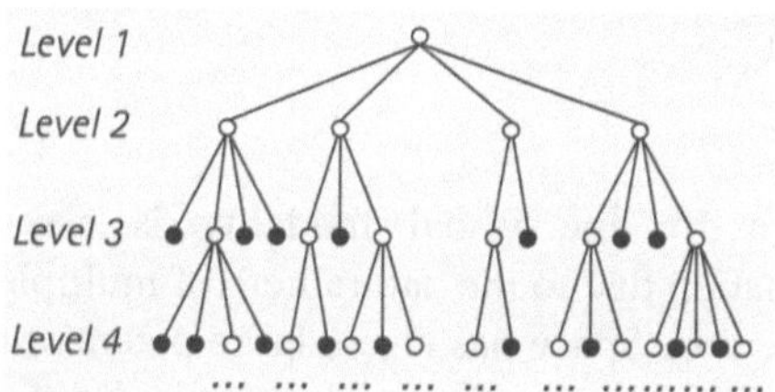

Fig. 1. Example of a hierarchical menu structure (open circles indicate choice nodes and full circles indicate terminal nodes, also called leaves).

Menu selection requires two user actions: navigating within the tree structure to steer the system toward a destination and selecting a menu item to execute a procedure or implement some action. User input can take different forms such as a mouse, a keyboard or numeric characters, but a simple two-way input means can also be considered, due to the two-dimensional nature of hierarchical menu structures. For the output modality, visual displays offer an efficient interaction mode and speech is also commonly used in non-visual interfaces. However, navigation in speech-based menus can be problematic due to a lack of explicit relation between menu levels when simple lists of words are used. Non speech-sounds have also been considered to support navigation in menu-based systems. Brewster *et al.* proposed earcons, i.e. abstract, synthetic sounds that can be used in structured combinations for menu sonification [3]. Finally, an attractive solution for auditory interfaces is to present structured menus with spatial sound, due to the analogy to visual representation of tree structures, as described in the next section.

3 Spatial Presentation and Navigation in a Menu Structure

3.1 Spatial Presentation of Sound Items

3-D sound has been used in different types of auditory displays where spatial sound separation is desired. HRTF processed sounds are normally heard outside the head when replayed over headphones and can be placed virtually at any position around the listener. However, limitations such as front-back confusions and elevation problems are common with non-individual HRTFs. For this reason, the audio interface considered here utilises a small number of sound positions, with a maximum of five sound items displayed along the horizontal plan. This strategy gives us only one dimension to work with, but it ensures that this spatial sound presentation will work for any user. A study on the discrimination of multiple non-speech sounds [4] showed that non-trained listeners can make a distinction between three or even five positions.

Fig. 2 presents three approaches for spatial sound separation in this restricted case. Standard stereo amplitude panning allows sound lateralisation (left picture), but this technique yields a very unpleasant listening experience for signals played at the extreme left or right. When using HRTFs (middle picture), the perceptual effect is more natural with a clear sound externalization at the sides. However, spatial discrimination at mid-positions (i.e. 320º and 40º azimuth) can be a problem for some users, as reported in [4]. The third approach consists in presenting the sound items along the axis of the ears (right picture). An optimal spatial separation and a natural perceptual effect can be obtained with a combination of stereo panning and 3-D sound processing.

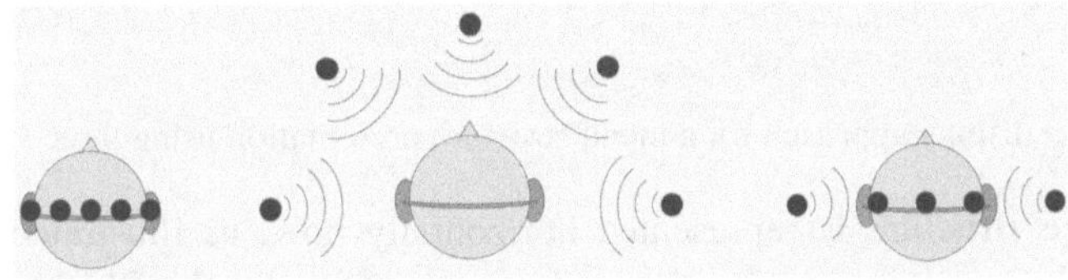

Fig. 2. Typical sound perception experienced over headphones for five sound positions obtained with stereo panning (left picture), HRTF processing (middle picture, 0º azimuth is the front source) and combined presentation (right picture).

3.2 Menu Navigation Using Spatial Item Presentation

With the horizontal spatialisation technique described earlier, several menu items can be presented, in a successive or a simultaneous mode, or using a time onset between items. Also, a horizontal scrolling technique can be used for displaying one dimension of the tree structure. This leaves us with two strategies for menu navigation, an *item display* or a *level display* approach.

Considering the tree structure vertically, as illustrated in fig. 3 (left picture), menu items from the same level can be presented at once. This item display approach allows rapid breadth searching and is therefore suitable for menu structures with long item lists. When a 3-item presentation is used (fig. 3, right picture), an intuitive input technique can be adopted using three buttons. Based on this spatial arrangement, the user can click towards the direction of the menu item of interest, i.e. left, middle or right. An additional button is needed to allow depth navigation in both ways (up and down navigation in the left picture, fig. 3). A roller would provide a natural input means for breadth searching.

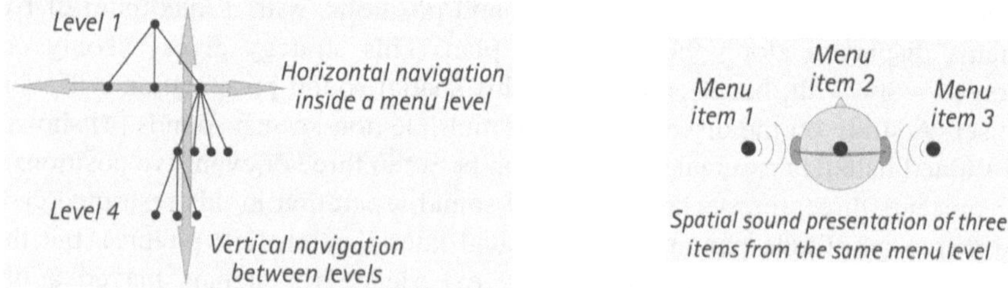

Fig. 3. Item display approach for a menu structure presentation using three sound positions.

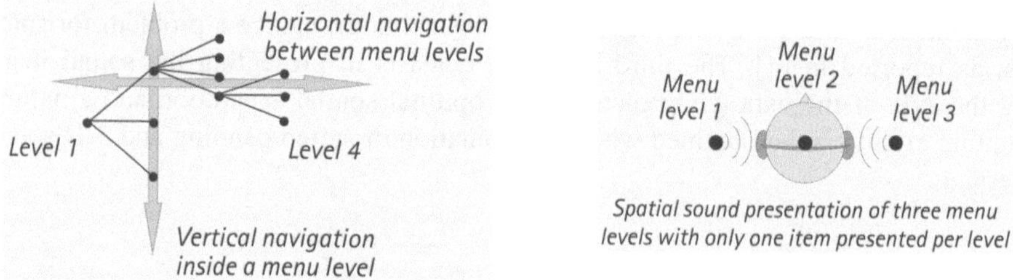

Fig. 4. Level display approach for a menu structure presentation using three sound positions.

If the tree structure is represented horizontally now, as illustrated in fig. 4 (left picture), only one item can be shown for each menu level with the present spatial configuration; however, several menu levels can be displayed at once (fig. 4, right picture). Using this level display approach, a relation between menu levels can be suggested, which gives explicit information about user's position in the menu. For instance, scrolling items of the menu level 2 will affect the item presented at sub-levels (to the right), which can be instantly displayed to the user. Therefore, this technique is more suitable for menu structures with a large depth. A different user input method is also necessary here. By using a 4-way arrows key for instance, menu navigation offers a clear spatial congruence between the input and the output. A problem with this level display approach, however, is its inefficiency for breadth searching, as simultaneous presentation of several items from long lists is not possible. Finally, a selection button is required at terminal nodes for the actual task implementation in both approaches.

4 Application to an Audio Player Interface

The spatial presentation technique described in this paper does not depend on the nature of sounds used to represent the menu items and can therefore be applied to different types of sonified menus. This includes for instance non-speech sounds for visual menu enhancement, speech for non-visual interfaces, or other audio items like audio clips, music tracks or speech streams with text-to-speech technology.

A case study on navigation in a large music collection is briefly described now, for the specific user task of creating a musical playlist. This type of interface, already considered by Pauws *et al.* [5], would be useful for a portable audio player due to the limited visual feedback existing on portable devices and the possibility of "eyes-free" interaction when using stereo headphones and a small remote control.

The process of selecting songs in a music collection requires two separate tasks: navigating in the collection to search songs, and selecting the chosen songs. A music database is usually organised as a hierarchical menu, which includes for instance the levels *music style*, *artist*, *album* and *song*. However, using a rigid tree structure to represent this list of items is not optimal because this would require the user to move back and forth between levels when scrolling all the songs from one artist. Direct breadth searching should therefore be allowed between different sub-menus of the same level. In this situation, user feedback about position in the menu would be useful and can be achieved best with the level display approach. In addition, by focusing the interaction on the middle position (i.e. only this item can be selected), feedback sounds can be played for the higher level to the left, and for the sub-level to the right. As a result, this interface can be described as a zooming function using a 4-way input key for navigation with a selection option for the center-item using an extra key. Finally, interaction feedback can be displayed to the user with speech, non-speech sounds and music. Speech is more appropriate to convey information about changes in menu levels, e.g. artist names. However, an automatic audio preview function of the center-item is also required, and non-speech sounds can enhance navigation feedback.

References

1. Ludwig, L., Pincever, N., Cohen, M., "Extending the Notion of a Window System to Audio", IEEE Computer (1990) 66-72.
2. Norman, K.L., "The psychology of menu selection: Designing cognitive control at the human/computer interface", Ablex Publishing Corporation (1991).

3. Brewster, S.A., Wright, P.C. & Edwards, A.D.N., "An evaluation of earcons for use in auditory human-computer interfaces", in Proc. of InterCHI'93, Amsterdam, NL: ACM Press, Addison-Wesley, (1993) 222-227.
4. Lorho, G., Marila, J., Hiipakka, J., "Feasibility of Multiple Non-Speech Sounds Presentation Using Headphones", in Proc. of ICAD'01, Espoo, Finland, (2001) 32–37.
5. Pauws, S., Bouwhuis, D., Eggen, E., "Programming and Enjoying Music with Your Eyes Closed", in Proc. of CHI2000, The Hague: ACM Press Addison-Wesley, (2000) 369-376.

"Do as I Say! … But Who Says What I Should Say - or Do?" On the Definition of a Standard Spoken Command Vocabulary for ICT Devices and Services

Bruno von Niman[1], Catriona Chaplin[2], Jose-Antonio Collado-Vega[3], Lutz Groh[4], Scott McGlashan[5], Wally Mellors[6], and David van Leeuwen[7]
(ETSI SPECIAL TASK FORCE STF182)[8]

[1] Ericsson Enterprise AB, Sweden
131 89 Stockholm, Sweden;
bruno.vonniman@ericsson.com

[2] Sony Ericsson Mobile Communications AB, Sweden

[3] Telefonica/ JACV Consult, Spain

[4] Siemens AG, Germany

[5] PipeBeach AB, Sweden

[6] WM Services, United Kingdom

[7] TNO Human Factors, Netherlands

[8] ETSI Specialist Task Force STF 182
http://portal.etsi.org/HF/STFs/STF182.asp;
hf_speech@list.etsi.fr

Abstract. This paper describes the development of a new ETSI Standard (ES): *Generic spoken command vocabulary for ICT devices and services.* It's basic approach focuses on simplifying the learning procedure for end-users, thereby allowing for reuse of basic knowledge between different terminal devices and services, leading to a faster and easier adoption of new technologies. The availability of common, basic interactive elements increases the transfer of learning between devices and services and improves the overall usability of the entire interactive mobile environment. Such a transfer becomes even more important in a world of ubiquitous devices and services. In particular, the paper discusses the importance of involving potential users of such products in this process, rather than relying on expert judgment alone to determine what the standard commands should be.

1 Introduction

Telecommunication, converging with information processing, and intersecting with mobility and the Internet, is leading to the development of new interactive applications and services, offering global access.

A technology enabling the most natural user interaction with these (often complex) systems and services is speech recognition. In recent years, speech recognition has

F. Paternò (Ed.): Mobile HCI 2002, LNCS 2411, pp. 425–429, 2002.
© Springer-Verlag Berlin Heidelberg 2002

become commercially viable on off-the-shelf devices and services- e.g. devices with telephone functionality (providing the dominant user interface in telecommunication). As the graphical user interface changed the way we interact with personal computers, voice user interfaces are shaping communication.

The results of this effort, an ETSI Standard (ES), will provide useful help to developers, leading to quicker, and more consistent and cheaper UI development.

ETSI, the European Telecommunications Standards Institute, ETSI, is involved in worldwide standardization open to all telecommunications players from around the globe. ETSI's Technical Committee Human Factors is responsible for human factors issues in all areas of telecommunications, producing standards, guidelines and reports that set the criteria necessary to ensure the widest possible accessibility of converging information and communications technology (ICT). It has a special responsibility to ensure that ETSI takes account of the needs of all users. This work is aligned with the European Commission's initiative ***eEurope***, a program for inclusive deployment of new, important, consumer-oriented technologies (http://europa.eu.int/information_society/eeurope).

2 Standardization of Spoken Commands

In early 2001, ETSI set up a 'Specialist Task Force' (STF) to develop a new ETSI Standard: ***Generic spoken command vocabulary for ICT devices and services***. The purpose of the standard is to simplify the learning procedure for end-users and to allow reuse of knowledge between different applications and devices. In all speech-controlled products and services adhering to the standard, one command will always mean the same thing to all users, even across the different products or services.

Specialist Task Force 182 is composed of seven individuals specializing in speech recognition, user interface design, telecommunications and standards. The scope of the work is to cover a minimum set of spoken commands to control the generic functions of ICT products and services that use speech recognition:

- Vocabularies for command, control and editing
- Applicable to navigation, information retrieval and basic call handling functions
- Specified in English, French, German, Italian and Spanish (see von Niman et al 2001 for further details).

2.1 Who Decides What Becomes a Standard?

Traditionally, ETSI has relied on experts to standardise telecommunications. Faced with three choices, we could:

- As the experts - decide the most suitable spoken commands, based on what would work best in a speech recogniser.
- Reuse spoken commands that are used in existing products and services (usually in English) and translate them to the other European languages.
- Embark on user-centred data collection in the countries in our scope.

We chose a methodology combining these elements, with the focus on data collection from native speakers of each language. However, our expertise on speech recognisers is still used to 'filter out' responses that cannot reasonably be used, and our knowledge of existing services and products is also applied where relevant.

3 Methodology for Collecting and Validating Spoken Commands

We wanted to elicit words that people would intuitively use, given the task that they want to complete. We considered two alternative types of stimulus material for our potential users: picture scenarios (storyboards) describing each function and written descriptions of each function.

3.1 Stage 1: Spontaneous Generation of Potential Command Words

Picture scenarios have been used previously in the development of a Swedish speech-controlled voice-mail service (MacDermid & Goldstein 1996). They have the advantage of being non-linguistic, so we could reuse them in all five data collection countries. However, we found from our pilot studies with pictures (see examples below) that they could be hard to interpret, especially for abstract concepts, and that the artist's choice of picture could itself bias the responses.

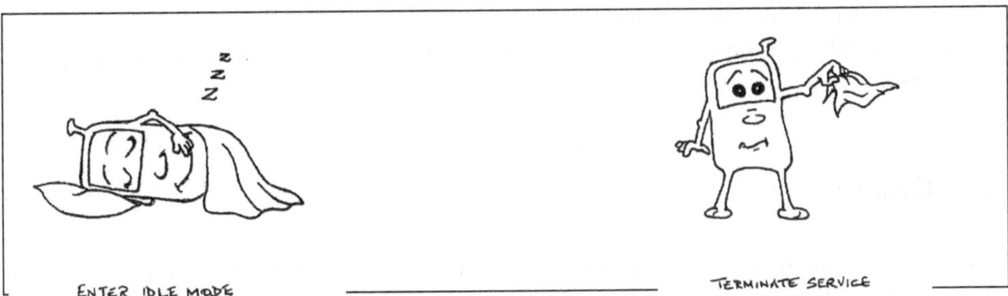

Fig. 1. Examples of picture scenarios (storyboards)

Written descriptions have been used in two US studies aiming to establish minimum sets of commands for a variety of speech applications (Cohen 2000, Guzman et al 2001). Descriptions proved easier than pictures to construct unambiguously, so we chose written stimuli, each text translated into the other target languages:

3.2 Stage 2: Phonetic Discrimination

By using phonetic dictionaries, we check if suggested command words that can be active simultaneously in a dialogue context could be recognised correctly by a speech recognition system. Unsuitable suggestions are rejected from the data collected.

It is possible to make 'Speak-to-me' suspend all activity and listen only for one special activation command from you, ignoring everything else you may say. (You do this when you want to use its services at intervals, but not switch it off altogether.) You now want to put 'Speak-to-me' into this suspended state.

You have finished what you wanted to do for today and want to leave 'Speak-to-me' altogether.

Fig. 2. Examples of written descriptions for the functions *Enter idle mode* and *Terminate*

3.3 Stage 3: Multiple-Choice Selection of Potential Command Words

The final user-centred data collection is also a web test containing the same function descriptions as in Stage 1, but this time a multiple-choice test with potential commands as the options. We obtained a multiple-choice set of options for each function by taking the spontaneous suggestions and removing command suggestions:

- problematic for a speech recogniser
- confusable with other functions
- with very low frequencies.

We also added commands we know to be used successfully in existing speech applications.

4 Results

The final results form the basis for and are thereby included in the ETSI Standard itself. They will be reported in detail and made available, together with copies of the ES, at the Conference.

Simplifying the learning procedure for end-users will allow for reuse of basic knowledge between different terminal devices and services and lead to a faster and easier adoption of new technologies.

5 Conclusion

By using this methodology, we feel confident that designers of spoken command-based products and services will want to conform to the ETSI standard. They will know that the standard is based on users' own terminology combined with expert judgement. This will make their products easier to use than if they invent their own commands, which requires users to learn a new vocabulary for each new product.

Furthermore, the availability of common, basic interactive elements increases the transfer of learning between devices and services and improves the overall usability of the entire interactive mobile environment. Such a transfer becomes even more important in a world of ubiquitous devices and services.

References

1. Cohen, Mike (2000). Universal Commands for Telephony-Based Spoken Language Systems. *SIGCHI Bulletin vol.32, no.2*, pp.25-30.
2. Guzman, Sandra, Robert Warren, Mark Ahlenius & David Neves (2001). Determining a set of acoustically discriminable, intuitive command words. *Proceedings of the Applied Voice Input/Output Society (AVIOS)*, pp.242-250, San José, USA.
3. MacDermid, Catriona & Mikael Goldstein (1996). The Storyboard method: establishing an unbiased vocabulary for keyword and voice command applications. *Adjunct proceedings of Human-Computer Interaction (HCI'96)*, pp.104-109, London, UK.
4. von Niman, Bruno, et al. (ETSI STF 182) (2001). Generic vocabulary for spoken commands. *Proceedings of Human Factors in Telecommunications (HFT'01)*, pp 305-306, Bergen, Norway.

Author Index